W9-CLO-442

## Selected Titles in This Series



(*Continued in the back of this publication*)

# Four-Dimensional Integrable Hamiltonian Systems with Simple Singular Points (Topological Aspects)

Translations of

# MATHEMATICAL MONOGRAPHS

Volume 176

# Four-Dimensional Integrable Hamiltonian Systems with Simple Singular Points (Topological Aspects)

L. M. Lerman
Ya. L. Umanskiy

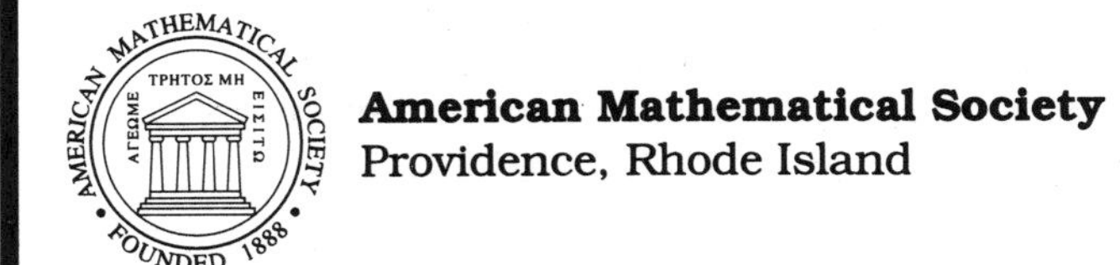


**American Mathematical Society**
Providence, Rhode Island

Л. М. Лерман, Я. Л. Уманский

ЧЕТЫРЕХМЕРНЫЕ ИНТЕГРИРУЕМЫЕ ГАМИЛЬТОНОВЫ СИСТЕМЫ С ПРОСТЫМИ ОСОБЫМИ ТОЧКАМИ (ТОПОЛОГИЧЕСКИЙ ПОДХОД)

Translated from the original Russian manuscript
by A. Kononenko and A. Semenovich

1991 *Mathematics Subject Classification.* Primary 58F05, 70Hxx.

Abstract. The main topic of this book is the isoenergetic structure of the Liouville foliation generated by an integrable system with two degrees of freedom and the topological structure of the corresponding Poisson action of the group $\mathbb{R}^2$. This is a first step towards understanding the global dynamics of Hamiltonian systems and applying perturbation methods. The main attention is paid to the topology of this foliation rather than to analytic representation. In contrast to books published before, the authors consistently use the dynamical properties of the action to achieve their results.

The book can be used by graduate students and researchers interested in studying dynamics of Hamiltonian systems. It can also be useful for people studying the geometric structure of symplectic manifolds.

---

**Library of Congress Cataloging-in-Publication Data**

Lerman, L. M. (Lev M.)
Four-dimensional integrable Hamiltonian systems with simple singular points (topological aspects) / L. M. Lerman, Ya. L. Umanskiy.
p. cm. — (Translations of mathematical monographs, ISSN 0065-3282 ; v. 176)
Translated from the original Russian manuscript.
Includes bibliographical references.
ISBN 0-8218-0375-1 (alk. paper)
1. Hamiltonian systems. 2. Four-manifolds (Topology) I. Umanskiy, Ya. L. (Yan L.) II. Title. III. Series.
QA614.83.L47 1998
514′.74—DC21 98-15030
CIP

---



∞ The paper used in this book is acid-free and falls within the guidelines
established to ensure permanence and durability.
Visit the AMS home page at URL: `http://www.ams.org/`

10 9 8 7 6 5 4 3 2 1 03 02 01 00 99 98

# Contents

# Introduction

The purpose of this book is to study a smooth integrable Hamiltonian system on a smooth four-dimensional symplectic manifold $M$ (a Hamiltonian system with two degrees of freedom in the terminology accepted in mechanics) in invariant domains containing singular points. The notion of integrability which appeared almost at the very beginning of the development of the theory of differential equations has undergone a significant evolution: from the attempts to obtain solutions of a differential equation in the form of elementary functions, their integrals, and inverse functions of integrals (integration in quadratures) in works of Johann, both Nicolai, and Daniel Bernoulli, Ricatti, Euler, Clairaut and other classics, via the discovery of the important particular cases of integrable systems in mechanics and geometry in the works of Euler, Lagrange, Jacobi, Neumann, Clebsch and many others, via the Liouville theorem, to the modern understanding of the integrability of a Hamiltonian system with $n$ degrees of freedom as the existence of $n$ almost everywhere independent integrals in the involution. Apparently, the Liouville theorem was the first general geometric result in the theory of integrable systems because this theorem depends not on a particular form of the system but only on the existence of $n$ independent integrals in the involution, and gives a description of the behavior of all trajectories of the system in the considered neighborhood. In [**1**] Arnold stated this theorem in the modern form and globalized the theorem's statement.

The modern breakthrough in the theory of finite-dimensional integrable systems occurred after the discovery of infinite-dimensional integrable Hamiltonian systems of the types of Korteveg–de Vries equations, nonlinear Schrödinger equation, sin-Gordon equation, Landau–Lifshits equation, and the development of the inverse scattering problem method and algebraic integration methods (Gardner, Green, Kruskal, Miura, Lax, Zakharov, Faddeev, Hirota, Toda, Calogero, Moser, Novikov, Dubrovin, Shabat, Mischenko, Fomenko, Bogoyavlensky and many others), which led to the discovery of new and rediscovery of already known finite-dimensional integrable Hamiltonian systems. These powerful analytic methods allow us to establish integrability (to be more precise, not to establish integrability of particular systems but to construct classes of integrable equations) and obtain explicit solutions. But these methods are poorly suited for obtaining a global description of a particular system or its phase portrait, its structure, etc., in the terminology used in the theory of dynamical systems. The first step in the development of geometric theory was made by Smale [**2**] and Marsden and Weinstein [**3**]. Smale formulated the approach to the study of a Hamiltonian system invariant with respect to some Lie group action, and Marsden and Weinstein [**3**] developed the notion of the reduction for a Hamiltonian system with symmetries. These works were among the most important ones which led the authors of this book to the idea of applying the dynamical systems methods to the study of Hamiltonian systems [**4, 5, 6, 7, 8, 73**].

Before we start describing our approach it is natural to put the following question. It is known that integrable systems form a pretty "thin" [**13**] subset in the set of all Hamiltonian systems on a symplectic manifold (in the space of their Hamiltonians), so is it worthwhile to study their structure? In our opinion, the answer is positive and it is motivated by the following considerations. In the first place, "general position" arguments, although quite fruitful, are not absolute. For example, they do not explain such a phenomenon as frequent occurrence of integrable models in different physical systems. Apparently, the symmetry considerations are always tacitly present in the constructions of such models, which often leads to their integrability, or, in the multidimensional case, to the existence of a sufficiently rich group of symmetries. In the second place, integrable systems often appear in local problems when the dynamics of a nonintegrable system in neighborhoods of singular points and periodic trajectories are studied. Moreover, when a singular point is degenerate, it becomes necessary to study the properties of families of the Hamiltonian systems in a neighborhood of this point [**14**]. It often leads to integrable normal forms depending on parameters, i.e., to a peculiar bifurcation theory in the class of integrable systems. It is worth mentioning that although a normal form transformation usually diverges, it nevertheless conveys a lot of information about the behavior of trajectories in a neighborhood of a singular point or a periodic orbit. The reason is that nonintegrability effects (for example, splitting of separatrix surfaces of singular points and periodic trajectories) are usually exponentially small when we are engaged in the local study of unfolding of a Hamiltonian system with a degenerate singular point under the condition of integrability of its normal form. This leads to a good asymptotic approximation of the real solutions by the solutions obtained from the normal form [**15**].

In our opinion, there is a third reason for the interest in integrable systems. It is known that the gradient systems played a significant role in the differential topology for the studies of the topology of smooth manifolds [**16**], in particular, for the solution of the famous Poincaré problem for $n > 4$. We believe that the study of integrable Hamiltonian systems may help in the study of the topology of symplectic manifolds. The future will show whether this is true or not.

In addition to the above reasons, one can mention another important argument in favor of the study of integrable systems: this is one of few classes of Hamiltonian systems for which it is possible, in principle, to carry out a complete study of the structure of the flow. Another well-known example of this type is the opposite case of complete nonintegrability—the case of the geodesic flows on manifolds of negative curvature (see [**17**] and the references therein). Intuition is sharpened on such classes.

The authors' interest in these problems was stimulated to a large extent by the desire to understand the structure of some model systems which appeared in the theory of the domain walls propagation in magnetic media [**18, 19**]. It is known that the main phenomenological equation of this theory is the Landau–Lifshits equation, which for one-dimensional medium (plane wave) has the form:

$$m_t = m \times m_{xx} + m \times Jm,$$

where $m(x,t)$ is a three-dimensional vector of magnetic momentum, $m^2 = 1$, $J = \operatorname{diag}(J_1, J_2, J_3)$, $x \in \mathbb{R}$. Stationary waves of the form $m(x,t) = v(\xi)$, $\xi = x - ut$ satisfy the equation which reduces, after a substitution, to an integrable Hamiltonian system whose phase space is the cotangent bundle $T^*S^2$ to the sphere

$S^2 : v_1^2 + v_2^2 + v_3^2 = 1$. This system depends on two parameters $u$ and $\epsilon = (J_3 - J_2)/(J_2 - J_1)$ and can be integrated in terms of Prim's $\theta$-functions [**20**]. However, it is still quite difficult to "see" the dynamics of all solutions from this fact. This led us to the problem of developing some version of a "qualitative theory of integrable Hamiltonian systems" [**4, 5**]. In the base of this theory we put the study of the orbit structure for the induced action of the group $\mathbb{R}^2$ generated by a pair of commuting Hamiltonian vector fields $X_H$, $X_K$, where $H$ is the Hamilton function of the Hamiltonian vector field, and $K$ is its additional integral. Such an approach was quite natural for us since it is typical for "Andronov's school of oscillation theory" to which the authors belong. From this point of view the first problem was to study the action in a neighborhood of the singular set of the action, i.e., in the union of orbits whose dimension is less than two. The theorem of Liouville–Arnold, which describes the structure of orbits in a neighborhood of a two-dimensional compact orbit, the torus, does not work here. We note that studying the structure of orbits in a neighborhood of a singular point of the action we are led to the necessity of the global study of their behavior, since usually there are orbits containing a singular point in their closure but deviating far from this singular point. Therefore, we are led to the notion of extended neighborhood of a singular point and to the study of the behavior of action orbits and Hamiltonian system trajectories in this neighborhood, which is invariant with respect to the action. This is the main topic of our book.

It should be noted that such a study is conducted in a four-dimensional extended neighborhood and cannot be reduced to the study of the dynamics on a smooth three-dimensional level set of the Hamilton function $H$. The constructed equivalence invariants are determined both by the behavior of the system on the level set of $H$ containing the singular point and by its behavior on nearby level sets. Clearly, for a global study of a system on a symplectic manifold one needs to know the behavior of the system on those level sets of the Hamilton function $H$ that do not contain singular points. This was done in the works of Fomenko and coauthors [**9, 10, 11**]. Since these results are presented in many publications and books, we will not repeat them here, referring the interested reader to the original works.

For the reader's convenience we list some nonstandard abbreviations used throughout the book:

IHVF – integrable Hamiltonian vector field,

PA – Poisson action,

SPT – singular periodic trajectory,

CSP – curve of singular points.

The exact meaning of this terms is explained in the appropriate sections of this book. Also we would like to mention the numeration of the formulas, definitions, statements (lemmas, propositions, theorems, corollaries) and remarks accepted in the book. Every chapter, except for small Chapter 4, is divided into sections, which have double numeration. For example, Section 3.2 denotes Section 2 of Chapter 3. Formulas are numbered consecutively within each chapter. Definitions, remarks and figures are numbered consecutively within each chapter, and their numbers are of the form m.n, where m is the number of the chapter and n is the number of the corresponding definition, remark, or figure. The statements are numbered consecutively within each section, so they have triple numbers p.q.r., where p is the number of the chapter, q, the number of the section, and r, the number of the corresponding statement.

We express our gratitude to V. M. Eleonsky, who brought to our attention integrable models in physics and discussed with us many questions related to integrable systems, and to L. P. Shilnikov whose lively interest has been of help to us in our work.

We acknowledge the partial financial support of the International Science Foundation (grant R97000), which allowed us to complete this book.

Nizhni Novgorod

September 1994

CHAPTER 1

# General Results of the Theory of Hamiltonian Systems

## 1.1. Hamiltonian systems on a symplectic manifold

In order to define a Hamiltonian system on a smooth[1] manifold $M$, this manifold should carry a special structure, called symplectic structure. In this section we review the relevant notions and definitions. The reader can find more detailed presentartion in the books [**21, 22, 23, 24, 25**].

Let $M$ be a smooth $2n$-dimensional manifold and $\Omega$ a smooth closed nondegenerate differential 2-form on $M$ (a symplectic structure). Recall that a form is said to be closed if its exterior derivative $d\Omega$ vanishes identically, and $\Omega$ is called nondegenerate if for every point $x \in M$ and every nonzero vector $\xi$ from the tangent space $T_xM$ there exists $\eta \in T_xM$ such that $\Omega(\xi, \eta) \neq 0$.

A nondegenerate 2-form may exist only on an even-dimensional manifold. The orientability of $M$ is another consequence of the existence of symplectic structure since there is a nondegenerate volume form $\Omega \wedge \cdots \wedge \Omega$ ($n$ times).

DEFINITION 1.1. The pair $(M, \Omega)$ is called a symplectic manifold.

Whenever there will be no ambiguity, we will denote a symplectic manifold simply by $M$.

A symplectic structure defines a natural isomorphism between the tangent bundle $TM$ and the cotangent bundle $T^*M$. Namely, let $x \in M$, $\xi \in T_xM$. Then $\Omega(\xi, \cdot)$ is a linear functional $\lambda$ on $T_xM$, that is, $\lambda \in T_x^*M$. The map $\xi \to \lambda$ is linear and, due to the nondegeneracy of $\Omega$, is an isomorphism between the spaces $T_xM$ and $T_x^*M$. We denote the inverse map $T^*M \to TM$ by $J$. Let us choose an arbitrary $C^k$-smooth function $H : M \to \mathbb{R}$. Its differential $dH$ is a 1-form. Thus, there is a well-defined $C^{k-1}$-smooth vector field $X_H$ on $M$,

$$X_H = JdH. \tag{1.1}$$

DEFINITION 1.2. The field $X_H$ is called a Hamiltonian vector field and the function $H$ is called its Hamilton function.

It follows from (1.1) that for all $\xi \in T_xM$ we have $dH_x(\xi) = \Omega(\xi, X_H(x))$, where $dH_x$ is the restriction of the form $dH$ to $T_xM$.

*Example* 1. Let $M = \mathbb{R}^{2n}$ with coordinates $(p, q) = (p_1, \ldots, p_n, q_1, \ldots, q_n)$, $\Omega = \sum_i dp_i \wedge dq_i = dp \wedge dq$. Then

$$dH = \sum_i \left[ \left( \frac{\partial H}{\partial p_i} \right) dp_i + \left( \frac{\partial H}{\partial q_i} \right) dq_i \right].$$

[1]Here and everywhere in the sequel smooth means $C^\infty$-smooth.

In the coordinates $(p, q)$ any vector field has the form

$$\sum_i \left[ X_i \frac{\partial}{\partial p_i} + Y_i \frac{\partial}{\partial q_i} \right].$$

Using (1.1) we can write the Hamiltonian field $X_H$ in the coordinates $(p, q)$ (recall that $dp_i \wedge dq_i(\xi, \eta) = dp_i(\xi)dq_i(\eta) - dp_i(\eta)dq_i(\xi)$):

$$dH_x(\xi) = \Omega(\xi, X_H(x)) = Y_i dp_i(\xi) - X_i dq_i(\xi).$$

Thus, $X_i = -\frac{\partial H}{\partial q_i}$, $Y_i = -\frac{\partial H}{\partial p_i}$.

The corresponding representation as a system of differential equations has the form:

$$\dot{p}_i = -\frac{\partial H}{\partial q_i}, \quad \dot{q}_i = -\frac{\partial H}{\partial p_i}, \tag{1.2}$$

or, in the abridged form, $\dot{x} = I\nabla H$, where the matrix $I$ is of the form

$$\begin{pmatrix} 0 & -E \\ E & 0 \end{pmatrix},$$

$E$ is the identity $(n \times n)$ matrix, and 0 is the zero $(n \times n)$ matrix.

*Example* 2. *Symplectic structure on the cotangent bundle: the phase space of a classical mechanical system.* Let $N$ be a smooth manifold and $T^*N$ its cotangent bundle. We define the 1-form $\omega$ on $T^*N$ as follows. Let $\pi : T^*N \to N$ be the canonical projection. A point $z \in T^*N$ is a point $x = \pi(z) \in N$ and a linear functional $l$ on $T_xN$. The differential $D\pi$ acts fiberwise from $T(T^*N)$ onto $TN$. Moreover, it acts linearly on each fiber of $T_z(T^*N)$, and maps it onto the fiber $T_xN$. Let us choose a vector $\xi \in T_z(T^*N)$. Define the value of the 1-form $\omega$ on $\xi$ by the formula: $\omega(\xi) = l(D\pi(\xi))$. Obviously, $\omega$ is a linear function of $\xi$ and depends smoothly on $z$. Thus, we have a differential 1-form. Its exterior derivative $d\omega$ is a closed nondegenerate 2-form, i.e., a symplectic structure.

Let us write $\omega$ in the local coordinates on $T^*N$. Let $U \subset N$ be a local chart and $(q_1, \dots, q_n)$ the local coordinates on $U$. Then, we have the coordinates $(p_1, \dots, p_n, q_1, \dots, q_n)$ on $T^*N|_U$, where $p_1, \dots, p_n$ are the coefficients of any 1-form in the basis made of the 1-forms $dq_1, \dots, dq_n$. Let us fix a point $\chi = (p_1^0, \dots, p_n^0, q_1^0, \dots, q_n^0)$. Then $x = (q_1^0, \dots, q_n^0)$, $l = p_1^0 dq_1 + \dots + p_n^0 dq_n$. The tangent vector $\xi \in T_\chi(T^*N)$ has a form $a_1 \frac{\partial}{\partial p_1} + \dots + a_n \frac{\partial}{\partial p_n} + b_1 \frac{\partial}{\partial q_1} + \dots + b_n \frac{\partial}{\partial q_n}$, and $D\pi(\xi) = b_1 \frac{\partial}{\partial q_1} + \dots + b_n \frac{\partial}{\partial q_n}$. Thus, $\omega(\xi) = l(D\pi(\xi)) = p_1^0 b_1 + \dots + p_n^0 b_n = p_1^0 dq_1(\xi) + \dots + p_n^0 dq_n(\xi)$. Since $\xi$ is an arbitrary vector, $\omega = p_1 dq_1 + \dots + p_n dq_n$ and the form $\Omega = dp_1 \wedge dq_1 + \dots + dp_n \wedge dq_n$ is obviously nondegenerate.

In the analytical mechanics $N$ is the configuration space of a mechanical system. Its Lagrange function (the Lagrangian) is a function on $TN$. The generalized momenta $p = \frac{\partial L}{\partial \dot{q}}$ are cotangent vectors. Thus, the phase space of the mechanical system is a cotangent bundle. If there is a fixed Riemannian mertic on $N$, then the classical mechanical system is defined as the system with the Lagrangian $L = T - V$, where $T(q, \dot{q})$ is the length of the tangent vector $\dot{q}$ at the point $q$, and $V$ is a smooth function on $N$.

Notice that in Examples 1 and 2, in a suitable coordinate system $(p_1, \dots, p_n, q_1, \dots, q_n)$, the form $\Omega$ was written locally as $\sum dp_i \wedge dq_i$. It turns out that this is a general fact. Namely, the following theorem holds.

THEOREM (Darboux). *Let $(M, \Omega)$ be a smooth symplectic manifold, and $x \in M$. There exist a neighborhood $U$ of the point $x$ and local coordinates $(p_1, \ldots, p_n, q_1, \ldots, q_n)$ such that $\Omega = \sum dp_i \wedge dq_i$ in these coordinates .*

Such coordinates are called symplectic and the neighborhood $U$ together with the symplectic coordinates in it is called a symplectic chart. For the proof of the Darboux theorem see [**21, 22, 23**].

DEFINITION 1.3. A diffeomorphism $f$ between symplectic manifolds $(M_1, \Omega_1)$ and $(M_2, \Omega_2)$ is called a symplectic diffeomorphism or a symplectomorphism if $\Omega_1 = f^*\Omega_2$, where $f^*$ is an induced mapping of forms.

Recall that if $f : M_1 \to M_2$, then $f^*\Omega_2(a, b) = \Omega_2(Df(a), Df(b))$, $a, b \in TM_1$. In particular, suppose that there are two symplectic charts: $(U; (p, q))$ on $M_2$ and $(V; (\xi, \eta))$ on $M_1$, and let $f : V \to U$ be a symplectic diffeomorphism. Then in the local coordinates $f$ has the form $p = p(\xi, \eta)$, $q = q(\xi, \eta)$ and the condition that $f$ is symplectic means that $dp \wedge dq = dp(\xi, \eta) \wedge dq(\xi, \eta)$. It follows that the Jacobi matrix $S$ of the transformation $x = \phi(\zeta)$, where $x = (p, q)$ and $\zeta = (\xi, \eta)$, satisfies the relation

$$SI(S)^T = I, \tag{1.3}$$

that is, the matrix $S = \frac{\partial \phi}{\partial \zeta}$ is symplectic. As we have seen in Example 1 above, every Hamiltonian vector field has a form $I\nabla H(x)$ in symplectic coordinates, where $\nabla H = (\frac{\partial H}{\partial p_1}, \ldots, \frac{\partial H}{\partial p_n}, \frac{\partial H}{\partial q_1}, \ldots, \frac{\partial H}{\partial q_n})^T$. Let us determine the form of the system in the coordinates $\zeta = (\xi, \eta)$ after a symplectic transformation (which is also called a canonical transformation). Differentiating the relation $x = \phi(\zeta)$ and using the equation $\dot{x} = I\nabla H(x)$ we have

$$\dot{\zeta} = S^{-1} I \nabla_x H(\phi(\zeta)),$$

where $\nabla_x$ denotes the gradient with respect to the variables $(p, q)$. Using (1.3) we obtain

$$\dot{\zeta} = IS^T \nabla_x H(\phi(\zeta)) = I\nabla_\zeta \hat{H}(\zeta),$$

where $\hat{H} = H(\phi(\zeta))$. Thus, the transformed system has the same form as the original one with the Hamilton function obtained from the original Hamilton function by plugging the new variables into it. The converse is also true: if a smooth transformation $x = \phi(\zeta)$ transforms a Hamiltonian system $\dot{x} = I\nabla H$, with an *arbitrary* Hamilton function $H(x)$, into a Hamiltonian system $\dot{\zeta} = I\nabla\hat{H}(\zeta)$ with the Hamilton function $\hat{H}(\zeta) = H(\phi(\zeta))$, then it is symplectic.

## 1.2. Poisson brackets and first integrals

A symplectic structure on a manifold $M$ defines an operation on the set $\mathcal{F}(M)$ of smooth functions on $M$ which makes $\mathcal{F}(M)$ a Lie algebra [**21, 22, 23, 24, 25**].

DEFINITION 1.4. The Poisson bracket $\{f, g\}$ of two smooth functions $f$ and $g$ is a smooth function defined by the formula

$$\{f, g\} = \Omega(X_g, X_f). \tag{1.4}$$

Let us find the explicit form of the Poisson bracket in the symplectic coordinates $(p, q)$. We have $\Omega = dp \wedge dq$, $X_f = I\nabla f$, $X_g = I\nabla g$. Therefore, $\{f, g\} = \Omega(X_g, X_f) = \sum_{i=1}[dp_i(X_g)dq_i(X_f) - dp_i(X_f)dq_i(X_g)] = \sum_i(-\frac{\partial g}{\partial q_i}\frac{\partial f}{\partial p_i} + \frac{\partial f}{\partial q_i}\frac{\partial g}{\partial p_i})$.

It follows easily from Definition 1.4 and the local representation of the Poisson bracket that the Poisson bracket is a bilinear skew-symmetric operation, which is also a derivation, that is, $\{fg,h\} = f\{g,h\} + g\{f,h\}$. In addition, the fact that $\Omega$ is closed gives us the following Jacobi identity:

$$\{f,\{g,h\}\} + \{h,\{f,g\}\} + \{g,\{h,f\}\} \equiv 0.$$

The notion of a Poisson bracket can be used as a basis for definition of a symplectic manifold. Namely, let $M$ be a smooth manifold, and suppose that a bilinear and skew-symmetric operation $(f,g) \to \{f,g\} \in \mathcal{F}(M)$ is defined on the set $\mathcal{F}(M)$ of smooth functions. In addition, suppose that for a fixed $g$ this operation is a derivation with respect to $f$ (see above) and satisfies the Jacobi identity. In this case $M$ is called a Poisson manifold and the operation is called a Poisson bracket.

In the local coordinates $(x_1,\dots,x_n)$ on the manifold $M$, the Poisson bracket can be written in the form

$$\{f,g\} = \sum_{i,j=1}^{n} a_{ij}\frac{\partial f}{\partial x_i}\frac{\partial g}{\partial x_j}, \tag{1.5}$$

where the matrix $A = (a_{ij})$ is skew-symmetric, $a_{ij} = \{x^i, x^j\}$. Formula (1.5) follows from the property of the Poisson bracket to be a derivation with respect to each argument (recall that a linear operator $L : f \to Lf$ which is a derivation of $\mathcal{F}(M)$, i.e., satisfies $L(fg) = fL(g) + gL(f)$, defines a vector field that has the form $\sum_{k=1}^{n}\frac{\partial}{\partial x_k}$ in the local coordinates [**27**]). The Jacobi identity is equivalent to the following identities for the coefficients $a_{ij}$ :

$$\sum_{n=1}^{n}(a_{ij}\frac{\partial a_{ks}}{\partial x_i} + a_{is}\frac{\partial a_{jk}}{\partial x_i} + a_{ik}\frac{\partial a_{sj}}{\partial x_i}) \equiv 0,$$

for all $i,j,k \in \{1,\dots,n\}$.

The Poisson bracket allows us to define the vector field $X_H$ corresponding to a fixed function $H \in \mathcal{F}(M)$. Recall that the transformation $L_H : \mathcal{F}(M) \to \mathcal{F}(M)$ defined by the formula $L_H(g) = \{g,H\}$ is a derivation and thus defines a vector field $X_H$. Namely, for each point $x \in M$ the vector $\xi = X_H(x)$ is a tangent vector such that for any function $g \in \mathcal{F}(M)$ the following relation holds:

$$\{g,H\}(x) = \frac{d}{dt}g(c(t))\,|_{t=0} = l_\xi(g), \tag{1.6}$$

where $c(t)$ is a smooth curve on $M$ that satisfies the conditions $c(0) = x$, $c'(0) = \xi$, and $l_\xi(g)$ is the derivative of $g$ in the direction of $\xi$. It is easy to check that this property does not depend on the choice of a curve satisfying these conditions.

Let us assume additionally that the Poisson bracket is nondegenerate, i.e., for any $x \in M$ we have $df_x = 0$ whenever $\{f,g\} = 0$ for any $g$. (In local coordinates this means that $\det(a_{ij}(x)) \neq 0$ for all $x \in M$. Moreover, since the matrix $a_{ij}$ is skew-symmetric, $\dim M = 2n$.) Then one can define a closed nondegenerate 2-form $\Omega$ on $M$, i.e., a symplectic structure. In order to do this we notice that for any vector $\xi \in T_xM$ there exists a function $f = f_\xi \in \mathcal{F}(M)$ such that $\xi = X_f(x)$. This follows from the nondegeneracy of the Poisson bracket. The easiest way to prove it is to show this in the local coordinates $(x_1,\dots,x_{2n})$. Then the vector $\xi$ can be written in the form $\xi = \sum b_i\frac{\partial}{\partial x_i}$, and the Poisson bracket has the form (1.5). For any $g \in \mathcal{F}(M)$ we have $l_\xi(g) = \sum b_i\frac{\partial g}{\partial x_i}$, and it must be equal to $\{f,g\}(x)$ with some, yet undetermined, function $f$. To find $f$ we have equations $b_i = \sum a_{ij}\frac{\partial f}{\partial x_j}$, or $\nabla f(x) = A^{-1}b$, where $b$ is a vector with components $b_i$. The function $f$ can be determined from this system. Any two such functions differ in the second or higher order terms (of course, we disregard the differences in the values of such functions at the point $x$). In this way we construct an equivalence class of functions $f_\xi$ corresponding to $\xi$.

Now the form $\Omega$ is defined by the formula

$$\Omega(\xi, \eta) = \{g_\eta, f_\xi\},$$

where $f_\xi$ and $g_\eta$ are any representatives of the above equivalence classes. Obviously, $\Omega$ is well defined. It is clear that the field $X_H$ determined by the function $H$ and a nondegenerate Poisson bracket is Hamiltonian vector field in the sense of Section 1.1.

The situation when the Poisson bracket is degenerate often occurs in applications. Let us assume, in addition, that the bracket has a locally constant rank, i.e., that $\operatorname{rank}(a_{ij})$ is constant in a neighborhood $U$ of the point $x$. Then there exist locally defined functions $f_1, \dots, f_s$ such that $\{f_i, g\} \equiv 0$ for all $g \in \mathcal{F}(M)$ and the functions $f_1, \dots, f_s$ are independent on $U$. Then the Poisson bracket is nondegenerate on their common level set $f_1 = C_1, \dots, f_s = C_s$, which is a smooth submanifold, and this level set is a local symplectic manifold [**24**].

*Example.* Let $M = \mathbb{R}^6 = T^*\mathbb{R}^3$ with the coordinates $(x_1, x_2, x_3, y_1, y_2, y_3)$. Define the Poisson bracket by the matrix [**28, 29**]

$$A = \begin{pmatrix} L(x) & L(y) \\ L(y) & 0 \end{pmatrix}, \quad L(x) = \begin{pmatrix} 0 & x_3 & -x_2 \\ -x_3 & 0 & x_1 \\ x_2 & -x_1 & 0 \end{pmatrix}.$$

It is easy to see that the functions $S = x_1^2 + x_2^2 + x_3^2$ and $K = x_1y_1 + x_2y_2 + x_3y_3$ have the property $\{S, F\} \equiv \{K, F\} \equiv 0$ for any smooth function $F$ on $\mathbb{R}^6$. The common level set of these two functions $S = s_0 > 0$, $K = k_0$ defines a smooth four-dimensional manifold $N$ diffeomorphic to $T^*S^2$, which is a symplectic submanifold.

Now let us assume that $M$ is symplectic and that there is a Hamiltonian vector field $X_H$ with the Hamilton function $H$ on it. Recall that a smooth function $f$ on $M$ is called a first integral of the field $X_H$ if $f$ is constant along the trajectories of $X_H$. The necessary and sufficient condition for the function $f$ to be a first integral is that $l_{X_H}(f) \equiv 0$, or, as follows from (1.6), $\{f, H\} \equiv 0$. We will say that the functions $f, g \in \mathcal{F}(M)$ are in involution if $\{f, g\} \equiv 0$. A set of functions $f_1, \dots, f_m$ is called involutive if $\{f_i, f_j\} \equiv 0$ for all $i, j \in \{1, \dots, m\}$. Thus the first integral of a Hamiltonian field is in involution with the Hamilton function. In particular, $H$ is an integral of the field $X_H$.

It is possible to obtain new integrals of the field $X_H$ from two integrals $f, g$ using the Poisson bracket. Indeed, from the Jacobi identity we have

$$\{\{f, g\}, H\} = -\{\{H, f\}, g\} - \{\{g, H\}, f\} \equiv 0,$$

i.e., $\{f, g\}$ is a first integral. However, this process does not always lead to independent integrals.

The independent involutive sets play a special role among the integrals. Let us denote the $n$-dimensional torus by $T^n = \mathbb{R}^n/\mathbb{Z}^n$.

THEOREM (Liouville; see, for example, [**14**]). *Suppose that there are $n$ smooth functions in involution $f_1, \dots, f_n$, $\{f_i, f_j\} \equiv 0$, $i, j \in \{1, \dots, n\}$, defined on the*

*$2n$-dimensional manifold $M$. Consider a level set for the functions $f_i$ :*

$$M_c = \{x \in M \mid f_i = c_i, i = 1, \dots, n\}.$$

*Suppose that the functions $f_1, \dots, f_n$ are independent on $M_c$, i.e., $df_1, \dots, df_n$ are independent at any point $x \in M_c$. Then*

1. *$M_c$ is a smooth submanifold diffeomorphic to $T^p \times \mathbb{R}^{n-p}$ and invariant with respect to the phase flow of the field $X_H$, $H = F(f_1, \dots, f_n)$. Moreover, $dF \neq 0$ when $f_i = c_i$, $i = 1, \dots, n$ (for example, $H = f_1$).*

2. *$M_c$ has a standard affine structure such that the phase flow can be rectified in this structure, i.e., we have $\dot{\varphi} = \text{const}$ in the affine coordinates $(\varphi_1, \dots, \varphi_n)$ on $M_c$.*

3. *If, in addition, $M_c$ is compact (i.e., $M_c = T^n$), then there exist symplectic coordinates "action-angle" $(I_1, \dots, I_n, \varphi_1, \dots, \varphi_n)$ in the neighborhood of $M_c$ in $M$ such that $H = H(I_1, \dots, I_n)$ in these coordinates.*

It follows from part 3 of the theorem that for a compact $M_c$ there exists a neighborhood $U$ in $M$ which is foliated by tori. Moreover, $U$ is diffeomorphic to $T^n \times D^n$, where $D^n$ is an $n$-dimensional ball. In $U$, the field $X_H$ has the form

$$\dot{I} = 0, \qquad \dot{\varphi} = \omega(I) = \frac{\partial H}{\partial I}.$$

In the general case, the frequency vector $\omega(I)$ changes from one torus to another which leads to the changes in the topology of the flows on the tori. *Example.* a) Let $(m, \omega) = \sum_{i=1}^{n} m_i \omega_i \neq 0$ at $I = I_0$, where $m_i \in \mathbb{Z}$, i.e., the frequencies $\omega_1, \dots, \omega_n$ are independent over the integers. Then any trajectory of the vector field $\dot{\varphi}$ on the $n$-dimensional torus is everywhere dense (transitive) [**21**].

b) Suppose that for $I = I_1$ there exists an integer vector $m$ such that $(m, \omega) = 0$. Then the set of such vectors forms a subgroup in $\mathbb{Z}^n$. Let $r$ be the rank of this subgroup. Then the torus $T^n$ splits into tori of dimension $n - r$, and the trajectory is transitive on every such torus.

## 1.3. Integrable systems and Poisson actions

The Liouville theorem shows that for a Hamiltonian system on a $2n$-dimensional symplectic manifold the presence of $n$ first integrals in involution that are independent in some neighborhood of a compact nondegenerate common level set of $n$ integrals, provides a complete information about the behavior of trajectories of this system. Therefore, such systems are called Liouville integrable or completely integrable. Generalizing slightly this notion we give the following definition.

DEFINITION 1.5. Let a Hamiltonian vector field $X_H$ be defined in a domain $D \subset M$ (in particular, $D$ may coincide with $M$), and let $\{F_1, \dots, F_n\}$, $F_1 \equiv H$, be a fixed involutive set of smooth functions on $D$ such that $dF_1, \dots, dF_n$ are independent almost everywhere in $D$. A vector field $X_H$ with the set of integrals $F_2, \dots, F_n$ is called an integrable Hamiltonian vector field (IHVF) and denoted by $(X_H, F_2, \dots, F_n)$.

The Liouville theorem is applicable in a region $D_0$ invariant with respect to the field $X_H$, in which $dF_1, \dots, dF_n$ are independent and the common level sets $F_1 = c_1, \dots, F_n = c_n$ are compact. Thus, such a region is foliated into invariant $n$-dimensional tori. However, this theorem says nothing about the structure of the foliation into the common level sets of the integrals near points at which the differentials $dF_1, \dots, dF_n$ are not independent. It turns out to be convenient to describe this foliation not in terms of functions $F_i$ but using the orbits of some naturally defined action of the group $\mathbb{R}^n$. Let us give the relevant definitions.

Let $N$ be a smooth manifold and $G$ a Lie group. A smooth map $h : G \times N \to N$ is called a (left) action of the group on the manifold if for any $g \in G$ the map $h(g, \cdot) = h_g$ is a diffeomorphism $h_g : N \to N$ and the following conditions are satisfied:

1. $h_e = \mathrm{id}_N$, where $e$ is the identity element of the group;
2. $h_{g_1 g_2} = h_{g_2} \circ h_{g_1}$, where $g_1 g_2$ is the product of elements $g_1$ and $g_2$ in $G$;
3. $h_{g^{-1}} = h_g^{-1}$, where $g^{-1}$ is the inverse of $g$.

An action $\Phi$ of a connected Lie group $G$ on a symplectic manifold $M$ is called a Poisson action [**21**] if

1. for all $g \in G$ the diffeomorphisms $\Phi_g$ are symplectic;
2. for any $\xi$ from a Lie algebra $\mathcal{G}$ of the group $G$ the action of the one-parameter subgroup $\exp(t\xi)$ has Hamilton function $H_\xi$ which depends linearly on $\xi$;
3. $H_{[\xi,\eta]} = \{H_\xi, H_\eta\}$, where $[\xi, \eta]$ is the commutator of the elements $\xi$ and $\eta$ in the algebra $\mathcal{G}$.

If the group $G$ is commutative, then the third condition means that for any $\xi, \eta \in \mathcal{G}$ the Hamilton functions $H_\xi$ and $H_\eta$ are in involution.

The set $\mathcal{O}_x = \{y \in M \mid h(g, x) = y \text{ for some } g \in G\}$ is called the orbit $\mathcal{O}_x$ of the group action $\Phi$ passing through the point $x \in M$.

In this book the most interesting case will occur when the acting group is $\mathbb{R}^n$. Then it is not difficult to show [**21**] that orbits of the action are subsets homeomorphic (in the intrinsic topology) to $\mathbb{R}^m \times T^k$, $m + k \leq n$, where $T^k$ is the $k$-dimensional torus. The orbits whose dimensions are less than $n$ are called singular.

Now let us return to the IHVF. Suppose that the Hamiltonian vector fields $X_{F_1}, \dots, X_{F_n}$ with Hamilton functions $F_1, \dots, F_n$ are complete, i.e., each solution of the corresponding equation can be extended to the whole $\mathbb{R}$. For example, the following condition is sufficient for this: the level sets of one of the functions $F_i$ are compact, or the common level sets of several functions $F_{i_1}, \dots, F_{i_k}$ are compact. Let us denote by $\varphi_i^{t_i}$ the flow corresponding to the field $X_{F_i}$. Then the following result holds (see, for example, [**21**]).

PROPOSITION. *Two smooth Hamiltonian vector fields $X_F$ and $X_G$ have commuting flows if and only if $\{F, G\} \equiv 0$.*

Commutativity of the flows $\varphi^t$ and $\psi^\tau$ means that $\varphi^t \circ \psi^\tau \equiv \psi^\tau \circ \varphi^t$ for any $t, \tau \in \mathbb{R}$.

It follows from this proposition that it is possible to define the action of the group $\mathbb{R}^n$ on $M$ by the formula

$$\Phi(t_1, \dots, t_n; x) = \varphi_n^{t_n} \circ \varphi_{n-1}^{t_{n-1}} \circ \dots \circ \varphi_1^{t_1}(x), \tag{1.7}$$

where $\varphi_i^{t_i}$ denotes the flow of the field $X_{F_i}$. It is not difficult to verify that this defines a Poisson action.

The action $\Phi$ generated by the functions $F_1, \dots, F_n$ will be called the induced Poisson action for the IHVF $(X_H, F_2, \dots, F_n)$.

Conversely, let $\Phi$ be a Poisson action of the group $\mathbb{R}^n$ on $M$. Let us choose any $n$ linearly independent vectors $\xi_1, \dots, \xi_n$ in $\mathbb{R}^n$, and denote the corresponding

Hamilton functions by $H_1, \dots, H_n$ (here we identify the Lie group $\mathbb{R}^n$ with its Lie algebra $\mathbb{R}^n$). Then each $H_i$ defines the IHVF in such a way that the remaining $H_j$, $i \neq j$, are its additional integrals. The vector fields $X_{H_i}$ are tangent to the orbits of the action $\Phi$. Therefore, each orbit is an invariant manifold of the vector field $X_{H_i}$. In particular, $H_i(\Phi(t,x)) \equiv H_i(x)$. Thus, the Hamilton function is invariant with respect to the action $\Phi$. This is also true for the action (1.7). Therefore, it is clear that the orbit foliation of the induced action is an important characteristic of the behavior of trajectories for the IHVF itself. Notice that the vector fields $X_{H_1}, \dots, X_{H_n}$ are independent at each point of a nonsingular orbit (of dimension $n$), whereas only $k$, $0 \leq k < n$, fields are independent on a singular orbit. Then the corresponding orbit will have dimension $k$. If $k = 0$, we shall call the corresponding orbit a singular point of the action.

On the other hand, the study of the IHVF cannot be completely reduced to the study of the action. For example, suppose that the action induced by the IHVF $X_H$ has a $k$-dimensional orbit $L$ and $x_0 \in L$. Then, there are two possibilities: 1) $X_H(x_0) = 0$; 2) $X_H(x_0) \neq 0$. In the first case the whole orbit $L$ consists of singular points of the field $X_H$. Indeed, if $x = \Phi(\tau, x_0)$ and $H(\Phi(t,x)) \equiv H(x)$, then $dH_x(\xi) \equiv dH_{\Phi^t(x)}(D_x\Phi^t(\xi))$. Let $t = -\tau$. Then we have $dH_x(\xi) = dH_x(\xi_0) = 0$, where $\xi_0 = D\Phi^{-\tau}(\xi)$, i.e., $X_H(x) \equiv 0$, $x \in L$. In the second case there is a $k$-dimensional invariant submanifold passing through the point $x_0$ and not containing singular points of $X_H$.

### 1.4. A local structure of the orbit foliation of an action

The local structure of the orbit foliation of an action near a point of a $k$-dimensional orbit $L$ may be described using the following construction. Let a point $x_0$ belong to the $k$-dimensional orbit of the action $\Phi$. Let us choose $n$ linearly independent vectors $\xi_1, \dots, \xi_n$ in $\mathbb{R}^n$ such that the first $k$ of the corresponding $n$ vector fields $X_{H_1}, \dots, X_{H_k}$ are linearly independent at the point $x_0$, and thus at every other point of the orbit $L$. Consider the auxiliary action $F$ of the group $\mathbb{R}^k$ generated by the commuting vector fields $X_{H_1}, \dots, X_{H_k}$. These fields are independent in a neighborhood of the point $x_0$ and the orbits of the original action $\Phi$ are foliated into the orbits of the action $F$. Therefore, there exists a neighborhood $U$ of the point $x_0$ which is foliated into local $k$-dimensional orbits ($k$-dimensional disks) of the action $F$. On the other hand, the neighborhood $U$ is also foliated into the $(2n-k)$-dimensional smooth disks defined by the equations $H_1 = c_1, \dots, H_k = c_k$. We will denote one of such disks by $M_c$, $c = (c_1, \dots, c_k)$. It is an integral manifold for the fields $X_{H_i}$, $i = 1, \dots, k$, hence is itself foliated into orbits of the action $F$. Let us denote by $N_c$ the quotient manifold obtained from $M_c$ and such that its points are orbits of the action $F$. Let $\pi : M_c \to N_c$ be the projection. Obviously, $N_c$ is a $C^\infty$-smooth $2(n-k)$-dimensional disk. Let us define a symplectic structure on $N_c$. Let $a \in N_c$, $\xi, \eta \in T_aN_c$, $\xi', \eta' \in T_xM_c$, where $\pi(x) = a$. With the point $x$ of the orbit $\mathcal{O}_x$ of the action $F$ fixed, the vectors $\xi', \eta'$ are defined not uniquely but up to some summands from the tangent space to the orbit $\mathcal{O}_x$ at the point $x$. The following statement holds [**3**].

THEOREM. *The value of a form $\Omega$ on a pair of vectors $\xi', \eta'$ does not depend on the choice of vectors $\xi', \eta'$, tangent to $M_c$ and such that $D\pi(\xi') = \xi, D\pi(\eta') = \eta$ and on the choice of vectors and a point on the orbit $\pi^{-1}(a)$. Thus, there is a* *well-defined symplectic structure $\Omega_c$ on $N_c$, and the pair $(N_c, \Omega_c)$ depends smoothly on $c$.*

PROOF. Let $\xi'', \eta'' \in T_x M_c$ and $D\pi(\xi'') = \xi, D\pi(\eta'') = \eta$. Then $D\pi(\xi'' - \xi') = 0$ and $D\pi(\eta'' - \eta') = 0$, that is, $r_1 = \xi'' - \xi', r_2 = \eta'' - \eta' \in \ker D\pi$. By the construction, the kernel $\ker D\pi$ at the point $x \in M_c$ coincides with the tangent space to the orbit $\mathcal{O}_x$, i.e., $r_1, r_2 \in T_x\mathcal{O}_x$. Therefore, $\Omega(\xi'', \eta'') = \Omega(\xi' + r_1, \eta' + r_2) = \Omega(\xi', \eta') + \Omega(\xi', r_2) + \Omega(r_1, \eta') + \Omega(r_1, r_2)$. The last three summands in this sum are equal to zero due to the following lemma (see [**21**]).

LEMMA. *Let $H_1, \dots, H_k$ be an involutive set of functions on a symplectic manifold $M$, $\dim M = 2n$, $k \le n$. Let $dH_1, \dots, dH_k$ be independent at a point $x \in M$. Let $\mathcal{O}_x$ be the orbit of the action $F$ generated by the functions $H_1, \dots, H_k$. Finally, let $N \subset M$ be the level set submanifold $H_1 = H_1(x), \dots, H_k = H_k(x)$. Then the tangent spaces $T_x\mathcal{O}_x$ and $T_x N$ are skew-orthogonal complements of each other in $T_x M$.*

Thus, $\Omega(\xi'', \eta'') = \Omega(\xi', \eta')$. The indepedence of the choice of the point on the orbit $\mathcal{O}_x$ follows from two facts. In the first place, $D\pi \circ DF_x(t; x)$ does not depend on $t$, $t \in \mathbb{R}^k$, since $\pi \circ F(t; x)$ does not depend on $t$. Therefore, if $D\pi(\xi') = \xi$, then $D\pi \circ DF_x(t; x)(\xi') = \xi$. In the second place, the fact that the action $F$ is symplectic implies that $\Omega(DF_x(t; x)\xi', DF_x(t; x)\eta') = \Omega(\xi', \eta')$. Thus, there is a well-defined closed 2-form $\Omega_c$ on $N_c$. It is easy to see from the definition of $\Omega_c$ that it is nondegenerate, thus it defines a symplectic structure on $N_c$. The smooth dependence of the pair $(N_c, \Omega_c)$ on $c$ follows from the smoothness and indepedence of functions $H_1, \dots, H_k$ in a neighborhood of the point $x$. The theorem is proved. □

Thus, we obtained a $2(n-k)$-dimensional reduced symplectic manifold $N_c$. On $N_c$ the action $R_c$ of the group $\mathbb{R}^{n-k}$ generated by the fields $X_{H_{k-1}}, \dots, X_{H_n}$ is defined. The action $R_c$ is well defined on $N_c$ since the fields $X_{H_j}$, $j = k+1, \dots, n$, commute with the action $F$. Notice that the point $\pi(L) \in N_c$ is a singular point of the action $R_c$. For future purposes it is useful to illustrate the reduction scheme in coordinates.

According to [**30**], there are symplectic coordinates $(p_1, \dots, p_n, q_1, \dots, q_n)$ in the neighborhood of the point $x_0 \in M$ such that $H_1 = p_1, \dots, H_k = p_k$ in these coordinates. Then the fact that the set $H_1, \dots, H_n$ is involutive implies that the remaining functions $H_{k+1}, \dots, H_n$ have the form

$$H_{k+j}(p_1, \dots, p_n, q_1, \dots, q_n) = h_j(p_1, \dots, p_k, p_{k+1}, \dots, p_n, q_{k+1}, \dots, q_n),$$

i.e., do not depend on $q_1, \dots, q_k$. Near the point $x_0$ the orbit $L$ is given by $L = \{x \mid x = \varphi_1^{t_1} \circ \dots \circ \varphi_k^{t_k}(x_0)\}$, where $\varphi_i^{t_i}$ are the flows with Hamilton functions $H_i$. In coordinates, the orbit $L$ is written in the form $p_i = p_i^0$, $i = 1, \dots, n$, $q_1 = q_1^0 + t_1, \dots, q_k = q_k^0 + t_k, q_{k+j} = q_{k+j}^0$, $j = 1, \dots, n-k$, where $x_0 = (p_1^0, \dots, p_n^0, q_1^0, \dots, q_n^0)$. In these coordinates the $2(n-k)$-dimensional submanifold $N_c$ is defined by the equations $H_1 \equiv p_1 = c_1, \dots, H_k \equiv p_k = c_k$, $q_1 = q_1^0, \dots, q_k = q_k^0$. The reduced form $\Omega_c$ is given by $\Omega_c = dp_{k+1} \wedge dq_{k+1} + \dots + dp_n \wedge dq_n$. We obtain a $2(n-k)$-dimensional symplectic submanifold with symplectic coordinates $(\hat{p}, \hat{q}) = (p_{k+1}, \dots, p_n, q_{k+1}, \dots, q_n)$. The reduced action $R_c$ of the group $\mathbb{R}^{n-k}$ is determined by the Hamilton functions $h_j$, $j = 1, \dots, n-k$, depending on the parameters $p' = (p_1, \dots, p_k)$. In the coordinates $(p_1, \dots, p_n, q_1, \dots, q_n)$, we have $dh_j = \sum_{s=1}^{n} \alpha_{js} dp_s$,

$j = k+1, \dots, n$, since $dH_{k+1}, \dots, dH_n$ are linear combinations of $dH_1, \dots, dH_k$ at the points of $L$. Therefore,

$$\frac{\partial h_j}{\partial p_{k+m}} = 0, \ \frac{\partial h_j}{\partial q_{k+m}} = 0, \ m = 1, \dots, n-k, \tag{1.8}$$

i.e., $p_{k+1} = p^0_{k+1}, \dots, p_n = p^0_n$, $q_{k+1} = q^0_{k+1}, \dots, q_n = q^0_n$ when $p_1 = p^0_1, \dots, p_k = p^0_k$. Therefore, under the reduction the image of the orbit $L$ is a singular point of the reduced action $R_c$. The reduction scheme that we have described is a simplified local version of the general scheme [**3**] for a symplectic action of a Lie group on a manifold.

The Hamilton functions $h_j$ depend on $k$ parameters $p'$. Thus, it is natural to expect that a singular point of the action, which exists for $p_i = p^0_i$, $i = 1, \dots, k$, is preserved if the parameters $p'$ vary. To clarify this question, let us consider the system of equations (1.8) in order to find the singular point. For a fixed $j$ it is a system of $2(n-k)$ equations with $2n-k$ unknowns $(p_1, \dots, p_k, p_{k+1}, \dots, p_n, q_{k+1}, \dots, q_n)$. For every $j$ let us write down the $(2n-2k) \times (2n-k)$ matrix $Q_j$ of the second derivatives of the functions $h_j$, evaluated at the point $x_0$,

$$Q_j = \begin{pmatrix} \frac{\partial}{\partial \hat{p}}\left(\frac{\partial h_j}{\partial p'}\right) & \frac{\partial}{\partial \hat{p}}\left(\frac{\partial h_j}{\partial \hat{p}}\right) & \frac{\partial}{\partial \hat{p}}\left(\frac{\partial h_j}{\partial \hat{q}}\right) \\ \frac{\partial}{\partial \hat{q}}\left(\frac{\partial h_j}{\partial q'}\right) & \frac{\partial}{\partial \hat{q}}\left(\frac{\partial h_j}{\partial \hat{p}}\right) & \frac{\partial}{\partial \hat{q}}\left(\frac{\partial h_j}{\partial \hat{q}}\right) \end{pmatrix}.$$

We denote by $N$ a smooth $(2n-k)$-dimensional disk $q_1 = q^0_1, \dots, q_k = q^0_k$ transversal to the orbit $L$.

PROPOSITION 1.4.1. *If for some $j \in 1, \dots, k$, $\operatorname{rank} Q_j = 2n-2k$, then there is a smooth $k$-dimensional submanifold $S$ in $N$, such that its intersection with $N_{p'}$ consists of singular points of the reduced action $R_{p'}$. Moreover, if the minor $\Gamma_0$ of $Q_j$ corresponding to the variables $(\hat{p}, \hat{q})$ is not equal to zero, then $S$ intersects $N_{p'}$ transversally at a single point for sufficiently small $\|p' - p'_0\|$. If $\Gamma_0 = 0$, but another minor of order $2n-2k$ is not equal to zero, then $S$ and $N_{p_0}$ are tangent at the point $x_0$.*

PROOF. Suppose that a minor corresponding to the derivatives with respect to the variables $(p_{i_1}, \dots, p_{i_l}, p_{k+s_1}, \dots, p_{k+s_m}, q_{k+r_1}, \dots, q_{k+r_t})$ (distinguished variables), $0 \le l \le k$, $0 \le m, t \le n-k$, $q+m+t = 2n-2k$, $1 \le i_\nu \le k$, $1 \le s_\nu \le n-k$, $1 \le r_\nu \le n-k$, does not vanish. By the implicit function theorem there exists a smooth solution of the system (1.8), explicitly solved for the distinguished variables. All points in $N$ determined by this solution form a $k$-dimensional submanifold $S$. Now we will show that every point in $S$ is singular with respect to the variables $(\hat{p}, \hat{q})$ for every Hamilton function $h_r(p', \hat{p}, \hat{q})$, $r \in 1, \dots, n-k$. In order to do this let us differentiate the involution condition $\{h_j, h_r\} \equiv 0$ with respect to the distinguished variables, and then substitute the obtained solution of the system (1.8). We obtain a system of homogeneous linear equations for $\frac{\partial h_r}{\partial \hat{p}}$ and $\frac{\partial h_r}{\partial \hat{q}}$. The determinant of this system is a nonzero minor of the matrix $Q_j$. Thus, it follows that the point $(\hat{p}, \hat{q})$ is singular for the reduced system with the Hamilton function $h_r$. Therefore, the point $(\hat{p}, \hat{q})$ is also singular for the reduced action of the group $\mathbb{R}^{n-k}$.

Now let us assume that the distinguished variables are $\hat{p}$ and $\hat{q}$, i.e., that $\Gamma_0 \ne 0$. Here $l = 0$, $m = t = n-k$. Then the solution of the system (1.8) has the form $\hat{p} = \varphi(p')$ and $\hat{q} = \psi(p')$. It implies that in $N$ the submanifold $S$ intersects every $N_{p'}$ transversally at a single point for $p'$ sufficiently close to $p_0$.

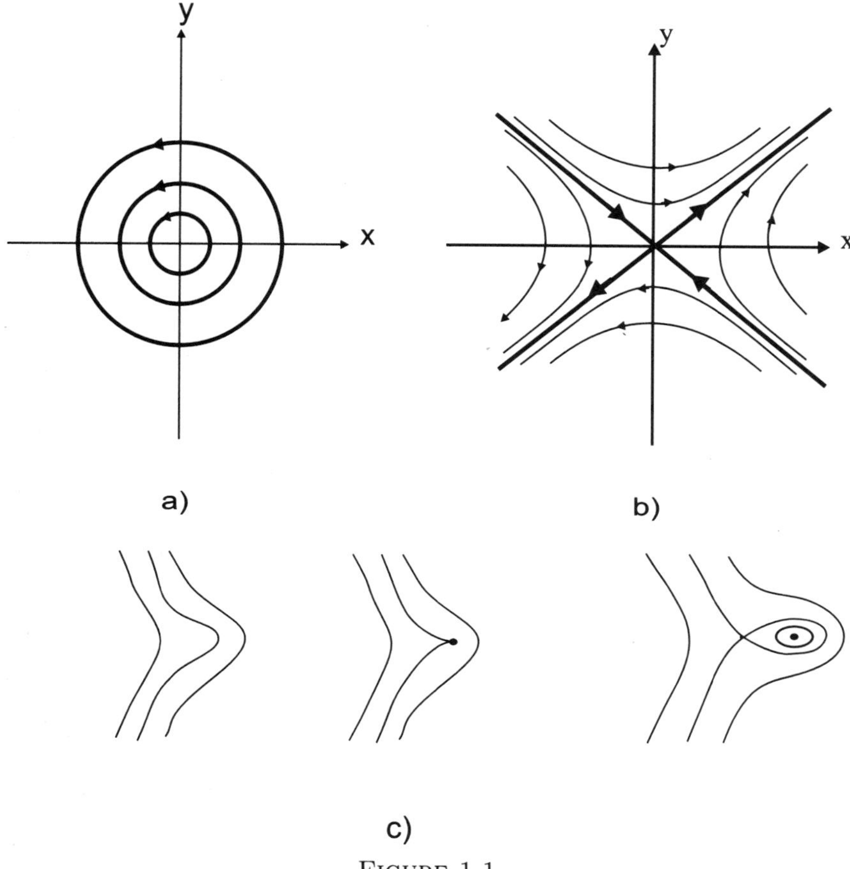


FIGURE 1.1

Now, let $\Gamma_0 = 0$, and suppose that another determinant of order $2(n-k)$ does not vanish. Without loss of generality we may assume that the free variables the solution depends on are $p^{(1)} = (p_1, \dots, p_l)$, $l \le k$, and $x^{(1)} = (x_{l+1}, \dots, x_k)$, where $x_{l+1}, \dots, x_k$ are certain $l-k$ variables from $(\hat{p}, \hat{q})$. The remaining variables are the distinguished variables. Let us denote them by $p^{(2)}$ and $x^{(2)}$. The equation for $S$ has the form

$$(1.9) \qquad p^{(1)} = p^{(1)},\ p^{(2)} = f(p^{(1)}, x^{(1)}),\ x^{(1)} = x^{(1)},\ x^{(2)} = g(p^{(1)}, x^{(1)}).$$

Let us show that a linear combination of $k-l$ vectors, obtained by differentiating the relations (1.9) with respect to $x^{(1)}$, belongs to the tangent space of the level set $N_{p_0}$. This condition means that 1-forms $dp_1, \dots, dp_l, dp_{l+1}, \dots, dp_k$ are equal to zero on these tangent vectors. It follows from (1.9) that the first $l$ forms are equal to zero on these vectors. Thus, a nontrivial linear combination of these vectors exists if the matrix $\left(\frac{\partial f}{\partial x^{(1)}}\right)$ is degenerate. Let us evaluate its determinant. In order to do this we recall that the functions (1.9) form a solution of the system (1.8). Let us substitute (1.9) into (1.8), differentiate the obtained identities with respect to $x^{(1)}$ and substitute the point $m$. We obtain a system of $2(n-k)$ nonhomogeneous linear equations for $\frac{\partial f}{\partial x^{(1)}}$ and $\frac{\partial g}{\partial x^{(1)}}$ with a nonzero determinant, coinciding with

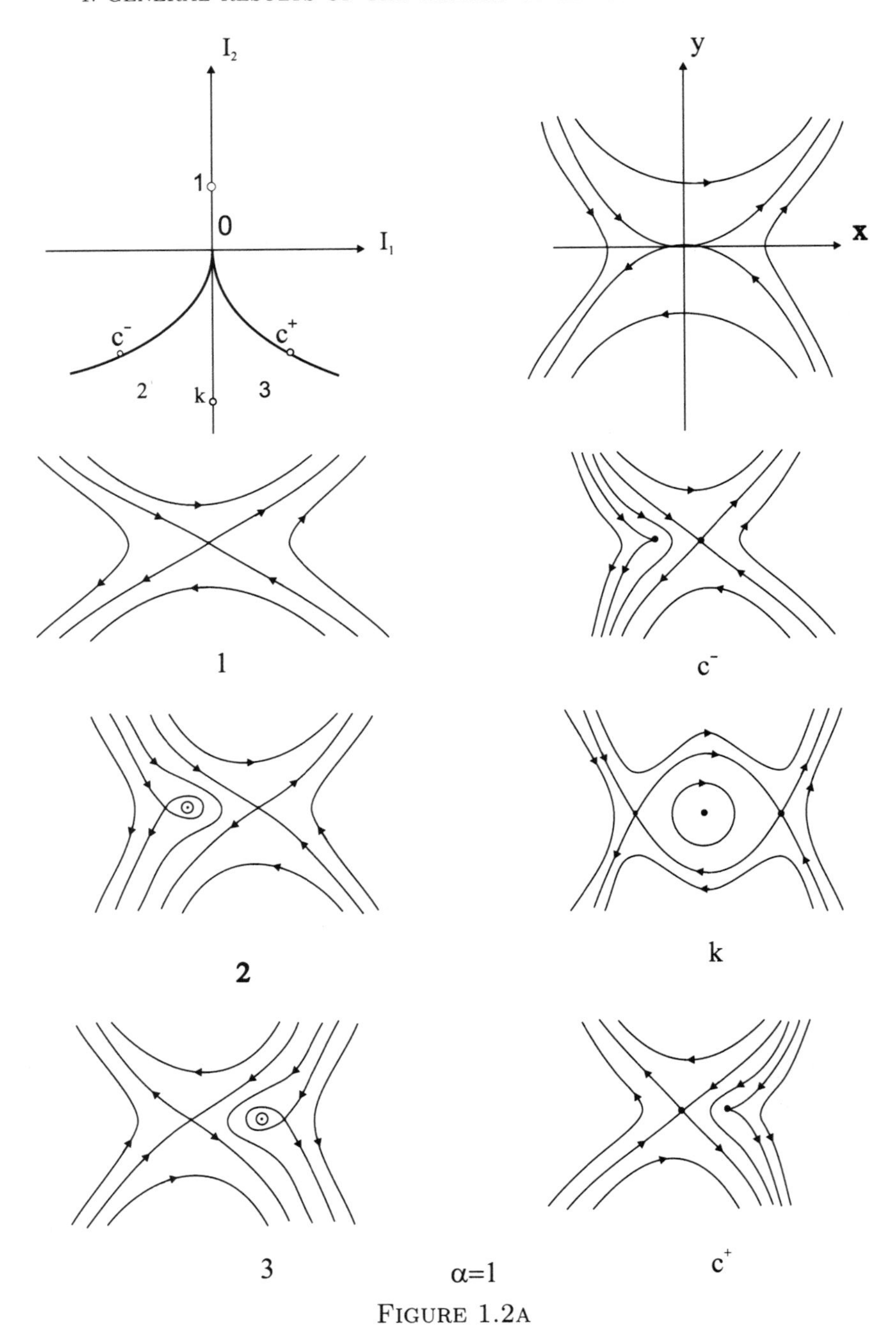


FIGURE 1.2A

the nonzero minor of the matrix $Q_j$. Solving it, we obtain

$$\det\left(\frac{\partial f}{\partial x^{(1)}}\right) = \frac{\Gamma_0}{\gamma_1} = 0.$$

The proposition is proved. □

Let us consider some examples of the application of the reduction.

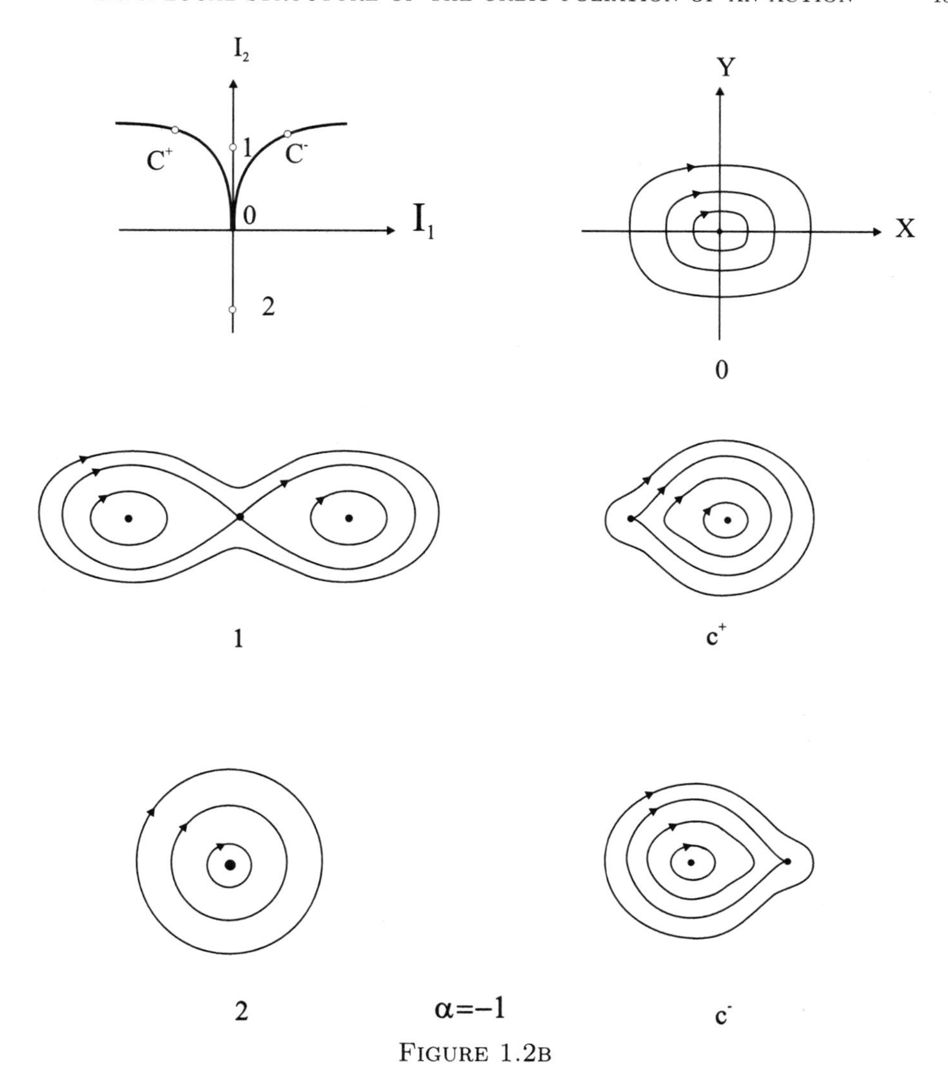


FIGURE 1.2B

*Example* 1. Let $n = 2$, $k = 1$. Then under the conditions of the proposition the reduction leads to a one-parameter family of two-dimensional Hamilton functions $h(I, x, y)$. Moreover, we may assume that for the base point $m$ we have $I = x = y = 0$, $h_x = h_y = 0$. It is natural to consider the case of a one-parameter family in general position. It is known from singularity theory [**31**] that there exist three versal families:

$$I + x^2 + y^2, \quad I + x^2 - y^2, \quad Ix - y^2 + x^3.$$

The first two cases correspond to the condition $\Gamma_0 \neq 0$. The structure of the Hamiltonian system does not change when $I$ changes in the neighborhood of $I = 0$. The third case corresponds to the condition $\Gamma_0 = 0$, $\Gamma_1 \neq 0$. The corresponding surface $S$ in $N$ is a smooth curve, which has the second order tangency with the plane $I = 0$. The curve itself is located on one side of the plane $I = 0$. The phase portraits of the system, for all three cases, are depicted in Figure 1.1.

*Example* 2. Let $n = 3$, $k = 2$. Then under the conditions of the proposition the reduction leads to the two-parameter family of two-dimensional Hamilton functions $h(I_1, I_2, x, y)$. In the four-dimensional cross-section $N$ the two-dimensional surface $S$ and the two-dimensional plane $N_m$ may either intersect transversally ($\Gamma_0 \neq 0$) or have points of contact of different degrees ($\Gamma_0 = 0$). In the case $\Gamma_0 \neq 0$ there are two types of families in general position, $(I_1 - x)^2 \pm (I_2 - y)^2$. If $\Gamma_0 = 0$, the families in general position have the form [**12**]: $I_1 x - x^3 - (I_2 - y)^2$ and $\pm x^4 + I_1 x^2 + I_2 x + 0.5y^2$. The phase portrait of the families in the case when $\Gamma_0 \neq 0$ is the same as in Figure 1.1. When $\Gamma_0 = 0$, the first family has the phase portrait as in Figure 1.1 c), and the parameter $I_2$ is insignificant. The phase portraits of the latter families for $\Gamma_0 = 0$ are depicted in Figure 1.2a for the minus sign and Figure 1.2b to the plus sign (see [**32**]).

## 1.5. Formulation of the classification problem for actions and IHVFs

When we study actions, a natural problem of "comparison" of two actions arises, i.e., we have to agree which actions we will consider identical and which ones different. This is a general question in dynamical systems theory [**33, 34, 35, 36**]. A similar question arises for two IHVFs.

Let $\Phi_1$ and $\Phi_2$ be two Poisson actions on smooth symplectic manifolds $M_1$ and $M_2$, respectively.

DEFINITION 1.6. The action $\Phi_1$ is said to be topologically equivalent to the action $\Phi_2$ if there exists a homeomorphism $h : M_1 \to M_2$ that transforms orbits of the action $\Phi_1$ into orbits of the action $\Phi_2$.

It is easy to see that this definition is a standard definition of the topological equivalence used in the theory of dynamical systems [**35**].

A similar definition may be given for the restrictions of the actions $\Phi_1$ and $\Phi_2$ to subsets $N_1 \subset M_1$ and $N_2 \subset M_2$ invariant with respect to these actions, i.e., the subsets consisting of whole orbits.

Definition 1.6 is insufficient for studying neighborhoods of singular orbits since in many cases the orbits leave any local neighborhood. That is, there is no neighborhood consisting of the whole orbits. Thus, it is necessary to generalize the notion of an invariant set.

A subset $G \subset M$ is called regular with respect to the action (or just regular) if $G$ is a closed $2n$-dimensional topological manifold such that its interior is a smooth open submanifold and its boundary $\partial G$ is a piecewise smooth $(2n-1)$-dimensional submanifold consisting of two piecewise smooth manifolds $R^{\perp}$ and $R^{\|}$, $R^{\perp}$ being transversal to the orbits of the action and $R^{\|}$ consisting of pieces of orbits.

The definition of the equivalence of actions on such sets must be modified accordingly.

DEFINITION 1.7. The action $\Phi_1$ on a regular set $G_1 \subset M_1$ is topologically equivalent to the action $\Phi_2$ on a regular set $G_2 \subset M_2$ if there exists a homeomorphism $h : G_1 \to G_2$ that maps the connected component of the intersection of the orbit of the action $\Phi_1$ with $G_1$ to a connected component of the intersection of the orbit of the action $\Phi_2$ with $G_2$, and maps the sets $R_1^{\|}$ and $R_1^{\perp}$ to the sets $R_2^{\|}$ and $R_2^{\perp}$.

A similar question about the equivalence arises for IHVFs as well. It would seem possible to use the standard definition of the topological equivalence of two IHVFs with respect to some homeomorphism transforming trajectories of one field into trajectories of the other. However, there is a certain problem here. Namely, it would be desirable if the equivalence relation produced a countable set of nonequivalent systems that could be distinguished by some invariants of combinatorial nature. However, it is known that the topological invariant for the flows on the Liouville tori is the Poincaire rotation numbers, which change continuously from torus to torus. Thus, the usual topological equivalence of vector fields leads to functional invariants (moduli) and does not allow us to get a countable number of equivalence classes (nevertheless, see [**37, 38**]). On the other hand, such a fine classification does not seem to be essential for the description of the topological structure of IHVFs. Therefore, we will adopt a weaker equivalence relation. Let $M_1$, $M_2$ be two smooth symplectic manifolds and $X_1$, $X_2$ two IHVFs on $M_1$, $M_2$ with Hamilton functions $H_1$, $H_2$, respectively. Also, let $\Phi_1$, $\Phi_2$ be the corresponding induced actions.

DEFINITION 1.8. The IHVF $X_1$ is isoenergetically equivalent to the IHVF $X_2$ if there exists a homeomorphism $h : M_1 \to M_2$ defining the equivalence of the actions $\Phi_1$ and $\Phi_2$ and, additionally, mapping every connected component of a level set of the function $H_1$ into a connected component of a level set of the function $H_2$.

An analogous definition may be given for the restrictions of the fields $X_1$, $X_2$ to sets $G_1$, $G_2$ regular with respect to the actions $\Phi_1$, $\Phi_2$.

The main goal of this book is to give a description of the structure of IHVFs and Poisson actions on a smooth four-dimensional symplectic manifold $M^4$ or on a subset $G$ regular with respect to the equivalence relations defined above.

CHAPTER 2

# Linear Theory and Classification of Singular Orbits

## 2.1. Linear Poisson actions

Let $\Phi : \mathbb{R}^n \times M \to M$ be a Poisson action of the group $\mathbb{R}^n$ on a $2n$-dimensional symplectic manifold $M$, and $p$ a singular point of the action $\Phi$, i.e., $\Phi(t,p) \equiv p$ for all $t \in \mathbb{R}^n$. The differentials $D\Phi^t : T_pM \to T_pM$ of the diffeomorphisms $\Phi^t \equiv \Phi(t,\cdot)$ at the point $p$ form a commutative subgroup $\Psi(t)$ of the Lie group $Sp(2n,\mathbb{R})$ of linear symplectic transformations of the linear symplectic space $V = (T_pM, \Omega_p)$. The family of transformations $\Psi(t)$ defines a linear Poisson action of the group $\mathbb{R}^n$ on the linear symplectic space $V$, which is called the linearization of the action $\Phi$ at the singular point $p$. Let $L$ be the subalgebra $L$ of the Lie algebra $sp(V)$, corresponding to the subgroup $\Psi(t)$. Then $L$ is commutative.

The above action $\Psi$ is a particular case of a general notion of a linear Poisson action (LPA) on a linear symplectic space. Namely, let $V$ be some $2n$-dimensional linear symplectic space and $[\cdot,\cdot]$ the skew-symmetric inner product in $V$. Let $\Psi : \mathbb{R}^n \times V \to V$ be a continuous action of the group $\mathbb{R}^n$ on $V$ such that for any fixed $t \in \mathbb{R}^n$ the transformation $\Psi^t = \Psi(t,\cdot)$ is a linear symplectic transformation of the space $V$. It is clear that the set of all transformations $\{\Psi(t) \mid t \in \mathbb{R}^n\}$ is a subgroup in the group $Sp_{2n}(V)$ of all symplectic transformations. An action of this type is called an LPA. To justify the terminology we should show that for any one-parameter subgroup $\Psi^{st}$, $s \in \mathbb{R}$, with $t \in \mathbb{R}^n$ fixed, there exists a uniquely defined Hamilton function $H_t$ which depends linearly on $t$. In order to do it we recall the method of constructing a linear Hamiltonian vector field on $V$. The skew-symmetric inner product defines the standard isomorphism $I : V \to V^*$ between the linear space $V$ and its dual (the space of linear functionals), $I(x) = [\cdot, x]$, $x \in V$. Let $B(x,y)$ be a symmetric bilinear form on $V$, and let $h(x) = B(x,x)/2$ be the corresponding quadratic form. Then for a fixed $x \in V$ the differential $dh(x) = B(x,\cdot)$ is a linear functional on $V$. Moreover, the family of these functionals depends linearly on $x$. The isomorphism $I$ generates a linear operator $x \to v(x)$, $v(x) = I^{-1}B(x,\cdot)$. Using the standard identification of $TV$ and $V \times V$ for a linear space $V$, we obtain a linear vector field on $V$, $\dot{x} = v(x)$. In other words, we have the formula

$$B(x,\xi) \equiv [\xi, v(x)] \quad \text{for all } \xi \in V.$$

Let us return to the action $\Psi$. Denote by $C : V \to V$ the infinitesimal operator of the one-parameter subgroup $\Psi^{st} : V \to V$, i.e., $Cx = \frac{\partial}{\partial s}\Psi^{st}|_{s=0}$. Thus, the equation $\dot{x} = CX$ determines the subgroup $\Psi^{st}$. Since the transformations $\Psi^t$ are symplectic, i.e. $[\Psi^t x, \Psi^t y] = [x,y]$ for any $t \in \mathbb{R}^n$, the operator $C$ satisfies the formula

$$[Cx, y] + [x, Cy] = 0. \tag{2.1}$$

It is clear that the operators $C_t$ depend linearly on $t \in \mathbb{R}^n$ and form a commutative Lie algebra of the subgroup $\{\Psi^t \mid t \in \mathbb{R}^n\}$. Let us introduce a bilinear functional $[x, Cy]$. This functional is symmetric due to the formula for $C$ and, therefore, $h(x) = [x, Cx]/2$ is a quadratic form. We have $dh(x) = [\cdot, Cx] = [\cdot, v(x)]$, where $v(x) = I^{-1}dh(x)$. Thus, the function $[x, Cx]/2$ is the Hamilton function of the one-parameter subgroup $\Psi^{st}$. By definition, the operators $C$ depend linearly on $t$, i.e., the Hamilton functions $h$ also depend linearly on $t$.

Let us formulate the corresponding statements in the coordinate form. In order to do it we choose a symplectic basis in $V$ [**21**]. In the corresponding symplectic coordinates $(p, q) = (p_1, \dots, p_n, q_1, \dots, q_n)$ the skew-symmetric inner product can be written as $[x, y] = (Ix, y)$. Here, $I$ is a $(2n \times 2n)$-matrix of the form

$$\begin{pmatrix} 0 & -E_n \\ E_n & 0 \end{pmatrix},$$

where $E_n$ is the identity matrix of order $n$, and $(\cdot, \cdot)$ is the Euclidean inner product:

$$(x, x') = \sum_i (p_i p_i' + q_i q_i'),$$

$x = (p, q)$, $x' = (p', q')$. The linear operators have matrix representations in coordinates. The condition $[Sx, Sy] = [x, y]$ for the operator $S$ to be symplectic can be written in the matrix form as $S^T I S = I$. By (2.1), the matrix $C$ of the infinitesimal operator $C$ satisfies the relation $IC = -C^T I$, i.e., since $I^T = I^{-1} = -I$, $IC$ is a symmetric matrix. In the standard notation of the theory of Lie groups and Lie algebras the matrices $\Psi^t$ belong to the group $Sp(2n, \mathbb{R})$ and the matrices $C$ belong to its Lie algebra $sp(2n, \mathbb{R})$. Let us denote by $A$ the symmetric matrix $-IC$. Then there is a quadratic Hamilton function $h(x) = [x, Cx]/2 = (Ix, Cx)/2 = (x, -ICx)/2 = (Ax, x)/2$. Here we use the standard notation $\dot{x} = IAx$, $Ax = \nabla h(x)$ with $\nabla$ being the gradient operator.

We return to the linear Poisson action $\Psi$ of the group $\mathbb{R}^n$. Let $\mathcal{A}$ be a commutative Lie subalgebra of operators $C$ corresponding to the commutative subgroup of symplectic transformations $\Psi^t : V \to V$, $t \in \mathbb{R}^n$. We only consider nondegenerate LPA which cannot be reduced to an action of a group of lower dimension. Therefore, $\mathcal{A}$ is an $n$-dimensional subalgebra.

DEFINITION 2.1. A linear Poisson action $\Psi$ is said to be simple if $\mathcal{A}$ is an $n$-dimensional subgroup containing an operator with simple eigenvalues.

In the symplectic basis the algebra $\mathcal{A}$ is represented by a commutative algebra of matrices of the form $IA$, $A^T = A$. The fact that $\mathcal{A}$ is $n$-dimensional implies the existence of $n$ linear independent over the field $\mathbb{R}$ matrices $IA_1, IA_2, \dots, IA_n$ such that $\mathcal{A} = \{\alpha_1 IA_1 + \dots + \alpha_n IA_n\}$, $(\alpha_1, \dots, \alpha_n) \in \mathbb{R}^n$. In particular, a basis of this type can be obtained by choosing a basis $t_1, \dots, t_n$ in the group $\mathbb{R}^n$ and taking the matrices of infinitesimal operators $\Psi^{st_1}, \dots, \Psi^{st_n}$ of $IA_1, \dots, IA_n$. How does one determine whether a given $n$-dimensional commutative subalgebra $\mathcal{A}$ contains an operator with simple eigenvalues? We may do the following: for any matrix $B = \alpha_1 IA_1 + \dots + \alpha_n IA_n$ we form its characteristic equation $\det(B - \lambda E) = 0$. It is a polynomial $\chi(\lambda)$ of degree $2n$, whose coefficients are polynomials in $\alpha_1, \dots, \alpha_n$. The discriminant $\Delta$ of this equation (i.e., the resultant of the polynomials $\chi(\lambda)$

and $\chi'(\lambda)$) is a polynomial in $(\alpha_1, \ldots, \alpha_n)$. If $\Delta(\alpha_1, \ldots, \alpha_n)$ is identically equal to zero, then every matrix in the algebra $\mathcal{A}$ has multiple eigenvalues (i.e., the LPA $\Psi$ is not simple). But if $\Delta(\alpha_1, \ldots, \alpha_n)$ is not identically zero, then there exists a set $(\alpha_1^0, \ldots, \alpha_n^0)$ of real numbers such that $\Delta \neq 0$. Indeed, if the polynomial $\Delta(\alpha_1, \ldots, \alpha_n)$ with real coefficients is zero for every set $(\alpha_1, \ldots, \alpha_n)$ of real numbers, then $\Delta \equiv 0$. Thus, all eigenvalues of the matrix $B$ are different for the set $(\alpha_1^0, \ldots, \alpha_n^0)$, i.e. the LPA $\Psi$ is simple.

In particular, if $n = 2$, the characteristic polynomial $\chi(\lambda)$ has a form $\chi(\lambda) = \lambda^4 + a\lambda^2 + b = 0$, where $b = \det(\alpha_1 A_1 + \alpha_2 A_2)$ is a homogeneous polynomial of degree 4, $a$ a homogeneous polynomial of degree 2, and $\chi'(\lambda) = 4\lambda^3 + 2a\lambda$. Thus, $\Delta = b(4b - a^2)/4$, i.e., the condition $\Delta \equiv 0$ is equivalent to the condition of $b$ or $a^2 - 4b$ being identically zero. Both $b$ and $a^2 - 4b$ are homogeneous polynomials of degree 4 in $(\alpha_1, \alpha_2)$. Let us note that $b = \lambda_1^2\lambda_2^2$, $4b - a^2 = -(\lambda_1^2 - \lambda_2^2)^2$, where $\lambda_1, \lambda_2, -\lambda_1, -\lambda_2$ are roots of the equation $\chi(\lambda) = 0$. In particular, it shows that the condition $b = 0$ implies that all operators in the algebra $\mathcal{A}$ have a pair of zero eigenvalues. If $b \neq 0$, $4b - a^2 = 0$, then all operators in $\mathcal{A}$ have multiple nonzero eigenvalues.

The classification of simple LPAs may be obtained from Definition 2.1. We denote by $C^d \in \mathcal{A}$ an operator with simple eigenvalues. Let us introduce a symplectic basis and the corresponding symplectic coordinates $(p_1, \ldots, p_n, q_1, \ldots, p_n)$ in $V$. Then the operator $C^d$ is represented by a matrix $IA^d$ and the corresponding Hamilton function has the form $H^d = (A^d z, z)/2$, $z \in V$. The matrix $IA^d$ has only simple eigenvalues and then, by the Williamson theorem [**40**], there exists a symplectic basis in $V$ and the corresponding coordinates $(x_1, \ldots, x_n, y_1, \ldots, y_n)$ such that in these coordinates the Hamilton function $H^d$ has the form

$$H^d = H^{(1)}(\xi_1) + H^{(2)}(\xi_2) + H^{(3)}(\xi_3),$$

where, up to the numbering of the dual pairs $(x_i, y_i)$, we have denoted

$$\begin{aligned} \xi_1 &= (x_1, y_1, \ldots, x_{k_1}, y_{k_1}), \\ \xi_2 &= (x_{k_1+1}, y_{k_1+1}, \ldots, x_{k_1+k_2}, y_{k_1+k_2}), \\ \xi_3 &= (x_{k_1+k_2+1}, y_{k_1+k_2+1}, \ldots, x_{k_1+k_2+2k_3}, y_{k_1+k_2+2k_3}), \\ & k_1 + k_2 + k_3 = n. \end{aligned}$$

Moreover,

$$\begin{aligned} H^{(1)}(\xi_1) &= \sum_{j=1}^{k_1} (\omega_j/2)(x_j^2 + y_j^2), \\ H^{(2)}(\xi_2) &= \sum_{j=1}^{k_2} \lambda_j x_{k_1+j} y_{k_1+j}, \\ H^{(3)}(\xi_3) &= \sum_{j=k_1+k_2+1}^{n} \alpha_{i-k_1-k_2}(x_j y_j + x_{j+1} y_{j+1}) + \beta_{j-k_1-k_2}(x_j y_{j+1} - x_{j+1} y_j). \end{aligned}$$

Here $\pm i\omega_1, \ldots, \pm i\omega_{k_1}$ are purely imaginary eigenvalues, $\pm\lambda_1, \ldots, \pm\lambda_{k_2}$ are real eigenvalues, and $\pm\alpha_1 \pm i\beta_1, \ldots, \pm\alpha_{k_3} \pm i\beta_{k_3}$ are complex-conjugate eigenvalues.

It follows from this form that the matrix $IA^d$ has a quasi-diagonal form:

$$IA^d = \operatorname{diag}(C_1, \ldots, C_{k_1}, D_1, \ldots, D_{k_2}, G_1, \ldots, G_{k_3}),$$

where

$$C_i = \begin{pmatrix} 0 & -\omega_i \\ \omega_i & 0 \end{pmatrix},$$

$$D_i = \begin{pmatrix} -\lambda_i & 0 \\ 0 & \lambda_i \end{pmatrix},$$

$$G_i = \begin{pmatrix} -\alpha_i & \beta_i & 0 & 0 \\ -\beta_i & -\alpha_i & 0 & 0 \\ 0 & 0 & \alpha_i & \beta_i \\ 0 & 0 & -\beta_i & \alpha_i \end{pmatrix}.$$

Since the eigenvalues of all matrices $C_i$, $D_i$, $G_i$ are different, all matrices $IA \in \mathcal{A}$ that commute with $IA^d$ have the same block form [**39**]. Moreover, the matrices of the same form commute with every block of the matrices $C_i$, $D_i$, $G_i$. Thus, the matrices $IA_1, \ldots, IA_n$ have the same form as the matrix $IA^d$. We denote the corresponding numbers for the matrix $IA_j$ by $\omega_{ij}$, $\lambda_{ij}$, $\alpha_{ij}$, $\beta_{ij}$.

From the form of the matrices $IA_1, \ldots, IA_n$ and the fact that $\mathcal{A}$ is $n$-dimensional (i.e., the linear indepedence of $IA_1, \ldots, IA_n$) we obtain the following result.

COROLLARY 2.1.1.

$$\det \Lambda = \begin{vmatrix} \omega_{11} & \cdots & \omega_{1n} \\ \cdots & \cdots & \cdots \\ \omega_{k_1 1} & \cdots & \omega_{k_1 n} \\ \lambda_{11} & \cdots & \lambda_{1n} \\ \cdots & \cdots & \cdots \\ \lambda_{k_2 1} & \cdots & \lambda_{k_2 n} \\ \alpha_{11} & \cdots & \alpha_{1n} \\ \beta_{11} & \cdots & \beta_{1n} \\ \cdots & \cdots & \cdots \\ \alpha_{1} & \cdots & \alpha_{1n} \\ \beta_{k_3 1} & \cdots & \beta_{k_3 n} \end{vmatrix} \neq 0.$$

From Corollary 2.1.1 we obtain the following corollary.

COROLLARY 2.1.2. *Any matrix of the form $IA$, $A = A^T$, commuting with $\mathcal{A}$ belongs to $\mathcal{A}$.*

The above results about the form of the matrices from the subgroup $\mathcal{A}$ allow us to introduce the following classification of simple LPAs. An LPA has a type $(k_1, k_2, k_3)$ if the operator $IA^d$ has $2k_1$ purely imaginary, $2k_2$ real, and $4k_3$ complex eigenvalues, $k_1 + k_2 + 2k_3 = n$.

Now let us turn to the case $n = 2$. There are four possible types of simple LPAs: $(2, 0, 0)$, $(1, 1, 0)$, $(0, 2, 0)$, $(0, 0, 1)$. The corresponding types will be called: center-center, saddle-center, saddle-saddle, focus-focus. We use the following definition for the classification of LPAs

DEFINITION 2.2. Two LPAs $\mathcal{A}$ and $\mathcal{A}'$ are called linearly equivalent if there exists a linear isomorphism $h : \mathbb{R}^4 \to \mathbb{R}^4$ which maps the orbits of the action $\mathcal{A}$ into orbits of the action $\mathcal{A}'$.

Now let us describe the orbit structure of a simple LPA. Let us choose linear symplectic coordinates $(x_1, x_2, x_3, x_4)$ in which the quadratic Hamilton function $H^d$ corresponding to the matrix $IA^d$ has the Williamson normal form $H = \lambda_1 \xi_1 + \lambda_2 \xi_2$, where

1. $\xi_1 = \frac{x_1^2+y_1^2}{2}$, $\xi_2 = \frac{x_2^2+y_2^2}{2}$ (center-center);
2. $\xi_1 = \frac{x_1^2+y_1^2}{2}$, $\xi_2 = x_2y_2$ (saddle-center);
3. $\xi_1 = x_1y_1$, $\xi_2 = x_2y_2$ (saddle-saddle);
4. $\xi_1 = x_1y_1 + x_2y_2$, $\xi_2 = x_1y_2 - x_2y_1$ (focus-focus).

Let $IB$ be another matrix in $\mathcal{A}$ linearly independent from $IA^d$. Let $K$ be a corresponding quadratic Hamilton function. Then $K$ has the same form $K = \mu_1\xi_1+\mu_2\xi_2$. Moreover, $\lambda_1\mu_2 - \lambda_2\mu_1 \neq 0$ (Corollary 2.1.1).

In particular, the set of solutions of the system $H = h$, $K = k$ is equivalent to the set of solutions of the system $\xi_1 = C_1$, $\xi_2 = C_2$ (in the corresponding parameter domains). It is easy to see that the set $\xi_1 = C_1$, $\xi_2 = C_2$ is invariant with respect to the LPA $\Psi$, i.e., consists of the whole orbits of the action. Conversely, any orbit of the LPA $\Psi$ is invariant with respect to both flows corresponding to the vector fields with Hamilton functions $\xi_1$ and $\xi_2$. Thus, the functions $\xi_1$ and $\xi_2$ are constant along the orbits.

This allows us to describe the structure of the common level set $L(C_1, C_2)$ of the functions $\xi_1 = C_1$, $\xi_2 = C_2$, and, thus, the possible orbit types of the linear action in each of the Cases 1–4.

PROPOSITION 2.1.1. *The set $L(C_1, C_2)$ is described as follows:*

A. *In Case 1:*

1. *a two-dimensional torus, for $C_1 > 0$, $C_2 > 0$;*
2. *a circle, for $C_1C_2 = 0$, $C_1^2 + C_2^2 \neq 0$;*
3. *a point, for $C_1 = C_2 = 0$.*

B. *In Case 2:*

1. *a union of two cylinders, for $C_1 > 0$, $C_2 \neq 0$;*
2. *a union of two nonclosed one-dimensional orbits, for $C_1 = 0$, $C_2 \neq 0$;*
3. *a circle, for $C_1 > 0$, $C_2 = 0$;*
4. *a "cross" consisting of a singular point and four one-dimensional orbits adjacent to it, for $C_1 = C_2 = 0$.*

C. *In Case 3:*

1. *four two-dimensional orbits homeomorphic to a plane, for $C_1C_2 \neq 0$;*
2. *the disjoint union of two copies of a product of a "cross"(see B.4) and a closed curve, for $C_1C_2 = 0$, $C_1^2 + C_2^2 \neq 0$. In this case every component consists of one one-dimensional nonclosed orbit and four two-dimensional orbits (planes), adjacent to it;*
3. *a direct product of two "crosses", for $C_1 = C_2 = 0$. In this case $L(C_1, C_2)$ consists of a singular point , 8 one-dimensional nonclosed orbits adjacent to it, and 16 two-dimensional orbits (planes), such that each of them contains a singular point and two one-dimensional orbits in its closure.*

D. *In Case 4:*

1. *a two-dimensional cylinder, for $C_1^2 + C_2^2 \neq 0$;*
2. *two two-dimensional planes intersecting transversally at a singular point, for $C_1 = C_2 = 0$. $L(0,0)$ consists of two orbits: cylinder and a singular point.*

PROOF. In cases 1–3 the solutions of the system $\xi_1 = C_1$, $\xi_2 = C_2$ are direct products of the solutions of each of the equations in the corresponding planes $L_1$ : $x_2 = y_2 = 0$ and $L_2 : x_1 = y_1 = 0$. It implies the proposition for these cases.

In case 4 the system of equations $\xi_1 = C_1$, $\xi_2 = C_2$ is linear with respect to, for example, $(x_1, x_2)$. Thus, for $y_1^2 + y_2^2 \neq 0$, the system has a single solution, namely, the cylinder defined by the equations

$$x_1 = \frac{C_1 y_1 + C_2 y_2}{y_1^2 + y_2^2}, \quad x_2 = \frac{C_1 y_2 - C_2 y_1}{y_1^2 + y_2^2}, \quad y_1^2 + y_2^2 \neq 0.$$

In particular, for $C_1 = C_2 = 0$ this cylinder coincides with the plane $x_1 = x_2 = 0$, punctured at the point zero. For $y_1 = y_2 = 0$ the solution exists only for $C_1 = C_2 = 0$ and coincides with the plane $y_1 = y_2 = 0$. Thus, for $C_1 C_2 \neq 0$ the solution is just one orbit, namely the cylinder. When $C_1 = C_2 = 0$ the solution consists of three orbits of the action: two cylinders (the planes $x_1 = x_2 = 0$ and $y_1 = y_2 = 0$ punctured at zero) and a singular point $x_1 = x_2 = y_1 = y_2 = 0$. The proposition is proved. □

REMARK 2.1. It follows from this proposition that in any of these four cases the common level set $H = h$, $K = k$ consists of the disjoint union of a finite number of two-dimensional orbits, provided it is a nondegenerate level set (does not contain a singular point). Every orbit is a connected component of the level set. A degenerate level set also consists of a finite number of orbits of different dimensions. Moreover, the set of points belonging to orbits of a given dimension (namely $0, 1, 2$) consists of a finite number of connected components. Every such component is a single orbit of the given dimension.

In particular, we obtain a structure of the set $\Sigma$ of one-dimensional orbits from this description. Namely, in cases 1–3, $\Sigma$ consists of two symplectic transversally intersecting planes. These planes are obtained as products of a singular point of the plane $L_1$ and the foliation of $L_2$, and vice versa, with interchanging $L_1$ and $L_2$. One-dimensional orbit foliations on these planes coincide with the foliations into trajectories of Hamiltonian vector fields on the corresponding planes $L_1$, $L_2$. In the case of focus-focus we have $\Sigma = \varnothing$.

Now we can give a linear classification of simple LPAs.

THEOREM 2.1.1. *Two simple LPAs $\Psi$ and $\Psi'$ are linearly equivalent if and only if their singular points are of the same type.*

PROOF. First we prove the "only if" part. Let $\Psi$ and $\Psi'$ be equivalent simple LPAs and $h : \mathbb{R}^4 \to \mathbb{R}^4$ the corresponding linear equivalence. Clearly, $h$ maps the set $\Sigma$ of one-dimensional orbits of the action $\Psi$ onto the set $\Sigma'$ of one-dimensional orbits of the action $\Psi'$. It follows immediately that the action $\Psi$ with a point of type focus-focus may be equivalent only to an action $\Psi'$ with a singular point of the same type. To distinguish the other three cases we notice that $h$ preserves the type of the orbit foliation of the action on the set $\Sigma$ of one-dimensional orbits of the action. Thus, the "only if" part follows from the fact that in the case of center-center there are singular points of the center type on both $L_1$ and $L_2$, in the case of saddle-center there is a center on one of the planes and a saddle on the other, and in the case of saddle-saddle the singular points on both planes are saddles.

Now we prove the "if" part. Assume that simple LPAs $\Psi$ and $\Psi'$ are of the same type. Let $\mathcal{A}$ and $\mathcal{A}'$ be the corresponding algebras. According to **[40]** there is a symplectic basis in $\mathbb{R}^4$ in which the matrix $IA^d \in \mathcal{A}$ has the standard form, i.e., its Hamilton function $H \doteq H^d = \lambda_1 \xi_1(x) + \lambda_2 \xi_2(x)$, $x \in \mathbb{R}^4$. Due to the simplicity of the LPA all matrices from $\mathcal{A}$ have the same form. On the other hand, there is a linear symplectic coordinate change $x = Sy$, which transforms the corresponding Hamilton function $'H^d$ into the same canonical form. Thus, for all Hamilton functions corresponding to the algebra $\mathcal{A}'$ we have $'H = 'H^d(x) = H^d(Sy) = \lambda_1'\xi_1(y)+\lambda_2'\xi_2(y)$. Let us choose another matrix in $\mathcal{A}$ linearly independent with $IA^d$, and let $K$ be the corresponding Hamilton function. Analogously, $H' = 'H^d$ and $K$ are independent Hamilton functions for $\mathcal{A}'$. Let us consider a system of equations $H(x) = h_0$, $K(x) = k_0$. Let $L$ be its solution set. It was mentioned above that $L$ consists of a finite number of whole orbits (see Proposition 2.1.1). Let us show that for the linear transformation $S$ defined in coordinates $(x)$ by the formula $\overline{x} = Sx$, the set $S(L)$ is a solution of the system $H'(x) = h_0'$, $K'(x) = k_0'$, with constants $h_0'$ and $k_0'$ to be defined below. Let $a \in L$. Then $h_0' = H(a) = \lambda_1\xi_1(a) + \lambda_2\xi_2(a)$, $k_0' = K(a) = \mu_1\xi_1(a) + \mu_2\xi_2(a)$. Hence, $H'(Sa) = H_1'(a) = \lambda_1'\xi_1(a)+\lambda_2'\xi_2(a)$, $K'(Sa) = \mu_1'\xi_1(a)+\mu_2'\xi_2(a)$. Let $b$ be any solution of the system $H(x) = h_0$, $K(x) = k_0$. Since $\Delta = \lambda_1\mu_1 - \lambda_2\mu_2 \neq 0$, this system is equivalent to the system $\xi_1(x) = \xi_1^0$, $\xi_2(x) = \xi_2^0$, where $h_0 = \lambda_1\xi_1^0+\lambda_2\xi_2^0$, $k_0 = \mu_1\xi_1^0+\mu_2\xi_2^0$. Therefore, $\xi_1(a) = \xi_1(b)$, $\xi_2(a) = \xi_2(b)$. Hence, $H'(Sb) = H_1'(b) = \lambda_1'\xi_1(b) + \lambda_2'\xi_2(b) = \lambda_1'\xi_1(a) + \lambda_2'\xi_2(a) = h_0'$. Similarly $K'(Sb) = k_0'$. We can prove the converse using the transformation $S^{-1}$ as a transformation in coordinates $(y)$. Thus, $S(L)$ is a set of solutions of the system $H'(x) = h_0'$, $K'(x) = k_0'$.

In coordinates $(x)$, the functions $H'$ and $K'$ have the form $H' = (1/2)(Px, x)$, and $K' = (1/2)(Qx, x)$. Thus, the Hamiltonian vector fields with Hamilton functions $H'$ and $K'$ have the form $IPSa$ and $IQSa$ at the point $a$. They are independent or dependent simultaneously with the pair $IS^TPSa$ and $IS^TQSa$ that corresponds to the Hamilton functions $H_1'(x) = H'(Sx)$ and $K_1'(x) = K'(Sx)$ at the point $a$. Due to the representation $H'(Sx) = \lambda_1'\xi_1(x) + \lambda_2'\xi_2(x)$ and $K'(Sx) = \mu_1'\xi_1(x)+\mu_2'\xi_2(x)$, the last two vectors are linear combinations of the pair of vectors $IAa$ and $IBa$ corresponding to the pair of Hamilton functions $H(x)$ and $K(x)$ at the point $a$. Therefore, an orbit of $\mathcal{A}$ passing through the point $a$ is mapped to a point $Sa$ which belongs to an orbit of the same dimension. Finally, since each orbit is a connected component of the set of points belonging to the orbits of a given dimension on the set $H = h_0$, $K = k_0$ (see Remark 2.1), the transformation $S$ maps orbits into orbits and defines the required mapping. The theorem is proved. □

The important feature of simple LPAs is their genericness. Namely, we will show that simple actions form an everywhere dense open set in an appropriately defined space of linear Poisson actions. First, let us define the topology on the space of LPAs. In order to do this we choose a symplectic basis in $V$. Then $V$ is canonically isomorphic to the standard linear symplectic space $\mathbb{R}^4$ with the skew-symmetric inner product $[x, y] = (Ix, y)$. In this case LPAs are identified with two-dimensional commutative subalgebras of the Lie algebra $sp(4, \mathbb{R})$ of Hamiltonian matrices of the form $IA$, $A^T = A$. The topology on $sp(4, \mathbb{R})$ is induced from the linear space of $(4 \times 4)$ matrices. Let $G_2$ be a two-dimensional Grassmannian over $sp(4, \mathbb{R})$ with the usual topology **[41]**. That is, $G_2$ is a manifold whose points are two-dimensional linear subspaces in $sp(4, \mathbb{R})$. There is a subset $CG_2$ in $G_2$ formed by two-dimensional commutative subalgebras. Let us choose a topology induced from $G_2$ in $CG_2$. The topological space $CG_2$ obtained is the space of LPAs. Now we will identify LPAs with points of $CG_2$.

THEOREM 2.1.2. *Simple LPAs form an open dense set in $CG_2$.*

PROOF. Let $\mathcal{A}$ be a simple LPA. Let $U$ be its neighborhood in $CG_2$ and $\mathcal{A}' \in U$. There is a matrix $IA^d$ in $\mathcal{A}$ with simple eigenvalues. Thus, if $U$ is a sufficiently small neighborhood of the point $\mathcal{A}$, then there exists a matrix close to $IA^d$ in $\mathcal{A}'$ with simple eigenvalues. Indeed, there exists a neighborhood $v$ of the point $IA^d$ in $sp(4,\mathbb{R})$ such that all matrices in $v$ have different eigenvalues of the same type as $IA^d$. For any point in $CG_2$ sufficiently close to $\mathcal{A}$ there is a corresponding two-dimensional linear subspace in $sp(4,\mathbb{R})$ intersecting $v$. Therefore, there exists a matrix $I(A^d)' \in v$ in $\mathcal{A}'$. Thus, $\mathcal{A}'$ is a simple LPA and the set of simple LPAs is open. Let us prove that it is dense. In order to do it we will use the list of all 15 normal Williamson forms of quadratic Hamilton functions and their centralizers (see Appendix A).

Let $\mathcal{A}$ be a nonsimple LPA. We have to show that there exists a simple LPA arbitrarily close to $\mathcal{A}$. We will prove more: there exist two nearby simple LPAs $\mathcal{A}$ and $\mathcal{A}'$ of different types. Later we will use this fact to prove the instability of nonsimple LPAs.

Let us choose two independent matrices $IA$ and $IB$ in $\mathcal{A}$. There exist symplectic coordinates in which $IA$ has a normal Williamson form. Then $IB$ belongs to the centralizer of the matrix $IA$. The corresponding quadratic forms for $IA$ and $IB$ belong to one of the existing 11 types $5-15$ in Appendix A. First, we consider Cases 6, 8, 11, 12, because the proof is identical for these cases. For example, consider Case 6. In this case, for a fixed subalgebra,

$$\Delta = -(\lambda^2\alpha_1^2 + \lambda\sigma\alpha_1\alpha_2 + d\alpha_2^2)^2\alpha_2^2(4d - \sigma^2),$$

where $\sigma = d_{11} + d_{22}$, $d = d_{11}d_{22} - d_{12}d_{21}$. Thus, the condition $\Delta = 0$ (see above) is equivalent to the condition $\delta = 4d - \sigma^2 = 0$ which singles out a three-dimensional submanifold of the four-dimensional centralizer. This submanifold is smooth everywhere except for the line $d_{11} = d_{22} = d_{12} = d_{21} = 0$ for which the action is degenerate. Obviously, changing the coefficients $d_{ij}$ (i.e., the matrix $IB$) slightly and remaining in the centralizer of the matrix $IA$, it is possible to obtain a nearby algebra $\mathcal{A}'$ such that $\Delta$ is not identically equal to zero for this algebra. That is, $\mathcal{A}'$ will be a simple LPA. It follows from Appendix A that the condition $\delta = 0$ in the centralizer separates two regions for the matrix $IB$: in one of them the pair $IA$, $IB$ generates an algebra corresponding to the saddle-saddle case, and in the other the pair $IA$, $IB$ generates an algebra corresponding to the focus-focus. Therefore, changing $\delta$ we may move from one region to the other, and thus obtain two different algebras $\mathcal{A}_1$ and $\mathcal{A}_2$ close to $\mathcal{A}$.

In Cases 5, 7, 9, 10, 13 the centralizers of the matrix $IA$ coincide with the algebras generated by the pair of matrices $(IA, IB)$. Therefore, the perturbation changing the type of a subalgebra must be constructed for the algebra $\mathcal{A}$, i.e., it is necessary to change both matrices $IA$ and $IB$. We will describe the perturbations of Hamilton functions for each of these cases.

Case 5.

$$H(\epsilon) = \lambda(p_1q_1 + p_2q_2) + p_1q_2 + \epsilon p_2q_1,$$
$$K(\epsilon) = \mu(p_1q_1 + p_2q_2) + \tau p_1q_2 + \epsilon\tau p_2q_1,$$
$$\lambda\tau - \mu \neq 0.$$

Case 7.

$$\begin{aligned} H(\epsilon) &= \frac{\kappa}{2}(q_1^2+q_2^2) + \omega(p_2q_1 - p_1q_2) + \frac{\epsilon}{2}(p_1^2+p_2^2), \\ K(\epsilon) &= \frac{\tau}{2}(q_1^2+q_2^2) + \sigma(p_2q_1 - p_1q_2) + \frac{\kappa\tau\epsilon}{2}(p_1^2+p_2^2), \\ &\kappa\sigma - \tau\omega \neq 0. \end{aligned}$$

Case 9.

$$\begin{aligned} H(\epsilon) &= \lambda p_1q_1 + \frac{\kappa}{2}q_2^2 + \frac{\epsilon}{2}p_2^2, \\ K(\epsilon) &= \mu p_1q_1 + \frac{\sigma}{2}q_2^2 + \frac{\sigma\epsilon}{\kappa}p_2^2, \\ \kappa &= \pm 1, \quad \lambda\sigma - \kappa\mu \neq 0. \end{aligned}$$

Case 10.

$$\begin{aligned} H(\epsilon) &= \frac{\omega}{2}(p_1^2+q_1^2) + \frac{\kappa}{2}q_2^2 + \frac{\epsilon}{2}p_2^2, \\ K(\epsilon) &= \frac{\nu}{2}(p_1^2+q_1^2) + \frac{\sigma}{2}q_2^2 + \frac{\sigma\epsilon\kappa}{2}p_2^2, \\ \kappa &= \pm 1, \quad \omega\sigma - \kappa\nu \neq 0. \end{aligned}$$

Case 13.

$$\begin{aligned} H(\epsilon) &= \frac{\kappa}{2}(p_1^2 - 2q_1q_2) - p_1q_2 + \frac{\epsilon}{2}p_2^2, \\ K(\epsilon) &= \frac{1}{2}[(\nu - \epsilon\kappa b)p_1^2 + 2\epsilon\kappa b p_1p_2 + \epsilon\kappa\nu p_2^2] \\ &\quad - p_1(\epsilon b q_1 + \kappa\nu q_2) - \frac{1}{2}(\epsilon\kappa b q_1^2 + 2\nu q_1q_2 + bq_2^2). \end{aligned}$$

In Case 14, since $\Delta \equiv 0$ along the centralizer, we also construct the perturbation of both matrices. The corresponding one-parameter family of perturbations of the Hamilton functions has the form

$$\begin{aligned} H(\epsilon) &= \frac{1}{2}(q_1^2 + \kappa q_2^2) + \frac{\epsilon}{2}(\kappa p_1^2 + p_2^2), \\ K(\epsilon) &= \frac{a}{2}q_1^2 + \frac{b}{2}q_2^2 + c(p_2q_1 - \kappa p_1q_2) + dq_1q_2 + \frac{\epsilon\kappa}{2}(ap_1^2 + 2\kappa d p_1p_2 + bp_2^2), \\ &\kappa = \pm 1. \end{aligned}$$

In Case 15 the quantity $\Delta$ also vanishes identically along the centralizer. However, this case reduces to Cases 9 and 10. Indeed, if $\delta = \nu_5^2 - \nu_4\nu_6 \neq 0$, then we have either Case 9 or 10 for the corresponding subalgebra. If $\delta = 0$, then we first perturb $IB$ inside the centralizer making $\delta \neq 0$, and then we get either Case 9 or 10.

It is easy to verify that in all constructed families of perturbed actions for a small $|\epsilon| \neq 0$ the algebras obtained have simple singular points of different types for different signs of $\epsilon \neq 0$. The theorem is proved. □

A useful characteristic of an LPA is its stability with respect to perturbations in the class of LPAs. Let us give the relevant definition.

DEFINITION 2.3. An LPA $\mathcal{A}$ is called linearly stable if there exists a neighborhood $U$ of the point $\mathcal{A}$ in $CG_2$ such that all LPAs in $U$ are linearly equivalent to the LPA $\mathcal{A}$.

THEOREM 2.1.3. *An LPA $\mathcal{A}$ is linearly stable if and only if $\mathcal{A}$ is a simple LPA.*

PROOF. Let $\mathcal{A}$ be a simple LPA. Then, by Theorem 2.1.2, there exists a neighborhood $U$ in $CG_2$ such that all LPAs in $U$ are simple and have the same type as $\mathcal{A}$, according to the proof of the theorem. Then, by the equivalence Theorem 2.1.1, all LPAs $\mathcal{A}' \in U$ are linearly equivalent to $\mathcal{A}$.

Conversely, let $\mathcal{A}$ be linearly stable. If $\mathcal{A}$ is not simple, then it follows from the proof of Theorem 2.1.2 that in any neighborhood of the point $\mathcal{A}$ there exist two simple LPAs of different types. By Theorem 2.1.1, this contradicts stability. The theorem is proved. □

## 2.2. Simple singular points of an action and IHVF (local theory)

In this section we study singular points of Poisson actions and IHVF. It was mentioned in Section 2.1 that if $\Phi$ is a Poisson action of the group $\mathbb{R}^4$ on $M$ and $p$ is its singular point, then there is a linear Poisson action defined in $T_pM$ which is called a linearized action.

DEFINITION 2.4. A singular point $p$ of a Poisson action $\Phi$ is called simple if the corresponding linearized action on $T_pM$ is a simple LPA.

By the type of a singular point of an action we will mean the type of the LPA corresponding to the simple point. Thus, in the case $n = 2$ there are simple points of the saddle-saddle, saddle-center, center-center, and focus-focus types.

Now, let $X_1$ be an IHVF with a Hamilton function $H_1$, and let $H_2, \dots, H_n$ be the additional integrals. Let us assume that $p$ is an isolated singular point of the field $X_1$.

LEMMA 2.2.1. *The point $p$ is singular for all vector fields $X_i$ with the Hamilton functions $H_i$, $i = 2, \dots, n$.*

PROOF. Denote the flow of the field $X_i$ by $f_i^t$, $i = 2, \dots, n$, and the flow of the field $X_1$ by $f_1^\tau$, $\tau \in \mathbb{R}$. Then $p \equiv f_1^\tau(p)$ for all $\tau \in \mathbb{R}$. Due to the commutativity of the flows $f_1^\tau$ and $f_i^t$ we have the formula

$$f_i^t(p) = f_i^t(f_1^\tau(p)) = f_1^\tau(f_i^t(p)),$$

which is true for all $(t, \tau) \in \mathbb{R}^2$. Therefore, the points $f_i^t(p)$ are singular points of the field $X_1$. Let $|t|$ be so small that $f_i^t(p)$ belongs to a neighborhood of the point $p$ which does not contain any other singular points of the field $X_1$. Then $f_i^t(p) \equiv p$. Thus, $p$ is a singular point of the vector field $X_i$. □

COROLLARY 2.2.1. *An isolated singular point of the IHVF $X_1$ is a singular point of the induced action.*

Notice that the point $p$ may be a nonisolated singular point for the fields $X_i$, $i \neq 1$.

*Example.* Let us consider $H_1 = \lambda_1 x_1 y_1 + \lambda_2 x_2 y_2$, $H_2 = x_2 y_2$, $\lambda_1 \neq \lambda_2$, $\lambda_1 \lambda_2 \neq 0$ in $(\mathbb{R}^4, dx \wedge dy)$. Then the point $O = (0, 0, 0, 0)$ is an isolated singular point of the field $X_1$. The field $X_2$ contains the whole plane $x_2 = y_2 = 0$ of singular points.

DEFINITION 2.5. A singular point $p$ of the IHVF $X_1$ is called simple if it is a simple singular point of the induced action $\Phi$, and the operator $L_1$ of linearization of the field $X_1$ at the point $p$ has distinct eigenvalues.

REMARK 2.2. By Definition 2.1, the algebra $\mathcal{A}$ generated by the operators $L_1$, $\dots, L_n$ of linearizations of the fields $X_1, \dots, X_n$ at the point $p$ is $n$-dimensional and contains an operator with simple eigenvalues. It is easy to see that $L_1$ may play a role of this operator for a simple singular point of the IHVF.

A special representation of a vector field in a neighborhood of a singular point, called its normal form [**42**], plays an important role in the local study of the singular point. For Hamiltonian vector fields, when symplectic coordinate transformations are used, it is more convenient to bring into normal form not the vector field but the Hamilton function itself.

It turns out that the normal form of a Hamilton function takes an especially simple form for simple singular points. Namely, they are functions of standard quadratic functions which appear in the Williamson normal form for the quadratic part of a Hamilton function (see Section 2.1). The first results about a formal normal form in a neighborhood of an elliptic point were obtained by Birkhoff [**43**]. The questions of divergence of normalizing transformations were studied by Siegel [**44**]. Further important results belong to Moser [**45**]. The condition of simplicity of a singular point for an integrable system with two degrees of freedom in the explicit form has appeared for the first time in [**46**], where it was used to obtain the convergence of the normalizing transformation in the integrable system. The normalizing form in a neighborhood of a simple singular point of an analytic IHVF with $n$ degrees of freedom was studied in the work of Vey [**47**]. However, the case when a linearization of a Hamilton function may have quadruples of complex roots $\pm\alpha \pm i\beta$, $\alpha\beta \neq 0$, $\alpha, \beta \in \mathbb{R}$, $i^2 = -1$, is not considered in his work. A similar result for a nonsimple (but resonance free) point of an analytic vector field was obtained in [**48**]. The $C^\infty$-smooth IHVF was studied in the Eliasson's thesis [**49**]. The case of an elliptic point is published in [**50**], where the validity of the result for other types of points is also claimed with the reference to [**49**]. Let us mention that for some problems a weaker form of the normalization of a Hamilton function is sufficient in the case of finite smoothness of the integrals. It was used in the case $n = 2$ in [**4**] for the study of the structure of invariant manifolds in a neighborhood of a singular point and the structure of one-dimensional orbits.

Let us proceed with precise formulations for the case of a four-dimensional symplectic manifold. Let $p$ be a simple singular point of an IHVF $X$, $H$ its Hamilton function, and $K$ an additional integral. It was stated above (see Section 2.1) that in a neighborhood of the point $p$ there exist symplectic coordinates $(x, y) = (x_1, x_2, y_1, y_2)$ such that in these coordinates the quadratic parts of the functions $H$ and $K$ have the forms

$$H_2 = \lambda_1\xi_1 + \lambda_2\xi_2,$$
$$K_2 = \mu_1\xi_1 + \mu_2\xi_2,$$

and $\Delta = \lambda_1\mu_2 - \lambda_2\mu_1 \neq 0$.

The above-mentioned results of Moser and Russmann for analytic $H$ and $K$, and the result of Eliasson for $C^\infty$-smooth $H$ and $K$ can be stated in a form convenient for us as folows.

THEOREM 2.2.1. *In a neighborhood of a simple singular point of a real analytic IHVF $X$ with two degrees of freedom there exist real analytic ($C^\infty$-smooth) coordinates in which $H = \lambda_1\xi_1 + \lambda_2\xi_2 + h(\xi_1, \xi_2)$, $K = \mu_1\xi_1 + \mu_2\xi_2 + k(\xi_1, \xi_2)$, where the functions $h$ and $k$ are real analytic ($C^\infty$-smooth) and their initial terms are of* *second degree at zero. The quadratic functions $\xi_1$ and $\xi_2$ depend on the type of the point and have the form:*

1. $\xi_1 = \frac{x_1^2+y_1^2}{2}$, $\xi_2 = \frac{x_2^2+y_2^2}{2}$ *(elliptic point),*
2. $\xi_1 = \frac{x_1^2+y_1^2}{2}$, $\xi_2 = x_2 y_2$ *(saddle-center),*
3. $\xi_1 = x_1 y_1$, $\xi_2 = x_2 y_2$ *(saddle),*
4. $\xi_1 = x_1 y_1 + x_2 y_2$, $\xi_2 = x_1 y_2 - x_2 y_1$ *(saddle-focus).*

We return to a simple singular point of a Poisson action $\Phi$ of the group $\mathbb{R}^2$. According to the definition of a Poisson action there exist two vectors $t_1$ and $t_2$ in $\mathbb{R}^2$ with the corresponding Hamilton functions $H_1$ and $H_2$ such that the Hamilton function of any other one-parameter subgroup $s(\alpha_1 t_1 + \alpha_2 t_2)$, $s \in \mathbb{R}$, has the form $\alpha_1 H_1 + \alpha_2 H_2$. Due to the simplicity of the singular point $p$ there exist symplectic coordinates such that, in a neighborhood of the point $p$, $H_1$ and $H_2$ have the form as in Theorem 2.2.1. But then Hamilton functions of all one-parameter subgroups have the same form.

This local representation of Hamilton functions allows us to prove easily the following result.

THEOREM 2.2.2. *Let $p$ be a simple singular point of a Poisson action $\Phi$ of the group $\mathbb{R}^2$ on $M$. Let $\Psi$ be a Poisson action on $T_pM$ linearized at $p$. There exist regular neighborhoods (see Chapter 1) $U \subset M$ of the point $p$, and $V \subset T_pM$ of the origin $0 \in T_pM$ such that the actions $\Phi$ and $\Psi$ are topologically equivalent in these neighborhoods.*

The proof of this theorem is similar to that of Theorem 2.1.1 for linear actions. The main point is that in the indicated symplectic coordinates the system of equations $H_1 = h_1$, $H_2 = h_2$ is equivalent to the system of equations $\xi_1 = C_1$, $\xi_2 = C_2$ in some neighborhood of the point $p$.

The next corollary follows from this theorem and Theorem 2.1.1.

COROLLARY 2.2.2. *Two Poisson actions $\Phi_1$ and $\Phi_2$ with simple singular points $p$ and $p'$ are topologically equivalent in regular neighborhoods of their simple singular points if and only if the singular points $p$ and $p'$ are of the same type.*

REMARK 2.3. In fact, the equivalence is a $C^\infty$-smooth (or analytic, in the case of analytic $H_1$ and $H_2$) diffeomorphism. Thus, this theorem is analogous to the result of [**51**].

REMARK 2.4. If, instead of actions, we consider a problem of the local isoenergetic classification of two IHVFs in neighborhoods of their singular points, then this problem is more difficult and will be solved later along with the global classification.

## 2.3. One-dimensional orbits of the action

In Chapter 1 we considered $k$-dimensional orbits of a Poisson action. Here we start a more detailed study of a Poisson action $\Phi$ of the group $\mathbb{R}^2$ on a four-dimensional symplectic manifold $M$ in a neighborhood of a one-dimensional orbit.

As usual, let us choose two independent vectors in $\mathbb{R}^2$. Let $H$ and $K$ be the Hamilton functions of one-parameter subgroups corresponding to these vectors. Let $\gamma$ be a one-dimensional orbit, and let a point $m \in \gamma$. Then one of the Hamiltonian fields $X_H$ or $X_K$ is nonsingular at the point $m$, i.e., $dH_m \neq 0$ or $dK_m \neq 0$ (or both of them are not equal to zero). Let $dH_m \neq 0$, for definiteness.

Let us apply the reduction described in Chapter 1 to this case. A neighborhood $U$ of the point $m$ is foliated into trajectories of the field $X_H$ on which $H$ and $K$ are constant (in Section 1.4 they will be orbits of the action $F$). We fix a value $h$ close to $H(m)$ such that the levels $H = h$ form a foliation of $U$. Each level $U \cap \{H = h\}$ is foliated by segments of trajectories of the field $X_H$. Identifying every such a segment with a point, we obtain the quotient manifold $D_h$, which is a two-dimensional symplectic manifold with respect to the induced 2-form $\Omega_h$ (see Section 1.4). Since flows for $X_H$ and $X_K$ commute on $U_h$, the translation of a segment of a trajectory of the field $X_H$ along the flow of the field $X_K$ gives a segment of a trajectory of the field $X_H$ as well. Therefore, after factorization, a local Hamiltonian system with one degree of freedom and a Hamilton function $\hat{K}_h$ is generated on $D_h$. The family of functions $\hat{K}_h$ possesses the property that the function $\hat{K}_0$ has a critical point. Generally speaking such a family may be quite complex, i.e., the critical point may split into several singular points or disappear, when $h$ varies. In singularity theory **[31]** one distinguishes the "simplest" families, which are called generic families. The meaning of this notion is that families sufficiently close to a given one (in a certain topology) must be equivalent to the original family.

Using the ideas of singularity theory we will formulate the notion of equivalent families applicable in our case. Let $\Lambda \subset \mathbb{R}$, $\Lambda' \subset \mathbb{R}$ be neighborhoods of zero in $\mathbb{R}$, $D$, $D'$ neighborhoods of zero in $\mathbb{R}^2$, and $f(\lambda, x)$, $g(\lambda', x')$ two smooth functions such that $D_x f(0,0) = D_{x'} g(0,0) = 0$, where $(\lambda, x) \in \Lambda \times D$, $(\lambda', x') \in \Lambda' \times D'$. We say that the families are equivalent if there exists a diffeomorphism $\kappa : \Lambda \to \Lambda'$, $\kappa(0) = 0$, and a smooth map $\chi : \Lambda \times D \to D'$ such that for each $\lambda \in \Lambda$ the map $\chi_\lambda = \chi(\lambda, \cdot)$ is a diffeomorphism satisfying the condition $g(\kappa(\lambda), \chi(\lambda, x)) \equiv f(\lambda, x)$. It is clear that this is an equivalence relation.

It is proved in singularity theory that a generic one-parameter family of functions of two variables containing a function with a critical point is equivalent to one of the following three standard families:

1. $h(\lambda, x, y) = \lambda + x^2 + y^2$;
2. $h(\lambda, x, y) = \lambda + x^2 - y^2$;
3. $h(\lambda, x, y) = \lambda x \pm x^2 + y^2$.

Now we can single out a general class of one-dimensional orbits.

DEFINITION 2.6. A one-dimensional orbit $\gamma$ of an action $\Phi$ of the group $\mathbb{R}^2$ on a four-dimensional symplectic manifold $M$ is called simple if a one-parameter family of two-dimensional Hamilton functions $\hat{K}_h$ is a generic one-parameter family.

Notice that choosing a different point $m'$ on the same orbit $\gamma$ we obtain a different family of Hamilton functions $\hat{K}'_h$. However, it is equivalent to the family $\hat{K}_h$. For the identity map $\kappa$, this equivalence is realized by a diffeomorphism determined by translations along trajectories of the field $X_H$.

The construction described above may be represented in the coordinate form. To this end let us choose symplectic coordinates $(x_1, x_2, y_1, y_2)$ in some neighborhood of the point $m \in \gamma$ such that the point $m$ has coordinates $(0,0,0,0)$ and $H(x_1, x_2, y_1, y_2) \equiv x_1$. Then the additional integral $K$ does not depend on the coordinate $y_1$ conjugate to $x_1$, $K = k(x_1, x_2, y_2)$. The disk $y_1 = 0$ is a cross-section to action orbits. The pieces of the trajectories of the field $X_H$ are $x_1 = x_1^0$, $x_2 = x_2^0$, $y_2 = y_2^0$, with the $y_1$ coordinate as a parameter, $|y_1| \leq d$. Let us apply the translation by the time $\tau$ along the trajectories of the field

$X_K : \dot{x}_1 = 0,\ \dot{y}_1 = k_{x_1},\ \dot{x}_2 = -k_{y_2},\ \dot{y}_2 = k_{x_2}$, to such a piece. We obtain a piece $x_1^\tau = x_1^0,\ x_2^\tau = \phi(\tau; x_1^0, x_2^0, y_2^0),\ y_2^\tau = \psi(\tau; x_1^0, x_2^0, y_2^0)$, shifted along $y_1$ by $\Delta^\tau = y_1^\tau - y_1 = \int_0^\tau k_{x_1}(x_1^0, x_2^\tau, y_2^\tau)d\tau$, where $\phi$ and $\psi$ are the solutions of the system $\dot{x}_2 = -k_{y_2},\ \dot{y}_2 = k_{x_2}$ with $x_1 = x_1^0$. For $\tau$ small enough this piece intersects $N$ at a unique point $(x^\tau, y^\tau)$ that belongs to the disk $N_h = N \cap \{x_1^0 = h\}$. Thus we have a local Hamiltonian flow $(x_2^0, y_2^0) \to (x_2^\tau, y_2^\tau)$ on $N_h$ generated by the Hamilton function $k(h, x_2, y_2)$ that depends on the parameter $h$. This is the family of reduced Hamilton functions $\hat{K}_h = k(h, x_2, y_2)$. Notice that the interval $x_1 = x_2 = y_2 = 0$ is a piece of the orbit $\gamma$, and the corresponding point $x_2 = y_2 = 0$ on the disk $N_0 : x_1 - y_1 = 0$ is a critical point of the Hamilton function $k(0, x_2, y_2)$. In general, all one-dimensional orbits of the action which intersect the neighborhood of the point $m$ are in one-to-one correspondence with the critical points of the functions $\hat{K}_h$ defined by the system of equations $k_{x_2} = 0,\ k_{y_2} = 0$. To find such points, consider (similarly to Chapter 1) the matrix

$$Q_1 = \begin{pmatrix} k_{x_2x_1} & k_{x_2x_2} & k_{x_2y_2} \\ k_{y_2x_1} & k_{y_2x_2} & k_{y_2y_2} \end{pmatrix}.$$

Later we will assume that $\operatorname{rank} Q_1 = 2$. Then it follows from Proposition 1.4.1 that the set of solutions of the system $k_{x_2} = 0,\ k_{y_2} = 0$ forms a smooth curve $l$ in $U_N$. For each point of the curve $l$ there is a piece of a one-dimensional orbit of the action passing through it. The union of such pieces forms a smooth two-dimensional surface $\Sigma$ in $U$. Denote by $\Gamma_0$, $\Gamma_1$, and $\Gamma_2$ the determinants of the following $2 \times 2$ matrices:

$$\begin{pmatrix} k_{x_2x_2} & k_{x_2y_2} \\ k_{y_2x_2} & k_{y_2y_2} \end{pmatrix}, \quad \begin{pmatrix} k_{x_2x_1} & k_{x_2x_2} \\ k_{y_2x_1} & k_{y_2x_2} \end{pmatrix}, \quad \begin{pmatrix} k_{x_2x_1} & k_{x_2y_2} \\ k_{y_2x_1} & k_{y_2y_2} \end{pmatrix}.$$

Along the curve $l$, these determinants are smooth functions of a point on $l$.

The assumption $\operatorname{rank} Q_1(m) = 2$ means that either $\Gamma_0(m) \neq 0$, or $\Gamma_1(m) \neq 0$, or $\Gamma_2(m) \neq 0$. If $\Gamma_0(m) \neq 0$, then the curve $l$ is transversal in $N$ to the planes $x_1 = \text{const}$ (Proposition 1.4.1), and, thus, $l$ is the graph of the mapping $x_2 = f(x_1)$, $y_2 = g(x_1)$. If $\Gamma_0(m) = 0$, then the curve $l$ is tangent to the plane $x_1 = 0$ at the point $m$ (Proposition 1.4.1). In this case $l$ can be parametrized as $x_1 = f(x_2)$, $y_2 = g(x_2)$ if $\Gamma_2(m) \neq 0$, or as $x_1 = f(y_2)$, $x_2 = g(y_2)$ if $\Gamma_1(m) \neq 0$. Along the curve $l$, the determinants $\Gamma_i$ are functions of $x_2$, $\gamma_i(x_2)$, $i = 1, 2, 3$, if $\Gamma_2(m) \neq 0$ (functions of $y_2$ if $\Gamma_1(m) \neq 0$).

To describe the location of the curve $l$ with respect to the plane $x_1 = 0$, which is tangent to it at the point $m$, we introduce the notion of the degree of tangency of a curve and a surface. Suppose we have a smooth surface $S$ and a smooth curve $l$ in $\mathbb{R}^3$ which are tangent at the point $m$, i.e., the vector $\gamma$ tangent to the curve $l$ at the point $m$ belongs to the plane $L$ tangent to $S$ at the point $m$. Let us introduce coordinates $(x, y, z)$ in $\mathbb{R}^3$ such that $m$ has coordinates $(0, 0, 0)$, $L$ is given by the equation $z = 0$, and the curve $l$ by the equations $z = f(x)$, $y = g(x)$, $f'(0) = 0$. Then the deviation of the curve from the plane $L$ is described by the function $f(x)$. By the degree $n$ of tangency between $l$ and $S$ we mean the order of the first nonzero derivative of the function $f(x)$ at the point $m$, i.e., $f^{(n)}(0) \neq 0$, $f^{(k)}(0) = 0$, $k < n$. Then the following assertion holds.

LEMMA 2.3.1. *If $\Gamma_0(m) = 0$, $\gamma_0'(m) \neq 0$, then the curve $l$ has quadratic tangency with the plane $x_1 = 0$. In particular, in some neighborhood of the point $m$ in $U_N$ the curve $l$ is located on one side of the plane $x_1 = 0$.*

PROOF. Suppose $\Gamma_2 \neq 0$. Then $x_1 = f(x_2)$, $y_2 = g(x_2)$ is the coordinate representation of $l$. These functions are solutions of the system $k_{x_2} = 0$, $k_{y_2} = 0$. It follows that

$$f'(x_2) = -\frac{\gamma_0(x_2)}{\gamma_2(x_2)}, \; g'(x_2) = -\frac{\gamma_1(x_2)}{\gamma_2(x_2)}.$$

Thus $f'(0) = 0$. Calculating $f''(0)$ we have

$$f''(0) = -\frac{\gamma_0'(0)}{\gamma_2(0)} \neq 0.$$

Therefore, the lemma is proved. □

The coordinate form of the simplicity condition for a one-dimensional orbit is given by the following theorem.

THEOREM 2.3.1. *A one-dimensional orbit $\gamma$ of the action $\Phi$ that passes through the point $m$ is simple if and only if* $\operatorname{rank} Q_1(m) = 2$, *and either $\Gamma_0 \neq 0$, or $\Gamma_0 = 0$, $\gamma_0'(0) \neq 0$.*

Proof. Let $\gamma$ be a simple orbit of the action. Then the corresponding family of Hamilton functions is equivalent to one of the families 1–3. That is, $k(h, x_2, y_2)$ is reduced by the substitution

$$h = a(\lambda), \; x_2 = u(\lambda, x, y), \; y_2 = v(\lambda, x, y),$$

$a(0) = 0$, $a'(0) \neq 0$, $(\det D(u,v)/D(x,y))|_{(0,0,0)} = d_0 \neq 0$, $u(0,0,0) = v(0,0,0) = 0$ to the form $\overline{k}(\lambda, x, y)$, where $\overline{k}$ is one of the families 1–3 above. It is easy to calculate that after such a substitution we have

$$\overline{\Gamma}_0(m) = \Gamma_0(m) d_0^2,$$
$$\overline{\Gamma}_1 = d_0\{a'(\Gamma_1 u_x + \Gamma_2 v_x) + \Gamma_0(u_\lambda v_x - u_x v_\lambda)\},$$
$$\overline{\Gamma}_2 = d_0\{a'(\Gamma_1 u_y + \Gamma_2 v_y) + \Gamma_0(u_\lambda v_y - u_y v_\lambda)\},$$

where

$$\overline{\Gamma}_0 = \overline{k}_{xx}\overline{k}_{yy} - \overline{k}_{xy}^2,$$
$$\overline{\Gamma}_1 = \overline{k}_{x\lambda}\overline{k}_{xy} - \overline{k}_{y\lambda}\overline{k}_{xx},$$
$$\overline{\Gamma}_2 = \overline{k}_{x\lambda}\overline{k}_{yy} - \overline{k}_{y\lambda}\overline{k}_{xy},$$

at $\lambda = x = y = 0$.

It follows that if $\overline{k}(\lambda, x, y)$ has the form 1 or 2, then $\overline{\Gamma}_0 \neq 0$, and therefore, $\Gamma_0 \neq 0$.

In Case 3 we have $\overline{\Gamma}_0(m) = 0$, $\overline{\Gamma}_1(m) = 0$, $\overline{\Gamma}_2(m) = \pm 2$. Therefore, $\Gamma_1(m) = \mp\frac{2v_x}{d_0^2 a'}$, $\Gamma_2(m) = \mp\frac{2u_x}{d_0^2 a'}$. Thus, either $\Gamma_1(m) \neq 0$, or $\Gamma_2(m) \neq 0$, since $d_0^2 a' \neq 0$ and $d_0 = u_x v_y - u_y v_x$, i.e., $u_x v_x \neq 0$. Therefore, $\operatorname{rank} Q_1(m) = 2$.

Now, we will show that $\gamma_0'(0) \neq 0$ in Case 3. For definiteness, let us assume that $\Gamma_2(m) \neq 0$, i.e., $u_x \neq 0$. Since $d_0 \neq 0$, in the new coordinates the curve $l$, the solution of the system $k_{x_2} = 0$, $k_{y_2} = 0$, satisfies the system of equations $k_x = 0$, $k_y = 0$. In Case 3 we have $\overline{\Gamma}_2(m) = \pm 2 \neq 0$, hence the curve $l$ has a representation $\lambda = \Lambda(x) = \pm 3x^2$, $y = G(x) \equiv 0$. It follows from the proof of Lemma 2.3.1 that $\gamma_0'(0) = -\Gamma_2(m) f''(0)$, where $h = f(x_2)$, $y_2 = g(x_2)$ is the representation of $l$ in the initial coordinates. In the new coordinates we have the same relation $\overline{\gamma}_0'(0) = -\overline{\Gamma}_2(m)\Lambda''(0) = \pm 12$. Let us find a connection between $\gamma_0'(0)$ and $\overline{\gamma}_0'(0)$. To do it we substitute the expressions $h = a(\lambda)$, $x_2 = u(\lambda, x, y)$, $y_2 = v(\lambda, x, y)$ into the relations $h = f(x_2)$, $y_2 = g(x_2)$. We obtain the implicit form of the curve

$l$ in the new coordinates: $a(\lambda) = f(u(\lambda, x, y))$, $v(\lambda, x, y) = g(u(\lambda, x, y))$. Then $\lambda = \Lambda(x)$, $y = G(x)$ is its solution. Substituting these expressions, differentiating twice in $x$, and taking into account that $\Lambda(0) = \Lambda'(0) = 0$, we have

$$\Lambda'' = \frac{f''(0)d_0}{a'(0)(v_y - g'u_y)} = \frac{f''(0)d_0\Gamma_2(m)}{a'(0)(\Gamma_1 u_y + \Gamma_2 v_y)} = \frac{f''(0)d_0\Gamma_2(m)}{a'(0)\overline{\Gamma}_2(m)},$$

where, according to Lemma 2.3.1, $g'(0) = -\frac{\Gamma_1(m)}{\Gamma_2(m)}$.

Thus, from these relations, we have a formula $\overline{\gamma}_0'(0) = \gamma_0'(0)d_0/a'(0)$. Therefore, $\gamma_0'(0) \neq 0$.

Conversely, let $\operatorname{rank} Q_1(m) = 2$, and either $\Gamma_0(m) \neq 0$, or $\Gamma_0(m) = 0$, $\gamma_0'(0) \neq 0$. In the case when $\Gamma_0(m) \neq 0$ we see that for $h = 0$, the critical point $x_2 = y_2 = 0$ for the function $k(h, x_2, y_2)$ is nondegenerate (its Hessian $\Gamma_0$ does not vanish). Therefore, for $|h|$ small enough the family $\hat{K}_h$ is equivalent to either family 1 (when $\Gamma_0(m) > 0$), or family 2 (when $\Gamma_0(m) < 0$). This follows from the parametric Morse lemma [**31**]. Now, let $\Gamma_0(m) = 0$, $\gamma_0'(0) \neq 0$. To be specific, we assume that $\Gamma_2(m) \neq 0$. Then the curve $l$ can be written in the form $x_1 = f(x_2)$, $y_2 = g(x_2)$. Let us normalize the family of functions $k(x_1, x_2, y_2)$ in a neighborhood of the critical point using the split function algorithm [**52**]. First we determine the type of the critical point $x_2 = y_2 = 0$ of the function $k(0, x_2, y_2)$. In this case $k^0_{x_2x_2}k^0_{y_2y_2} - (k^0_{x_2y_2})^2 = \Gamma_0(m) = 0$, and $k^0_{x_2x_2} + k^0_{y_2y_2} \neq 0$, since otherwise $k^0_{x_2x_2} = k^0_{y_2y_2} = k^0_{x_2y_2} = 0$, which contradicts the condition $\operatorname{rank} Q_1(m) = 2$. Let us assume for definiteness that $k^0_{y_2y_2} \neq 0$. Here and below superscript zero in the notation for a second derivative means that the derivative is calculated at the point $(0, 0, 0)$. By a nondegenerate linear coordinate change $x_2 = \alpha_{11}u + k^0_{y_2y_2}v$, $y_2 = \alpha_{21}u + k^0_{x_2y_2}v$, the function $k(0, x_2, y_2)$ reduces to the form $k_1(u, v) = (1/2)au^2 + O_3(u, v)$, where $\alpha_{11}k^0_{x_2y_2} + \alpha_{21}k^0_{y_2y_2} \neq 0$ (the determinant of the coordinate change) and $O_3(u, v)$ denotes the terms of the third order at the origin. Moreover, due to the nondegeneracy of the coordinate change one has $a = k^0_{x_2x_2}\alpha_{11}^2 + 2k^0_{x_2y_2}\alpha_{11}\alpha_{21} + k^0_{y_2y_2}\alpha_{21}^2 \neq 0$. On the next step [**52**], a nonlinear coordinate change $U_1 = u + U_2(u, v)$, $v_1 = v$, with $U_2$ a quadratic homogeneous polynomial, reduces the function to the form

$$k_2(u_1, v_1) = (1/2)au_1^2 + (1/6)Av_1^3 + O_4(u_1, v_1),$$

where $O_4(u_1, v_1)$ denotes the terms of fourth and higher order, and the coefficient $A$ has the form

$$k_{x_2x_2x_2}k^3_{y_2y_2} - 3k_{x_2x_2y_2}k^2_{y_2y_2}k_{x_2y_2} + 3k_{x_2y_2y_2}k_{y_2y_2}k^2_{x_2y_2} - k_{y_2y_2y_2}k^3_{x_2y_2},$$

where all the derivatives are calculated at the point $(0, 0, 0)$. Differentiating the function $\gamma_0(x_2)$ at the point $x_2 = 0$ and using the relations

$$\Gamma_2(m)k^0_{x_2y_2} - \Gamma_1(m)k^0_{y_2y_2} = 0 \text{ and } \Gamma_0(m) = 0,$$

we obtain

$$A = k^0_{y_2y_2}\gamma_0' \neq 0.$$

Thus, in a neighborhood of the origin the function $k(0, x_2, y_2)$ reduces to the form

$$(1/2)au^2 + (1/6)Av^3 + O_4, \quad aA \neq 0.$$

Now we return to the family $k(x_1, x_2, y_2)$. Carrying out the preceding calculations we obtain the following formula:

$$\begin{aligned} k(x_1, u_1, v_1) = {} & k(x_1, 0, 0) + a_{10}(x_1)u_1 + a_{01}(x_1)v_1 \\ & + (1/2)[(a + a_{20}(x_1))u_1^2 + 2a_{11}(x_1)u_1v_1 + a_{02}(x_1)v_1^2] \\ & + (1/6)[(a_{30}(x_1)u_1^3 + 3a_{21}(x_1)u_1^2v_1 \\ & + 3a_{12}(x_1)u_1v_1^2 + (A + a_{03}(x_1))v_1^3] + O_4(u_1, v_1), \end{aligned}$$

where $a_{ij}(0) = 0$, $a'_{01}(0) = k^0_{x_1y_1} = k^0_{x_1x_2}k^0_{y_2y_2} - k^0_{x_1y_2}k^0_{x_2y_2} = \Gamma_2(m) \neq 0$.

Using the fact that $a$ and $A$ are not equal to zero, the translations $u_1 \to u_1 + c(x_1)$, $c(0) = 0$, transformations similar to those used above, and the parameter change $x_1 \to \lambda$, we obtain that

$$k(\lambda, u_2, v_2) - \alpha_0(\lambda) = \lambda v_2 \pm (1/2)u_2^2 + v_2^3 + O_4(u_2, v_2),$$

where the coefficients in $O_4$ depend on $\lambda$. This means that the family $\lambda v_2 \pm (1/2)u_2^2 + v_2^3$ is versal [**31**] for the family $k(\lambda, u_2, v_2)$, and, thus, for the family $k(x_1, x_2, y_2)$.

These results allow us to give the following definition.

DEFINITION 2.7. A one-dimensional orbit is called elliptic, hyperbolic, or parabolic if the corresponding function $k$ has as a versal family the family 1, 2, or 3, respectively.

The level sets of the foliations for each of these families are well known, and are depicted in Figure 1.1 in Chapter 1.

Thus, for each $x_1 = c$, we have obtained a description of the foliation of the disk $N$ into the level sets of the function $k(x_1, x_2, y_2)$.

From that, one can easily get a description of the local structure of the action orbits in a neighborhood of the point $m$. Namely, the orbit foliation is a product of the foliated disk $N$ and the interval $|y_1| \leq d$.

Let us describe the structure of a neighborhood of an elliptic or hyperbolic simple closed one-dimensional orbit. Let $\gamma$ be such an orbit. Choose two independent one-parameter subgroups in $\mathbb{R}^2$. Let $H$ and $K$ be the corresponding Hamilton functions. Without loss of generality we may assume that the fields $X_H$ and $X_K$ are nondegenerate along $\gamma$, and that $H(\gamma) = K(\gamma) = 0$. Since a symplectic manifold is orientable, a tubular neighborhood of a closed smooth curve is diffeomorphic to the direct product $D^3 \times S^1$, where $D^3$ is a smooth three-dimensional ball. However, in our case this neighborhood possesses additional stratifications related with the foliation into action orbits. We will construct a regular neighborhood (see Chapter 1) of the orbit $\gamma$. To this end, we take a point $m \in \gamma$ and construct (as before) the section $N$.

1. *The case of an elliptic orbit* $\gamma$. We choose a neighborhood of the point $m$ in $N$ as follows. On each disk $N_h = N \cap \{H = h\}$, $|h| \leq h_*$, where $h_* > 0$ is small enough, we take a closed curve $c_0(h)$ such that it is a level curve for the function $K$, it encloses the point $l(h) = l \cap N_h$, and is close enough to it. In addition, it is possible to choose the curves $c_0(h)$ in such a way that their union over $h$ defines a smooth cylinder $C$ transversal to the disks $N_h$ in $N$. The top curve $c_0(h_*)$ and the bottom curve $c_0(-h_*)$ define two disks which we will call the top lid $B_t$ and the bottom lid $B_d$. The union of the cylinder $C$ and the two lids $B_t$ and $B_d$ forms a topological sphere which bounds a neighborhood of the point $m$ in $N$. Let us, also, denote this neighborhood by $N$ and preserve the notation $N_h$ for the disk

$N \cap \{H = h\}$. The neighborhood $N$ contains a part of the curve $l$ (see above) whose end-points belong to $B_t$ and $B_d$. On each of the disks $N_h$ there is a foliation into the level curves $K = k$ of the function $K$, where $k$ varies between $K(l(h))$ and $K(c_0(h))$. This is a foliation into the closed curves $c(h, k)$ which tend to the point $l(h)$. Consider trajectories of the field $X_H$ through each point of $N$. Since $h_*$ is small and $c_0(h)$ and $l(h)$ are close to each other, every such trajectory comes back to $N$. It follows from the invariance of $H$ and $K$ along the flow that trajectories that begin on the closed curve $c(h, k)$ come back to it. Since the mapping of the disk $N_h$ into itself is symplectic with respect to the restriction of the symplectic form $\Omega$, it preserves the orientation. Therefore, each curve $c(h, k)$ defines a Liouville torus, and a trajectory originating at $l(h)$ defines a closed one-dimensional orbit of the action. The union of all such tori and the one-dimensional orbits forms a regular neighborhood $U$ of the orbit $\gamma$. Notice that in this case $\mathbb{R}^{\perp} = \varnothing$ and $\mathbb{R}^{\parallel} = \partial U$ (see the definition of a regular neighborhood). We will call the regular neighborhood of an elliptic orbit constructed above a canonical neighborhood.

PROPOSITION 2.3.1. *Let $\gamma_1$ and $\gamma_2$ be two elliptic one-dimensional orbits of Poisson actions $\Phi_1$ and $\Phi_2$, and let $U_1$ and $U_2$ be their canonical neighborhoods. Then the action $\Phi_1$ on $U_1$ is topologically equivalent to the action $\Phi_2$ on $U_2$.*

PROOF. Below, the objects related to $\Phi_1$ or $\Phi_2$ will be marked by the subscripts 1 or 2, respectively. Let us consider the lateral face of the canonical neighborhood $U_1$, which is formed by the trajectories of the field $X_{H_1}$ passing through the points of the cylinder $C_1 \subset N_1$. This set $\Xi$ is diffeomorphic to the direct product of a torus $T^2$ and the interval $I = [-h_1^*, h_1^*]$ and $\Xi$ bears a smooth foliation into the Liouville tori. Also, there is a smooth foliation $B$ of each such torus into closed curves. These curves are the boundaries of the two-dimensional disks, being stable (or unstable) manifolds of the singular points for the auxiliary gradient system (see Appendix 2). It is easy to see that this foliation into closed curves is orientable and may be defined by a periodic vector field. Therefore, it is not difficult to construct a smooth foliation into two-dimensional cylinders transversal to the closed curves and the Liouville tori in $\Xi$. This allows one to define the coordinates $(\phi_1, \theta_1, h_1)$ in $\Xi$, where $(\phi_1, \theta_1)$ are the cyclic coordinates on the Liouville tori $H = h$. Moreover, $\phi_1 = \text{const}$ defines a leaf of the foliation $B$ on the torus, and $\theta_1 = \text{const}$ defines a fiber of the foliation into cylinders that was constructed above. The coordinates $(\phi_2, \theta_2, h_2)$ on $U_2$ are defined analogously. By $\Sigma_i$ we will denote the smooth cylinder consisting of one-dimensional orbits contained in $U_i$.

Next, let us consider the moment maps $\mu_i : U_i \to \mathbb{R}^2 = \{(H_i, K_i)\}$ defined by $\mu_i(m) = (H_i(m), K_i(m))$, $m \in U_i$. Then it follows from the construction of the canonical neighborhood that $\sigma_i = \mu_i(U_i)$ is a region on the $(H_i, K_i)$-plane bounded by the intervals $H_i = \pm h_i^*$ and by the two curves $K_i = \lambda_i(H_i)$, which is the image of $\Xi$, and $K_i = \nu_i(H_i)$, which is the image of the cylinder $\Sigma_i$. Let us define a fiberwise map $q : \sigma_1 \to \sigma_2$ in the triangular form $q(H_1, K_1) = (q^{(1)}(H_1), q^{(2)}(H_1, K_1))$. Moreover, we define it in such a way that the curve $(H_1, \lambda_1(H_1))$ is mapped into the curve $(H_2, \lambda_2(H_2))$ and the curve $(H_1, \nu_1(H_1))$ is mapped into the curve $(H_2, \nu_2(H_1))$. In addition, let us define two circle homeomorphisms $g_\phi$ and $g_\theta$. Then there is a homeomorphism $g_0 : \Xi_1 \to \Xi_2$ defined coordinatewise by $g_0(\phi_1, \theta_1, h_1) = (g_\phi(\phi_1), g_\theta(\theta_1), q^{(1)}(h_1))$.

Now let a point $m_1 \in U_1 \backslash \Sigma_1$ be fixed. Then the numbers $h_1 = H_1(m_1)$, $k_1 = K_1(m_1)$, and a trajectory $L_1(m_1)$ of the auxiliary gradient vector field $Y_1$ are determined. The trajectory $L_1(m_1)$ crosses $\Xi_1$ at a unique point $m_1'$.

Let $m_2' = g_0(m_1')$ and let $L_2(m_2')$ be the trajectory of the field $Y_2$ passing through $m_2'$. The function $K_2$ varies monotonically along this trajectory. Thus, there exists a unique point $m_2 = g(m_1)$ on the trajectory $L_2(m_2')$ such that $K_2(m_2) = q^{(2)}(H_1(m_1), K_1(m_1))$. It is obvious that this map is a homeomorphism in $U_1 \backslash \Sigma_1$ and extends uniquely to the whole $\Sigma_1$ since the trajectories of all the points of the closed curve from the foliation $B_1$ of $\Xi_1$ converge to the unique point on $\Sigma_1$. The same arguments are valid for $Y_2$. The proposition is proved. □

2. *The case of a hyperbolic orbit* $\gamma$. Consider a closed hyperbolic orbit $\gamma$ and a point $m$ on $\gamma$. As before, let $N$ be a three-dimensional section transversal to $\gamma$. It was shown that there is a smooth curve $l$ in $N$ passing through the point $m$ and transversal to the disks $N_h$. For each point $l(h) = l \cap N_h$ there exists a closed one-dimensional hyperbolic orbit $\gamma(h)$ passing through $l(h)$, $\gamma = \gamma(0)$ and the union of these curves for all $|h| \leq h^*$ is a smooth two-dimensional symplectic cylinder $\Sigma$. There are two smooth curves $W^s(h)$ and $W^u(h)$ in $N_h$ which intersect transversally at the point $l(h)$ and form a "cross", the level set $K = K(l(h))$ of the function $K$ on $N_h$. The cross divides the neighborhood of the point $l(h)$ in $N_h$ into four sectors. Moreover, in each pair of the opposite sectors the signs of the function $K - K(l(h))$ are the same, and in the adjacent sectors they are opposite.

Let us consider the Poincaré diffeomorphism $f_h : N_h \to N_h$ generated by trajectories of the field $X_H$ near the closed trajectory $\gamma(h)$. Since this is a symplectic diffeomorphism, it preserves the orientation of $N_h$, and, moreover, it preserves the level sets of the function $K$. Therefore, each of the curves $W^s(h)$ and $W^u(h)$ is mapped into itself. Then there are the following two possibilities:

1. the restriction of $f_h$ to $W^s(h)$ and $W^u(h)$ preserves the orientation;
2. the restriction of $f_h$ to $W^s(h)$ and $W^u(h)$ changes the orientation.

We call the first case orientable and the second nonorientable. We call $\gamma(h)$ orientable or nonorientable hyperbolic orbit, respectively. It is clear that the property of being orientable or nonorientable is preserved for all orbits close enough to $\gamma(h)$. It is easy to see that for an orientable hyperbolic orbit, locally near $\gamma(h)$, the common level set $H = h$, $K = K(l(h))$ is a union of two smooth cylinders intersecting transversally at $\gamma(h)$. Similarly, for a nonorientable hyperbolic orbit this set is the union of two smooth Möbius bands intersecting transversally at $\gamma(h)$. Notice that locally each cylinder consists of the one-dimensional orbit $\gamma(h)$ (see Figure 6.1 in Chapter 6) and two two-dimensional cylindrical orbits, and the Möbius band consists locally of the one-dimensional orbit $\gamma(h)$ and one two-dimensional cylindrical orbit. It is clear that the cylinders (correspondingly, Möbius bands) depend smoothly on $h$ and form a smooth three-dimensional submanifold diffeomorphic to the direct product of a cylinder (Möbius band) and the interval $I = [-h_*, h_*]$. The manifolds $W^s$ and $W^u$ intersect transversally in $M$ along $\Sigma$.

To construct the canonical neighborhood of the orbit $\gamma(h)$ we proceed as follows. On each of the cylinders $W_h^s$ and $W_h^u$ forming the level set $H = h$, $K = K(l(h))$ we choose two smooth curves such that they are located on different sides of $\gamma(h)$ and do not intersect $\gamma(h)$. Let us denote these curves by $\sigma_i^s(h)$ and $\sigma_i^u(h)$, $i = 1, 2$. Such curves may be chosen so that their union over $h$ forms two smooth cylinders transversal to $W_h^s$, or $W_h^u$, respectively. Let us continue each curve $\sigma_i^s(h)$ and $\sigma_i^u(h)$

along the trajectories of the auxiliary system $Y$, assuming that a direct product Riemannian metric is chosen in a neighborhood of $\gamma(h)$. Let us continue these trajectories so that the function $K - K(l(h))$ vary from $-\delta_*$ to $\delta_*$ on them, where $\delta_*$ is a small positive number. The last step in the construction of the neighborhood is the choice of action orbits defined by the conditions $H = h$, $K = K(l(h)) \pm \delta_1$, $0 < \delta_1 < \delta_*$. Since on the level $H$ the trajectories of the field $Y$ are transversal to the orbits (i.e., the level submanifolds of the function $K$), we obtain a neighborhood which possesses all the properties of a canonical neighborhood. A cross-section of it is shown in Figure 2.1. In the nonorientable case we do the same, except that, on the Möbius band, we construct only one simple closed curve $\sigma^s(h)$, $\sigma^u(h)$ such that each of the curves is homotopic to the double covering of the curve $\gamma(h)$ and does not intersect it.

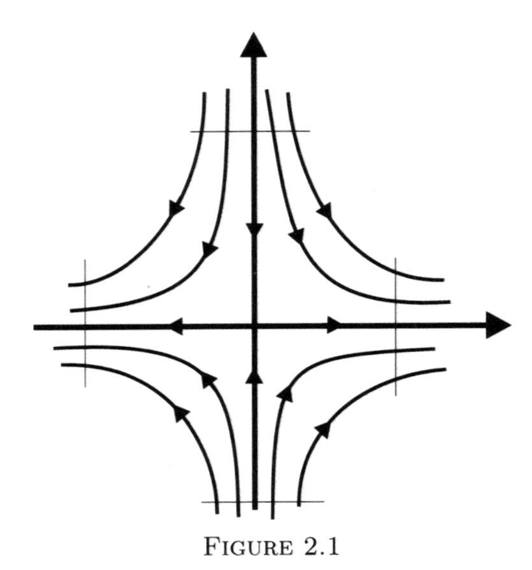

FIGURE 2.1

Now, let us assume that we have two Poisson actions $\Phi_1$, $\Phi_2$ on four-dimensional symplectic manifolds $M_1$, $M_2$ with hyperbolic one-dimensional orbits $\gamma_1$, $\gamma_2$ and canonical neighborhoods $U_1$, $U_2$.

PROPOSITION 2.3.2. *The action $\Phi_1$ on $U_1$ is topologically equivalent to the action $\Phi_2$ on $U_2$ if and only if the orbits $\gamma_1$ and $\gamma_2$ are of the same orientation type.*

PROOF. The "only if" part is obvious. Indeed, under the equivalence, $W_1^s$ and $W_1^u$ are transformed into the corresponding sets for the second system, but their topologies are different in the orientable and nonorientable cases.

Now we prove the "if" part. In order to construct the conjugating homeomorphism of canonical neighborhoods we use the $Y$-invariant continuous foliation constructed in Appendix 2, whose leaves are two-dimensional disks. We introduce coordinates $(\theta, h)$ on $\Sigma$ (for example, action-angle variables), $0 \leq \theta \leq 2\pi$, $|h| \leq h_*$. Then this foliation defines the structure of a fiber bundle over $\Sigma$ with a two-dimensional disk as a fiber. Moreover, this bundle is isomorphic to a direct product in the orientable case, and is a nontrivial bundle with the structure group $\mathbb{Z}_2$ in the nonorientable case. We denote by $D(\theta, h)$ the fiber over a point $(\theta, h)$. Such a disk, together with the trajectories of the field $Y$, is depicted in Figure 2.1.

Notice that, according to the construction of the canonical domain, the boundary of each such disk contains two pairs of smooth curves $\nu_i^+(\theta,h)$, $\nu_i^-(\theta,h)$, $i=1,2$. The pair $\nu_1^+(\theta,h)$, $\nu_2^+(\theta,h)$ is the intersection of the disk $D(\theta,h)$ with the level set $K-K(l(h))=\delta_*$, and the pair $\nu_1^-(\theta,h)$, $\nu_2^-(\theta,h)$ is the intersection of the disk $D(\theta,h)$ with the level set $K-K(l(h))=-\delta_*$. In the orientable case the union over all $\theta$ of the curves $\nu_i^+(\theta,h)$, with $i$ fixed, forms an annulus $Q^+(h)$ (part of the two-dimensional orbit of the action), which belongs to the boundary of the canonical domain. Thus, there are two annuli $Q_i^+(h)$, $i=1,2$, in the orientable case. In the nonorientable case the union over $\theta$ of the curves $\nu_1^+(\theta,h)$ and $\nu_2^+(\theta,h)$ forms a unique annulus $Q(h)$, which intersects a fixed disk $D(\theta,h)$ along two curves $\nu_1^+(\theta,h)$ and $\nu_2^+(\theta,h)$. Each of the annuli $Q_i^+(h)$ or $Q^+(h)$ contains a simple smooth closed homotopically nontrivial curve $N_i^s(h)$, the trace of the stable manifold of the saddle curve $\gamma(h)$ of singular points of the field $Y$ (see Appendix B). Analogous annuli $Q_i^-(h)$ and $Q^-(h)$ are obtained from the curves $\nu_i^-(\theta,h)$.

Consider a smooth foliation of $Q_i^+(h)$ (or $Q^+(h)$) into closed curves, containing $N_i^s(h)$ as fibers and transversal to the curves $\nu_i^+(\theta,h)$. Obviously, such a foliation can be chosen to depend smoothly on $h$. In the orientable case the existence of two foliations of $Q_i^+(h)$ allows us to introduce two coordinates on it, $(\theta,n_i)$, $|n_i|\le n_i^*$. In addition we assume that $n_i=0$ on $N_i^s(h)$. In the nonorientable case we introduce the coordinates $(\psi,n)$ in $Q^+(h)$, $0\le\psi\le 4\pi(\mathrm{mod}\,4\pi)$ on $N^s(h)$. Moreover, we introduce the coordinate $\psi$ so that the points $(\psi,n_0)$ and $(\psi+2\pi(\mathrm{mod}\,4\pi),n_0)$ belong to the same disk $D(\theta,h)$ for a fixed $n_0$. (On $D(\theta,h)$ these points lie on the different curves $\nu_1^+(\theta,h)$ and $\nu_2^+(\theta,h)$.) Then, the curve $\gamma(h)$ is doubly covered by the curves $n=n_0$.

Now we construct the conjugating homeomorphism in the orientable case. The curves $N_i^s(h)$ divide each annulus $Q_i^+(h)$, $i=1,2$, into two semiannuli. We shall say that a semiannulus contained in $Q_1^+(h)$ is incidental with a semiannulus contained in $Q_2^+(h)$ if the trajectories of the field $Y$ emanating from these semiannuli arrive at the same annulus $Q_j^-(h)$ (see Figure 2.2).

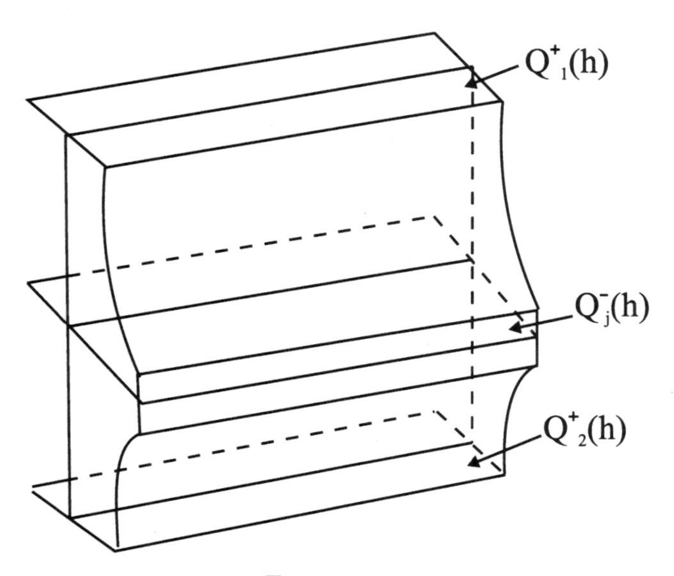


FIGURE 2.2

We define the following homeomorphisms:

1. The cylinder homeomorphisms $\phi_\Sigma : \Sigma \to \Sigma'$, preserving the foliation into one-dimensional orbits.

2. The homeomorphisms $\phi_n^i : [-n_i^*, n_i^*] \to [-'n_i^*, 'n_i^*]$, $\phi_n^i(0) = 0$.

These homeomorphisms define a correspondence between the semiannuli contained in $Q_1^+(h)$, $Q_2^+(h)$ and the corresponding semiannuli of the second system. Obviously, these homeomorphisms can be chosen in such a way that incidental semiannuli of the first system correspond to incidental semiannuli of the second. From now on we assume this to be so.

3. The homeomorphisms $\phi_k : [-\delta_*, \delta_*] \to [-\delta'_*, \delta'_*]$, $\phi_k(0) = 0$, $\phi_k(\delta_*) = \delta'_*$. These homeomorphisms define correspondence between the points of the trajectories of the field $Y$. We recall that the function $K$ is monotone decreasing along such a trajectory.

Let $x$ be an arbitrary point of $U_1$ not in the unstable manifold $W^u(\Sigma)$ of the set $\Sigma$ of singular points of the field $Y$. Let $h = H(x)$, $k = K(x)$, $|h| \le h_*$, $|K(x) - K(l(h))| \le \delta_*$, and let $D(\theta, h)$ be the corresponding disk. As time decreases, the trajectory of the field $Y$ passing through such a point will transversally intersect the level set $K = K(l(h)) + \delta_*$ at some point $x_1$ in the intersection of $Q_1^+(h)$ or $Q_2^+(h)$ with the disk $D(\theta, h)$.

The homeomorphism $\phi_\Sigma$ maps the point $(\theta, h)$ into the point $(\theta', h')$.

The homeomorphism $\phi_n^i$ defines the mapping of the coordinate $n_i(x)$ into the corresponding coordinate $n'_i$ on a certain annulus $'Q_i^+(h')$ of the second system. Thus, $D'(\theta', h')$ uniquely defines a point such that there is a unique trajectory of the field $Y$, not contained in the unstable manifold of the set of singular points $\Sigma'$ of the field $Y'$, passing through this point.

The homeomorphism $\phi_k$ uniquely defines a point on this trajectory. Namely, this is the point $x'$ for which

$$K'(x') = K'(l(h')) + \phi_k(k - K(l(h))).$$

Obviously, the map constructed above is a homeomorphism everywhere except for the points of $W^u(\Sigma)$. However, this homeomorphism can be extended uniquely to the points of $W^u(\Sigma)$, since for $x_n \to x_* \in W^u(\Sigma)$ there is defined a point $y = (\theta(x_*), h(x_*)) \in \Sigma$ and a curve $\Gamma$ consisting of the point $y$, the stable separatrix of the field $Y$ coming to the point $y$, and the unstable separatrix of $Y$ emanating from $y$. Those trajectories of the field $Y$ passing through the points $x_n$ that do not belong to $W^u(\Sigma)$ converge to the curve $\Gamma$. However, the points of $\Gamma$ are separated by the function $K$, which is monotonic on $\Gamma$ and decreases from $K(l(h(x_*))) + \delta_*$ to $K(l(h(x_*))) - \delta_*$. It is easy to see that the homeomorphism constructed above preserves the level sets of the functions $H$ and $K$ and, therefore, maps parts of the orbits of the action $\Phi_1$ that belong to $U_1$ into the corresponding parts of the orbits of the action $\Phi_2$ that belong to $U_2$.

In the nonorientable case the construction is analogous. The only difference is that we define one homeomorphism $\phi_n : [-n_*, n_*] \to [-n'_*, n'_*]$, and the pair of points on $Q^+(h) \cap D(\theta, h)$ corresponding to a closed curve $n = n_0$, is mapped into the pair of points on $Q^+(h') \cap D(\theta', h')$, corresponding to the curve $n' = n'_0 = \phi_n(n_0)$, where $(\theta', h') = \phi_\Sigma(\theta, h)$. In fact, this correspondence defines a homeomorphism between the annuli $Q^+(h)$ and $Q^+(h')$. From this point on, the conjugating homeomorphism is constructed exactly as in the orientable case. The proposition is proved. □

## 2.4. Singular trajectories of IHVF

Now we turn to the study of trajectories of IHVF. There are trajectories of two types. Trajectories of the first type are at the same time one-dimensional orbits of the induced action. Such trajectories will be called singular trajectories. Trajectories of the second type lie on two-dimensional integral manifolds which are two-dimensional orbits of the induced action. In this section we will study only the singular trajectories.

DEFINITION 2.8. A singular trajectory $\gamma$ of the field $X_H$ is called simple if it is a simple one-dimensional orbit of the induced action.

According to the classification of simple one-dimensional action orbits we will distinguish simple singular trajectories of three types: elliptic, hyperbolic, and parabolic. As in the case of one-dimensional orbits, singular trajectories of the field $H$ may be closed (singular periodic trajectories or SPTs) or nonclosed. Let us present examples of nonclosed singular trajectories of elliptic and hyperbolic types.

*Example* 1. Suppose $(M,\Omega) = (\mathbb{R}^4, dx_1 \wedge dy_1 + dx_2 \wedge dy_2)$ and $H = \lambda x_1 y_1 + (\omega/2)(x_2^2 + y_2^2)$, $\lambda\omega \neq 0$, $K = \gamma x_1 y_1 + (\nu/2)(x_2^2 + y_2^2)$, $\Delta = \lambda\nu - \omega\gamma \neq 0$. IHVF $(X_H, K)$ has an invariant symplectic plane $x_2 = y_2 = 0$ such that all trajectories on it, except for the singular point $x_1 = y_1 = 0$, are nonclosed elliptic singular trajectories. Let us show this. Let us choose a cross-section $N : x_1 = d$. The level sets $H = c$ on $N$ are defined by the equations $y_1 = \frac{c-(\omega/2)(x_2^2+y_2^2)}{\lambda d}$, and the restriction of the function $K$ to $N_c = N \cap \{H = c\}$ has the form

$$k_c(x_2, y_2) = (\gamma/\lambda)c + (\Delta/2)(x_2^2 + y_2^2).$$

The family $k_c$ is exactly the one-parameter family of the reduced Hamilton functions which defines the type of a one-dimensional action orbit, and therefore, the type of a singular trajectory. Note that the trajectories of the field $X_H$ on the symplectic plane $x_1 = y_1 = 0$ are hyperbolic SPTs.

*Example* 2. As in Example 1, let $(M,\Omega) = (\mathbb{R}^4, dx_1 \wedge dy_1 + dx_2 \wedge dy_2)$. Consider $H = \lambda_1 x_1 y_1 + \lambda_2 x_2 y_2$, $\lambda_1\lambda_2 \neq 0$, $K = \mu_1 x_1 y_1 + \mu_2 x_2 y_2$, $\lambda_1\mu_2 - \lambda_2\mu_1 \neq 0$. Then, as in Example 1, we obtain that all trajectories, except for the singular point, are nonclosed hyperbolic singular trajectories on the invariant symplectic two-dimensional planes $x_1 = y_1 = 0$ and $x_2 = y_2 = 0$.

We now turn to singular periodic trajectories of IHVF. The following natural question arises immediately: is there any connection between the types of periodic trajectories of the Hamiltonian vector field $X_H$ and their types as SPT for IHVF? We recall [**22**] that for an autonomous Hamiltonian vector field $X_H$ a periodic trajectory $\gamma$ always has two unit multipliers. Moreover, the linear type of the periodic trajectory is defined by the set of remaining multipliers. In the case of two degrees of freedom there are two more multipliers $\mu_1, \mu_2$ satisfying the relation $\mu_1\mu_2 = 1$. Therefore, in the general case, either $\bar{\mu}_1 = \mu_2 = \exp(i\alpha)$, $\alpha \neq 0, \pi$ (elliptic trajectory), or $\mu_1^{-1} = \mu_2 \in \mathbb{R}$, $\mu_1 \neq \pm 1$ (hyperbolic trajectory).

Therefore, one should distinguish the notions of elliptic (hyperbolic) SPT for IHVF $(X_H, K)$ and elliptic (hyperbolic) trajectory of the field $X_H$. Generally speaking, these notions are independent and there is no rigid connection between them.

*Example.* The pair $H = \lambda_1 x_1 y_1 + \omega(x_2^2 + y_2^2)$, $K = x_1^2 y_1^2$ defines the IHVF $(X_H, K)$ in $(\mathbb{R}^4, dx_1 \wedge dy_1 + dx_2 \wedge dy_2)$. The plane $x_1 = y_1 = 0$ is invariant and is filled in by SPTs which are hyperbolic trajectories of the field $X_H$. However, for IHVF $(X_H, K)$ they are not hyperbolic SPTs, since on the cross-section $y_2 = 0$ the one-parameter family of the reduced Hamilton functions has the form $K = x_1^2 y_1^2$, i.e., it is not a generic family.

Nevertheless, the following statement is true.

PROPOSITION 2.4.1. *Suppose $\gamma$ is an elliptic (hyperbolic) SPT for the IHVF $(X_H, K)$, and the vector field $X_K$ does not have singular points on $\gamma$. Then*

1. *if $\gamma$ is a hyperbolic SPT, then $\gamma$ is a hyperbolic periodic trajectory of either the field $X_H$, or $X_K$;*
2. *if $\gamma$ is an elliptic SPT, then the absolute values of all multipliers of the trajectory $\gamma$ for the fields $X_H$ and $X_K$ are equal to unity.*

The proof of this proposition is based on the following lemma.

LEMMA 2.4.1. *Let $\gamma$ be an elliptic or oriented hyperbolic SPT for the IHVF $(X_H, K)$ such that $X_K \neq 0$ along $\gamma$. Then there exists a neighborhood $U$ of the trajectory $\gamma$ and $C^\infty$-smooth symplectic coordinates $(I, \theta, x, y)$ such that in these coordinates $\Omega = dI \wedge \theta + dx \wedge dy$ and*

$$H = H_0(I) + \xi H_1(I, \theta, x, y), \quad K = K_0(I) + \xi K_1(I, \theta, x, y), \tag{2.2}$$

*where, in the elliptic case, $\xi = x^2 + y^2$ and, in the hyperbolic case, $\xi = xy$. For the nonoriented hyperbolic trajectory this statement is true in the double covering of a neighborhood of the trajectory $\gamma$.*

PROOF. The proof of this lemma for the hyperbolic case is given in [**5**]. The proof for the elliptic case for analytic IHVF is given in [**48**] (as a particular case of a more general statement), and the smooth case can be proved using the reduction procedure [**3**].

Now let us prove Proposition 2.4.1. First we consider the hyperbolic case. Using the form (2.2) for the integrals $H$ and $K$, one can easily see that the condition of hyperbolicity of SPT, $\Gamma_0 < 0$, takes the form

$$-[H_0'(0)K_1(0, \theta, 0, 0) - K_0'(0)H_1(0, \theta, 0, 0)]^2 / (H_0'(0))^2 < 0,$$

i.e., since $H_0'(0)K_0'(0) \neq 0$,

$$\frac{K_1(0, \theta, 0, 0)}{K_0'(0)} \neq \frac{H_1(0, \theta, 0, 0)}{H_0'(0)} \tag{2.3}$$

for any $\theta$. Now, let us find multipliers of the periodic solution $x = y = I = 0, \theta = H_0'(0)t$ of the vector field $X_H$. In order to do this we describe the system linearized at this solution,

$$\dot{J} = 0, \quad \dot{\tau} = H_0'' J, \quad \dot{\xi} = -\xi H_1(0, H_0'(0)t, 0, 0), \quad \dot{\eta} = \eta H_1(0, H_0'(0)t, 0, 0),$$

where the linearized 2-form is $dJ \wedge dr + d\xi \wedge d\eta$. Then the multipliers different from unity are

$$\mu = \exp\left[\frac{\int_0^{2\pi} H_1(0, s, 0, 0)ds}{H_0'(0)}\right] \text{ and } \mu^{-1}.$$

Analogously, the multipliers of the periodic solution $x = y = I = 0$, $\theta = K_0'(0)t$ of the vector field $X_K$ (the set described by these functions gives the same one-dimensional orbit as the above-mentioned periodic solution of the vector field $X_H$) are

$$\nu = \exp\left[\frac{\int_0^{2\pi} K_1(0,s,0,0)ds}{K_0'(0)}\right] \quad \text{and } \nu^{-1}.$$

If $\mu \neq 1$, then $\gamma$ is a hyperbolic periodic trajectory of the field $X_H$. By integrating the inequality (2.3) we get $\mu \neq \nu$; therefore, if $\mu = 1$, then $\nu \neq 1$, i.e., $\gamma$ is a hyperbolic periodic trajectory of the field $X_K$. Now suppose that $\gamma$ is an elliptic SPT. Due to (2.2), the condition of SPT being elliptic leads to the inequality (2.3). The system linearized along the periodic solution $x = y = I = 0, \theta = K_0'(0)t$, has the form

$$\dot{J} = 0, \quad \dot{\tau} = H_0''J, \quad \dot{\xi} = -\eta H_1(0, H_0'(0)t, 0, 0), \quad \dot{\eta} = \xi H_1(0, H_0'(0)t, 0, 0).$$

After the change of variables $u = \xi + i\eta$, the last two equations become one complex equation

$$\dot{u} = iH_1(0, H_0'(0)t, 0, 0)u.$$

Therefore, the multipliers are

$$\mu = \exp\left[\frac{i\int_0^{2\pi} H_1(0,s,0,0)ds}{H_0'(0)}\right] \quad \text{and } \mu^{-1},$$

i.e., $|\mu| = 1 = |\mu^{-1}|$. The proposition is proved. □

We now turn to the description of the structure of the IHVF in a neighborhood of SPT $\gamma$. The theorems on the structure of orbits of the action in the canonical neighborhood of a simple one-dimensional orbit proved in the previous section also allow one to describe the structure of a neighborhood of SPT with respect to the isoenergetic equivalence.

Suppose that $\gamma$ is an SPT for the IHVF $(X_H, K)$. By a canonical neighborhood of $\gamma$ we understand the canonical neighborhood constructed above of a one-dimensional orbit of the induced action $\Phi$. Recall that we actually have four types of SPTs: elliptic, oriented hyperbolic, nonoriented hyperbolic, and parabolic.

PROPOSITION 2.4.2. *Suppose that $\gamma$, $\gamma'$ are SPTs for the IHVFs $(X_H, K)$, $(X_{H'}, K')$, and $V, V'$ are their canonical neighborhoods. Then IHVF $(X_{H'}, K')$ in $V'$ is isoenergetically equivalent to IHVF $(X_H, K)$ in $V$ if and only if $\gamma$ and $\gamma'$ are of the same type.*

The proof of this proposition follows from the constructions presented in the proofs of the corresponding statements for one-dimensional orbits of the induced actions, since the conjugating homeomorphisms constructed there preserve the foliation into the level sets of the functions $H$ and $K$, respectively.

CHAPTER 3

# IHVF and Poisson Actions of Morse Type

The goal of this chapter is to describe a class of IHVFs and Poisson actions that will be called the IHVFs and actions of Morse type. The reason for introducing this class is that, as we have pointed out earlier, the structure of IHVFs and Poisson actions may be quite complex. For example, the reduction in a neighborhood of a one-dimensional orbit reduces the study of the structure of the foliation into orbits to the study of a foliation into level curves of a one-parameter family of the reduced Hamilton functions, and this family may have degenerate singular points unless we impose some additional restrictions. The problem becomes even more complicated if one passes to many degrees of freedom. In this case critical points requiring continuous invariants (moduli) for their classification may appear during the reduction [**34**]. On the other hand, the selected class of systems must be general enough in a sense that IHVFs and actions close to the given one must have the same structure. The question about genericness of the IHVFs under consideration has not been solved yet (see, however, [**37**]). Therefore, we will restrict ourselves to some examples, illustrating the "naturality" of the selected class of systems.

## 3.1. Limit set of an orbit of the action

Let $\Phi$ be an action of the group $\mathbb{R}^n$ on a manifold $M^k$, $k = \dim M^k > n$. At this moment, $\Phi$ is not assumed to be Poisson. For $x \in M^k$, we denote by $L_x$ an orbit of the action passing through $x$, i.e., $L_x = \{y \in M^k \mid y = \Phi(t;x),\ t \in \mathbb{R}^n\}$. Let $O_x$ be the stationary subgroup of the point $x$, $O_x = \{t \in \mathbb{R}^n \mid \Phi(t;x) = x\}$. Then $L_x$ is the image, under an injective immersion, of the manifold $\mathbb{R}^n/O_x$, which is a quotient group of $\mathbb{R}^n$ with respect to the subgroup $O_x$ and is isomorphic (as a Lie group) to the product $T^p \times \mathbb{R}^q$, $p+q \le n$ (see Chapter 1). Let us denote the quotient map by $\pi : \mathbb{R}^n \to \mathbb{R}^n/O_x$. We will call the orbit $L_x$ noncompact if $q > 0$, and compact if $q = 0$. Let us introduce the notion of a limit point and the limit set of the orbit $L_x$. Let $\{t_m\}$ be a sequence in $\mathbb{R}^n$ that has no limit points and satisfies the condition that the sequence $\pi(\{t_m\})$ in $\mathbb{R}^n/O_x$ does not have limit points either. Such a sequence is called completely divergent with respect to the orbit $L_x$. Nevertheless, the sequence $y_m = \Phi(t_m;x)$ may have limit points in $M^k$. The point $z \in M^k$ is called a limit point of the orbit $L_x$ if there exists a sequence $t_m$ completely divergent with respect to the orbit $L_x$ and such that $\Phi(t_m;x) \to z$. The set of all limit points of the orbit $L_x$ will be called its limit set. Note that our definition slightly differs from the conventional one [**33**]. In particular, the limit set of a compact orbit is empty by our definition. As in the case of regular dynamical systems [**33**], the following statement holds.

PROPOSITION 3.1.1. *If the closure of an orbit $L_x$ is compact, then the limit set of the orbit $L_x$ is nonempty, closed, and invariant (i.e., consists of whole orbits).*

In the case when $L_x$ belongs to its limit set we say that the orbit $L_x$ is Poisson stable. Another feature of a Poisson stable orbit is local disconnection of a set formed by an intersection of the orbit with a neighborhood (in $M^k$) of an arbitrary point on this orbit.

Now let us turn to the study of a Poisson action of $\mathbb{R}^n$ on a symplectic manifold $M^{2n}$. The following statement is true.

LEMMA 3.1.1. *Any $n$-dimensional orbit of a Poisson action of the group $\mathbb{R}^n$ is a locally connected set in $M^{2n}$.*

PROOF. Suppose that $x \in M$ and $L_x$ is $n$-dimensional. Then one can choose $n$ linearly independent one-parameter subgroups in $\mathbb{R}^n$ such that the corresponding Hamilton functions $H_1, \ldots, H_n$ are independent at the point $x$. Therefore, there exists a neighborhood $U$ of the point $x$ such that in $U$ the functions $H_1, \ldots, H_n$ generate $n$ transversal $(2n-1)$-dimensional foliations $H_1 = h_1, \ldots, H_n = h_n$. Moreover, the values of each of the functions $H_1, \ldots, H_n$ vary monotonically in the direction transversal to the corresponding foliation. Therefore, there exists a neighborhood $U_1 \subset U$ of the point $x$ such that there is only one fiber of the $n$-dimensional foliation $H_1 = H_1(x), \ldots, H_n = H_n(x)$ in $U$ that locally coincides with the orbit of the action. The lemma is proved. $\square$

This lemma immediately implies the following result.

COROLLARY 3.1.1. *An $n$-dimensional orbit of a Poisson action of the group $\mathbb{R}^n$ on a symplectic manifold $M^{2n}$ is Poisson stable.*

The following property of the dimension holds for the Poisson actions of the group $\mathbb{R}^n$.

PROPOSITION 3.1.2. *The limit set of an $n$-dimensional orbit is either empty or consists of orbits of dimension less than $n$.*

PROOF. Suppose that a point $y$ belongs to the limit set of an $n$-dimensional orbit $L_x$ ($y$ does not belong to $L_x$ due to Corollary 3.1.1) and an $n$-dimensional orbit $L_y$ passes through this point. Since $y$ is a limit point, there exists a sequence $\{t_m\}$ in $\mathbb{R}^n$ completely divergent with respect to $L_x$ such that the sequence $x_m = \Phi(t_m, x) \in L_x$ converges to $y$ in $M^{2n}$. Choosing Hamilton functions $H_1, \ldots, H_n$ as in Lemma 3.1.1 and using the continuity of $\Phi$ we see that the functions $H_1, \ldots, H_n$ take equal values $h_1^0 = H_1(x_m) = H_1(y), \ldots, h_n^0 = H_n(x_m) = H_n(y)$ on the orbits $L_x$ and $L_y$. Thus, we have a contradiction, since there is only one fiber of the $n$-dimensional foliation $H_1 = h_1, \ldots, H_n = h_n$ with the values $h_1 = h_1^0, \ldots, h_n = h_n^0$ in the neighborhood of the point $y$. The proposition is proved. $\square$

The situation with orbits of a dimension $k < n$ is essentially different. Let us give an example of Poisson action of the group $\mathbb{R}^2$ in $M^4$ which has a Poisson stable one-dimensional orbit. We will also demonstrate a possibility of the destruction of this structure by a small perturbation in the class of Poisson actions.

*Example.* Let us consider a geodesic flow on a flat two-dimensional torus $T^2$, i.e., a Hamiltonian system on the cotangent bundle $T^*T^2$ with coordinates $(p_1, p_2, \phi_1 (\bmod 2\pi), \phi_2 (\bmod 2\pi))$ and the form $\Omega = dp_1 \wedge d\phi_1 + dp_2 \wedge d\phi_2$ defined by the Hamilton function $H = (1/2)(p_1^2 + p_2^2)$. This system is integrable with the additional integral $K = \alpha p_1 + \beta p_2$ (for arbitrary $\alpha, \beta$). Consider one of the invariant tori $T_{hk}$ defined by the condition $p_1^2 + p_2^2 = h^2$, $\alpha p_1 + \beta p_2 = k$. The field $X_K$

defines a linear flow $\dot{\phi}_1 = \alpha, \dot{\phi}_2 = \beta$ on it. If $\alpha/\beta$ is irrational, then the trajectories of the field $X_K$ are everywhere dense on $T_h$. For $k_0 = h\sqrt{\alpha^2+\beta^2} > 0$ two tori $H = h$, $\alpha p_1 + \beta p_2 = k_0 > 0$ and $H = h$, $\alpha p_1 + \beta p_2 = -k_0$ are the sets of maximum and minimum points of the restriction of the function $K$ to the level $H = h^2/2$, respectively, and therefore, consist of one-dimensional orbits of the induced action $\Phi$. These orbits coincide with the trajectories of the field $X_K$ (and $X_H$ as well). Therefore, if $\alpha/\beta$ is irrational, they converge to themselves. Let us show that by slightly perturbing the action in the compact part of the phase space, one can destroy the tori of one-dimensional orbits, obtaining instead a finite number of isolated one-dimensional orbits of the action. Let us choose a sequence of rational numbers $\frac{m_s}{n_s}$ converging to the irrational number $\frac{\alpha}{\beta}$ as $s \to \infty$. Then the sequence of $\epsilon_s$ is determined with $\epsilon_s$ being a unique solution of the equation $\frac{\alpha+\epsilon}{\beta+\epsilon} = \frac{m_s}{n_s}$ for $s$ large enough. Then, as is easily checked, $\epsilon_s \to 0$ and the Hamilton function

$$H_\epsilon = (1/2)(p_1^2 + p_2^2) + \epsilon \sin(n\phi_1 - m\phi_2)$$

for $\epsilon = \epsilon_s$ defines IHVF $(X_H, K)$ with

$$K_\epsilon = (\alpha+\epsilon)p_1 + (\beta+\epsilon)p_2.$$

In the bounded region $p_1^2 + p_2^2 \leq h^2$ this IHVF is a small perturbation of the initial IHVF. It is evident that the level set $H_\epsilon = h/2$, being a slight deformation of the torus $H = h/2$, also is a three-dimensional torus. The set $\Sigma$ where the fields $X_{H_\epsilon}$ and $H_{K_\epsilon}$ are collinear, is defined by the equations

$$(\alpha+\epsilon)p_2 - (\beta+\epsilon)p_1 = 0, \qquad \cos(n\phi_1 - m\phi_2) = 0.$$

In $T^*T^2 = \mathbb{R}^2 \times T^2$ this set consists of a product of two closed curves on the torus $T^2 = \{(\phi_1, \phi_2) \,(\mathrm{mod}\, 2\pi)\}$ and a straight line on the plane $(p_1, p_2)$. On the level $H_\epsilon = h^2/2 = (1/2)(p_1^2 + p_2^2) + \epsilon \sin(n\phi_1 - m\phi_2)$ the set $\Sigma$ cuts out four closed curves defined by the equations

$$(\alpha+\epsilon)p_2 - (\beta+\epsilon)p_1 = 0, \qquad p_1^2 + p_2^2 = h^2 \pm \epsilon, \qquad n\phi_1 - m\phi_2 = \pi/2, 3\pi/2.$$

It is easy to show that two curves are elliptic (maximum and minimum of the restriction of $K$ to the level $H = h^2$) and the other two are saddle.

## 3.2. Class of Poisson actions and IHVFs of Morse type

Let $\Phi$ be a Poisson action of the group $\mathbb{R}^2$ on a four-dimensional symplectic manifold $M$. Let us introduce the following notation for distinguishing a class of actions. Two orbits of dimension less than two will be called *coupled* if they belong to the closure of one and the same noncompact orbit. A set of one-dimensional and zero-dimensional orbits that contains, along with every orbit, all orbits coupled with it, is called a *garland*. The union of all the orbits that contain the elements of a garland in their closures, is called the separatrix set of the garland. From now on we only consider connected separatrix sets.

Let us turn to the decription of the class of Poisson actions which we will consider later.

DEFINITION 3.1. A Poisson action $\Phi$ of the group $\mathbb{R}^2$ will be called a Morse action if the following conditions hold:

1. The closure of any orbit is compact (Lagrange stability).
2. The set of singular points is finite, and they are all simple.

3. All one-dimensional orbits are simple. There is a finite number of parabolic orbits.
4. Any garland with connected separatrix set contains at most one singular point.

Simultaneously with Definition 3.1 we will formulate the related definition for an IHVF $(X_H, K)$.

DEFINITION 3.2. An IHVF $(X_H, K)$ will be called a Morse IHVF if the following conditions hold:

1. The Poisson action generated by $(X_H, K)$ is a Morse action.
2. Singular points of the field $X_H$ do not have multiple eigenvalues.

From now on we will use the abbreviation PA for a Poisson action of the group $\mathbb{R}^2$ of Morse type on a four-dimensional symplectic manifold. Analogously, by IHVF we mean an integrable Hamiltonian vector field of Morse type.

## 3.3. General properties of PA of Morse type

Let us establish the main properties of PA. Recall (Proposition 3.1.2) that for any Poisson action of the group $\mathbb{R}^2$ the boundary set of a two-dimensional orbit, provided it is not empty, consists of orbits of lower dimension. As we saw in the previous section, the property that the dimension decreases in passing to the boundary does not necessarily hold for one-dimensional orbits. For PA the situation is different.

PROPOSITION 3.3.1. *For PA the limit set of a noncompact orbit consists of orbits of lower dimension.*

PROOF. Let $L_x$ be a noncompact orbit. It follows from the Lagrange stability and Proposition 3.1.1 that the limit set of the orbit $L_x$ is not empty and consists of the whole orbits. For two-dimensional orbits Proposition 3.3.1 follows from Proposition 3.1.2. For a one-dimensional orbit we use part 3 of Definition 3.1 and the results on local structure of orbits close to a given simple one-dimensional orbit (see Section 2.3). It follows from this description that the set of points of the orbit is locally connected for all types of simple orbits. Therefore, one-dimensional orbits of a Morse action are not Poisson stable. Suppose now that $y \in M$ is a limit point for $L_x$ (then $y$ does not belong to $L_x$) and $L_y$ is the orbit passing through the point $y$. The orbit $L_y$ cannot be a two-dimensional orbit, since a neighborhood of a point on a two-dimensional orbit is foliated into two-dimensional orbits. Also $L_y$ cannot be a one-dimensional orbit. Indeed, in this case one can choose two independent one-parameter subgroups in $\mathbb{R}^2$ such that one of the corresponding Hamilton functions, for example $H$, does not have singular points on $L_y$. Then we can apply the reduction scheme (see Section 2.3) and it follows from this scheme that there are no other one-dimensional orbits except for $L_x$ on the level $H = H(y)$ in the neighborhood of the point $y$. On the other hand, since $y$ is a limit point for $L_y$, there are points of the orbit $L_x$ in any neighborhood of the point $y$ on the level $H = H(y)$. This contradiction proves the proposition. □

Now let us describe the local structure of the set of orbits of PA containing a simple singular point $p$ in their limit set (local separatrix set of a singular point). Let us choose two independent one-parameter subgroups in $\mathbb{R}^2$ and let $H$, $K$ be the corresponding Hamilton functions. As indicated in Theorem 2.2.1, there exists a neighborhood of a simple singular point $p$ such that $H$, $K$ have the form $H = h(\xi, \eta)$, $K = k(\xi, \eta)$ in this neighborhood and the homogenous quadratic functions $\xi$, $\eta$ depend on the type of the point. Suppose that an orbit $L$ contains the point $p$ in its limit set. Then in a sufficiently small neighborhood of the point $p$ the points of this orbit (to be more precise, the part belonging to the neighborhood) are solutions of the system $H = H(p)$, $K = K(p)$. By a local $p$-orbit of the action we will mean a connected component of the intersection of an action orbit with the neighborhood under consideration and containing $p$ in its closure.

PROPOSITION 3.3.2. *The local separatrix set of a simple singular point $p$ consists of:*

1. *The point $p$ itself in the case of elliptic point; the point $p$ and four one-dimensional local $p$-orbits forming a "cross", in the case of saddle-center.*
2. *The point $p$, eight one-dimensional local $p$-orbits, and 16 two-dimensional local $p$-orbits in the case of saddle-saddle. There exists a local symplectic diffeomorphism taking this set of local $p$-orbits into the union of four Lagrange disks in the linear symplectic space $(\mathbb{R}^4, dx_1 \wedge dy_1 + dx_2 \wedge dy_2)$ defined by the system of equations $x_1y_1 = 0$, $x_2y_2 = 0$. Under this diffeomorphism coordinate semiaxes correspond to one-dimensional $p$-orbits, and the sectors on the Lagrange planes obtained by removing axes correspond to two-dimensional $p$-orbits.*
3. *The point $p$ and two two-dimensional $p$-orbits (homeomorphic to an open punctured disk) in the case of focus-focus. Their union forms two smooth Lagrange disks intersecting transversally at the point $p$.*

The proof of this statement follows from the local representation of the functions $H$, $K$ and from the types of the functions $\xi$, $\nu$ in each of the four cases (Theorem 2.2.1). Now let us establish a connection between the type of a simple singular point and the type of one-dimensional orbits passing through a neighborhood of this point. In order to do this we will describe the degeneracy set $\Sigma$ in a neighborhood of a simple singular point.

PROPOSITION 3.3.3. *In a sufficiently small neighborhood of a singular point of the focus-focus type the set $\Sigma$ coincides with this point. There exist neighborhoods of singular points of the saddle-saddle, saddle-center, and center-center types such that the set $\Sigma$ is the union of two symplectic two-dimensional disks $\Sigma_1$, $\Sigma_2$, intersecting transversally at a singular point. The foliation into one-dimensional orbits of the action on these disks is diffeomorphic to the foliation into level curves of either the function $x^2 + y^2 = c$ (center) or the function $x^2 - y^2 = c$ (saddle) on the plane $\{(x, y)\}$. In the center-center case both foliations on $\Sigma_1$, $\Sigma_2$ are of the center type, in the saddle-center case the foliation is of the center type on one of the disks and of the saddle type on the other, and in the saddle-saddle case both foliations are of the saddle type.*

PROOF. Let us choose, as usual, two Hamilton functions $H$, $K$ and symplectic coordinates such that $H = h(\xi, \eta) = \lambda_1\xi + \lambda_2\nu + \cdots$, $K = k(\xi, \eta) = \nu_1\xi + \nu_2\nu + \cdots$ in these coordinates. The degeneracy set is determined by the condition of linear dependence of the vector fields $X_H$, $X_K$. For the functions $H$, $K$ of the above type, the condition of simplicity of a singular point (Corollary 2.1.1) given by $\Delta = \lambda_1\nu_2 - \lambda_2\nu_1 = D(h, k)/D(\xi, \eta)\,|_{(\xi=\eta=0)} \neq 0$ implies that the linear dependence of the vector fields $X_H$, $X_K$ is equivalent to the linear dependence of the vector fields $X_\xi$, $X_\eta$ (possibly in a smaller neighborhood). Therefore, we have

1. Center-center: $\xi = (x_1^2 + y_1^2)/2$, $\eta = (x_2^2 + y_2^2)/2$, $X_\xi = (-y_1, 0, x_1, 0)^\top$, $X_\eta = (0, -y_2, 0, x_2)^\top$, $\Sigma_1 = \{x_2 = 0,\ y_2 = 0\}$, $\Sigma_2 = \{x_1 = 0,\ y_1 = 0\}$. The foliation on $\Sigma_1$ coincides with the trajectories of the field $X_\xi$, and that on $\Sigma_2$ with the trajectories of the field $X_\eta$, i.e., they are both centers.

2. Saddle-center: $\xi = (x_1^2 + y_1^2)/2$, $\eta = x_2 y_2$, $X_\xi = (-y_1, 0, x_1, 0)^\top$, $X_\eta = (0, -x_2, 0, y_2)^\top$, $\Sigma_1 = \{x_2 = 0,\ y_2 = 0\}$, $\Sigma_2 = \{x_1 = 0,\ y_1 = 0\}$. The foliation on $\Sigma_1$ coincides with the trajectories of the field $X_\xi$, and that on $\Sigma_2$ with the trajectories of the field $X_\eta$, i.e., one is a center, and the other is a saddle.

3. Saddle-saddle: $\xi = x_1 y_1$, $\eta = x_2 y_2$, $X_\xi = (-x_1, 0, y_1, 0)^\top$, $X_\eta = (0, -x_2, 0, y_2)^\top$, $\Sigma_1 = \{x_2 = 0, y_2 = 0\}$, $\Sigma_2 = \{x_1 = 0, y_1 = 0\}$. The foliations on $\Sigma_1$ and $\Sigma_2$ are of the saddle type.

4. Focus-focus: $\xi = x_1 y_1 + x_2 y_2$, $\eta = x_1 y_2 - x_2 y_1$, $X_\xi = (-x_1, -x_2, y_1, y_2)^\top$, $X_\eta = (x_2, -x_1, y_2, -y_1)^\top$.

Therefore, we can see that $X_\xi$ and $X_\eta$ are only dependent at the point $x_1 = x_2 = y_1 = y_2 = 0$. The proposition is proved. □

Now we can define the types of one-dimensional orbits on the disks $\Sigma_i$, $i = 1, 2$.

PROPOSITION 3.3.4. *One-dimensional orbits belonging to the disks $\Sigma_i$, $i = 1, 2$, are simple and have the following form:*

1. *elliptic on both disks $\Sigma_i$, $i = 1, 2$, for the center-center;*
2. *hyperbolic on $\Sigma_1$, elliptic on $\Sigma_2$ for the saddle-center;*
3. *hyperbolic on both disks $\Sigma_i$, $i = 1, 2$, for the saddle-saddle.*

PROOF. Let us consider, for example, the disk $\Sigma_1$. Without loss of generality we can assume that $\lambda_1 \neq 0$. Suppose that $m_0 = (x_1^0, 0, y_1^0, 0) \in \Sigma_1$, $x_1^0 y_1^0 \neq 0$. Let us introduce the following notation: $c = h(\xi^0, 0)$, $\xi^0 = x_1^0 y_1^0$ or $\xi^0 = ((x_1^0)^2 + (y_1^0)^2)$. We will work in a small neighborhood of the point $p = (0,0,0,0)$ such that the equation $h(\xi, \eta) = c$ has a unique solution for $\xi$, $\xi = \varphi(\eta, c) = c/\lambda_1 - (\lambda_2/\lambda_1)\eta + \cdots$ in this neighborhood. Then the reduced Hamilton function $K$ can be expressed in the form $\hat{k}(x_2, y_2, c) = k(\varphi(\eta, c), \eta) = \nu_1 c_1/\lambda_1 + (\Delta/\lambda_1)\eta + \cdots$ on the level $H = c$. Therefore, the point $x_2 = y_2 = 0$ corresponding under the reduction to the point $m_0$ is a critical point for the function $\hat{k}$ with respect to variables $(x_2, y_2)$. Depending on the form of $\eta$ we have a point of the center type if $\eta = (x_2^2 + y_2^2)/2$, or a point of the saddle type if $\eta = x_2 y_2$; this can be verified by a direct calculation. This proves all statements of our proposition. □

COROLLARY 3.3.1. *PA has a finite number of noncompact one-dimensional orbits, which can be of the elliptic and hyperbolic type only, and form loops (i.e., their closures contain one singular point). The limit set of a noncompact elliptic orbit consists of one singular point of the saddle-center type; the limit set of a noncompact hyperbolic orbit consists of one singular point of the saddle-saddle type. All parabolic orbits are compact.*

PROOF. It follows from Proposition 3.3.4 that the limit set of a noncompact one-dimensional orbit contains one singular point, which is a simple singular point. It follows from Proposition 3.3.4 that all such orbits are elliptic or hyperbolic, i.e., parabolic orbits are closed. The finiteness of the set of noncompact orbits follows from the finiteness of the number of singular points and the finiteness of the set of local $p$-orbits (see Proposition 3.3.2). Finally, it follows from condition 4 of

Definition 3.1 that a noncompact orbit contains one singular point in its closure, and the type of this point is determined by Proposition 3.3.4. □

We now start studying the structure of the limit set of a noncompact two-dimensional orbit. As we know, these orbits can be of two types: a cylinder and a plane.

PROPOSITION 3.3.5. *Assume that a Poisson action satisfies conditions 1–3 of Definition 3.1. Then the limit set of a two-dimensional orbit of the cylinder type consists of at most two components such that each of them is either a singular point of the focus-focus type or a compact one-dimensional orbit of the parabolic or hyperbolic type. The limit set of a two-dimensional orbit of the plane type is a contour consisting of singular points of the saddle-saddle type and noncompact one-dimensional orbits of the hyperbolic type.*

PROOF. Let $L$ be a one-dimensional orbit of the cylinder type and $x \in L$. A stationary subgroup $O_x$ of the point $x$ has the form $n\vec{e}$, $\vec{e} \in \mathbb{R}^2$, $n \in \mathbb{Z}$. Let us consider a one-parameter subgroup $t\vec{e}$, $t \in \mathbb{R}$, and the corresponding Hamilton function $K$. Then the trajectories of the field $X_K$ are closed and nonhomotopic to zero on $L$. Let us fix one of these closed trajectories and denote it by $\gamma$. This closed curve divides $L$ into two open cylinders. It is clear then that the limit set of the orbit $L$ consists of at most two connected components. Let us consider one of the components $\Gamma$ of the limit set and a cylinder $L_1 \subset L$ adjacent to it such that $\gamma$ is one of its boundary curves. Suppose that $\Gamma$ contains a singular point $p$. It follows from Proposition 3.3.2 that this point cannot be of the center-center or saddle-center type, i.e., there are no two-dimensional local $p$-orbits adjacent to these types of singular points. In order to prove the proposition we have to exclude the saddle-saddle case. Suppose that the proposition is not true and $p$ is of the saddle-saddle type. Then $L$ contains one of local two-dimensional $p$-orbits (a sector $\sigma$ between two local one-dimensional $p$-orbits, which also obviously belong to $\Gamma$). Let $H$ be the second Hamilton function, corresponding to another one-parameter subgroup independent of $t\vec{e}$. It is clear that the trajectories of the field $X_H$ on $L$ are transversal to all (closed) trajectories of the field $X_K$ for any choice of the second independent subgroup in $\mathbb{R}^2$.

Let us add to $\sigma$ all adjacent one-dimensional local $p$-orbits and the point $p$. Let us denote this set by $\hat{\sigma}$. It is always possible to choose a one-parameter subgroup in $\mathbb{R}^2$ and, consequently, the Hamilton function $H$ so that the point $p$ be a saddle of the field $X_H$ and $\hat{\sigma}$ belong to the stable manifold. It follows that there exist coordinates in a neighborhood of the point $p$ such that $H = h(\xi, \eta) = \lambda_1\xi + \lambda_2\eta + \cdots$, $K = k(\xi, \eta) = \nu_1\xi + \nu_2\eta + \cdots$, $\Delta = \lambda_1\nu_2 - \lambda_2\nu_1 \neq 0$ (simplicity condition for the point $p$). Changing $H$ to $H_1 = aH + bK$ (which is equivalent to choosing a new one-parameter subgroup in $\mathbb{R}^2$) we obtain that $\lambda_1' = a\lambda_1 + b\nu_1$, $\lambda_2' = a\lambda_2 + b\nu_2$, i.e., it is possible to choose $a, b$ such that $\lambda_1', \lambda_2'$ have any fixed set of signs. In these coordinates the local $p$-orbit belongs to one of the four Lagrange planes. For example, let it be the plane $y_1 = y_2 = 0$ and the sector $\sigma$ be $x_1 > 0, x_2 > 0$. Then it is a sector on the stable manifold of the point $p$, provided $\lambda_1' > 0, \lambda_2' > 0$. In this case there exists a point $m$ on $\gamma$ such that a trajectory of the field $X_H$ which tends to $p$ as $t \to \infty$ passes through the point $m$. Therefore, all trajectories of the filed $X_H$ passing through the points of $\gamma$ belong to $L_1$ for increasing $t$, and their $\omega$-limit set belongs to $\Gamma$.

First, let us show that all points of $\gamma$ belong to the stable manifold of the point $p$. Let $m', m' \neq m$ be an arbitrary point of $\gamma$. Since $\gamma$ is a trajectory of the filed $X_K$, there exists $s$ such that $\Phi(s, 0; m') = m$, where $\Phi(t_1, t_2; x)$ is the image of the point $x$ under the action of the element $(t_1, t_2) \in \mathbb{R}^2$. Here $\Phi(t, 0; x)$ is the flow of the field $X_K$, and $\Phi(0, t; x)$ is the flow of the field $X_{H_1}$. If $t_n \to \infty$, then $\Phi(s, t_n; m') = \Phi(0, t_n; \Phi(s, 0; m')) = \Phi(0, t_n, m) \to p$, since $p$ is the only limit point of the trajectory $\Phi(0, t, m)$. Therefore, $\Phi(s, 0, y) = p$. But $p$ is a singular point for the action, hence $y = p$. Thus, $\gamma$ belongs to the stable manifold of the point $p$. Since there are no one-dimensional orbits of the action on $L$, all points on $\gamma$ reach $\sigma$ in finite time, which contradicts the fact that $\gamma$ is not homotopic to zero on $L$. Therefore, either $\Gamma$ coincides with a point of the focus-focus type or $\Gamma$ does not contain any singular points. In the latter case $\Gamma$ is compact and connected, i.e., it coincides with a one-dimensional compact orbit. Such an orbit cannot be elliptic, since there are no two-dimensional compact orbits adjacent to it.

Now we consider an orbit of the plane type. It is clear that the limit set $\Gamma$ of the orbit $L$ is connected and consists of finitely many one-dimensional orbits and singular points. Let us show that $\Gamma$ contains a singular point of the saddle-saddle type. Assume the contrary. The set $\Gamma$ cannot contain singular points of the saddle-center, center-center types and one-dimensional elliptic orbits, since there are no two-dimensional orbits adjacent to them (in other words, these singular points and orbits do not belong to the $\omega$-limit set of the two-dimensional orbits). This fact and the absence of points of the saddle-saddle type in $\Gamma$ imply that there are no one-dimensional hyperbolic noncompact orbits in $\Gamma$. Therefore, $\Gamma$ can only contain a singular point of the focus-focus type or closed one-dimensional orbits of parabolic or hyperbolic type. The fact that $\Gamma$ is connected, closed, and one-dimensional implies that it coincides with one of these orbits.

In the case where $\Gamma$ consists of the focus-focus type singular point $p$, the closure of the set $\Gamma$ is a smooth two-dimensional sphere $S$. It is possible to choose one-parameter subgroups of $\mathbb{R}^2$ so that the corresponding vector field $X_H$ have a singular point of the saddle-focus type at the point $p$, and its sphere $S$ coincide with its local stable manifold locally in a neighborhood of the point $p$. Since the Euler characteristic of the sphere $S$ is two and the index of the singular point $p$ for the restriction of the filed $X_H$ to the sphere $S$ is one, there exists another singular point in $S \backslash \{p\}$.

In the other case the closure of $L$ is a smooth two-dimensional disk ($\Gamma$ is a parabolic or an orientable hyperbolic one-dimensional orbit) or a smooth projective plane ($\Gamma$ is a nonorientable hyperbolic one-dimensional orbit). It is possible to choose the Hamilton function $H$ so that $\Gamma$ be a closed trajectory of the field $X_H$. Then there should be a singular point of the filed $X_H$ on $L$, since the Euler characteristic of a disk and a projective plane is not equal to zero. In all cases the existence of a singular point of the filed $X_H$ in $L$ is in contradiction with the fact that $\dim L = 2$.

Therefore, $\Gamma$ contains only singular points of the saddle-saddle type. Therefore, it contains one-dimensional hyperbolic orbits adjacent to these points and belonging to the boundaries of local two-dimensional $p$-orbits belonging to $L$. The statement is proved. □

CHAPTER 4

# Center-Center Type Singular Points of PA and Elliptic Singular Points of IHVF

The results of the previous chapters allow us to turn to the main goal of this book: obtaining invariants of the topological conjugacy of PA and isoenergetic equivalence of IHVFs in some special neighborhoods of separatrix garland sets which contain a unique simple singular point (in the extended neighborhoods of a singular point), and to the proof of the equivalence theorems. We will consider each type of singular points separately.

Let us first study the IHVF $(X_H, K)$ with an elliptic singular point $p$. It has been pointed out already that there is a neighborhood $U$ of the point $p$ such that in this neighborhood $H$ and $K$ have the form

$$(4.1)\quad H = h(\xi_1, \xi_2) = \omega_1\xi_1 + \omega_2\xi_2 + \cdots, \qquad K = k(\xi_1, \xi_2) = \nu_1\xi_1 + \nu_2\xi_2 + \cdots,$$

where $\xi_i = (x_i^2 + y_i^2)/2$.

The simplicity condition of a singular point means that $\Delta = \omega_1\nu_2 - \omega_2\nu_1 \neq 0$ and $\omega_1\omega_2 \neq 0$. Without loss of generality we can assume that $\omega_1 > 0$.

Let us construct a neighborhood of the point $p$ which is invariant under the action. In this simplest case the garland of the point $p$ and its separatrix set coincide with the point $p$. Therefore, the extended neighborhood $V$ is just a properly chosen neighborhood of the point $p$ in $M$.

We will distinguish between two cases: A ($\omega_1\omega_2 > 0$) and B ($\omega_1\omega_2 < 0$). Under the assumption $\omega_1 > 0$, Case A implies that the quadratic form $\omega_1(x_1^2 + y_1^2)/2 + \omega_2(x_2^2 + y_2^2)/2$ is positive definite. Case B implies that $\omega_2 < 0$.

To construct the neighborhood $V$, let us consider the positive quadrant $\xi_1 \geq 0$, $\xi_2 \geq 0$ of the plane $(\xi_1, \xi_2)$. In Case A we consider the curve $l:\ h(\xi_1, \xi_2) = h_* > 0$ for sufficiently small $h_*$. This curve cuts off a curvilinear triangle $\delta$ with sides that coincide with the axes $\xi_1 = 0$, $\xi_2 = 0$ and the curve $l$ from the quadrant. The triangle $\delta$ is smoothly foliated by the curves $h(\xi_1, \xi_2) = \epsilon$, $0 < \epsilon \leq h_*$, and every such curve transversally intersects the sides $\xi_1 = 0$, $\xi_2 = 0$ of the triangle (Figure 4.1a).

In Case B let us choose a line $l$: $\alpha\xi_1 + \beta\xi_2 = c$, $\alpha > 0$, $\beta > 0$ in the positive quadrant $\xi_1 \geq 0$, $\xi_2 \geq 0$ of the plane $(\xi_1, \xi_2)$ in such a way that $\beta\omega_1 - \alpha\omega_2 \neq 0$. Then, for sufficiently small $c > 0$ and for sufficiently small $|\epsilon|$ this line cuts off from the positive quadrant a triangle $\delta$ with sides transversal to the curve $h(\xi_1, \xi_2) = \epsilon$ for $|\epsilon|$ small enough. For $\epsilon = -h_2^* = h(0, c/\beta) < 0$ and $\epsilon = h_1^* = h(c/\alpha, 0)$, these curves pass through the vertices of the triangle lying on the axis $\xi_1 = 0$ and $\xi_2 = 0$, respectively. The curve $h(\xi_1, \xi_2) = 0$ passes through the point $(0, 0)$ (Figure 4.1b).

Let us denote by $\mu : x \to (\xi_1(x), \xi_2(x))$ the modified moment transformation. Then we choose $V = \mu^{-1}(\delta)$.

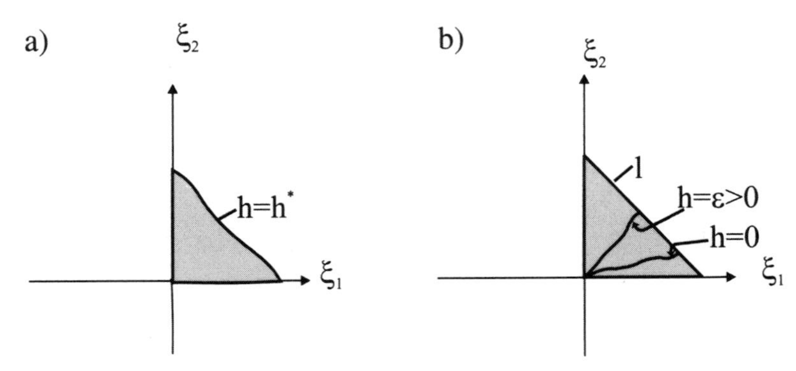


FIGURE 4.1

LEMMA 4.1.1. *$V$ is invariant under the induced action and homeomorphic to a four-dimensional ball $D^4$. $\partial V$ is homeomorphic to a three-dimensional sphere $S^3$.*

PROOF. In Case A our lemma follows from the Morse lemma [**31**], since $V$ is the set of smaller values of a smooth function $H$ in a neighborhood of a nondegenerate singular point $p$ such that the quadratic part of $H$ in $p$ is positive definite.

Let us consider Case B. The preimage of the interval $\alpha\xi_1 + \beta\xi_2 = c$, $\xi_1 \geq 0$, $\xi_2 \geq 0$ is a three-dimensional ellipsoid $\alpha(x_1^2+y_1^2)/2+\beta(x_2^2+y_2^2)/2 = c$, and $\mu^{-1}(\delta)$ is a region whose boundary is this ellipsoid. The boundary is contracted to a point as $c \to +0$. Therefore, $\mu^{-1}(\delta)$ is a closed four-dimensional ball $D^4$. Since $\xi_1$ and $\xi_2$ are local integrals of IHVF $(X_H, K)$, $\mu^{-1}(\delta)$ is invariant under the induced action. □

Now we describe the partition of the neighborhood $V$ by levels of the integral $H = h$. We denote $V_\epsilon = \{x \in V | H(x) = \epsilon\}$.

PROPOSITION 4.1.1. *In Case A, $V_\epsilon$ is diffeomorphic to $S^3$ for $\epsilon > 0$, $V_0 = \{p\}$. In case B, the sets $V_\epsilon$ are diffeomorphic to a solid torus $D^2 \times S^1$ for $\bar{\epsilon}\{0, h_1^*, h_2^*\}$. $V_0$ is homeomorphic to a cone on a torus, and $V_{h_1^*}$ and $V_{h_2^*}$ are closed curves.*

PROOF. In Case A our proposition follows from the Morse lemma [**31**]. Let us consider Case B. For example, let $\epsilon > 0$. We may assume that $\epsilon, \xi_1, \xi_2$ are so small that the equation $h(\xi_1, \xi_2) = \epsilon$ has a unique solution of the form $\xi_2 = \phi(\xi_1, \epsilon) = \epsilon/\omega_2 - (\omega_1/\omega_2)\xi_1 + O(\epsilon^2 + \xi_1^2)$. Therefore, $V_\epsilon$ is the preimage of the piece of the curve $\xi_2 = \phi(\xi_1, \epsilon)$ between the sides $\alpha\xi_1 + \beta\xi_2 = c$ and $\xi_2 = 0$ of the triangle $\delta$ (see Figure 4.1). For $\xi_2 = 0$ we obtain the closed curve $x_2 = y_2 = 0$, $x_1^2 + y_1^2 = 2\xi_1^*(\epsilon)$, where $\xi_1^*(\epsilon)$ is the unique solution of the equation $h(\xi_1, 0) = \epsilon$. For $\xi_2 > 0$ and fixed $\xi_1$, $\xi_1^*(\epsilon) \leq \xi_1 \leq \hat{\xi}_1(\epsilon)$, we obtain a torus $T(\xi_1) = \{x_1^2+y_1^2 = 2\xi_1, x_2^2+y_2^2 = 2\phi(\xi_1, \epsilon)\}$ in $V_\epsilon$, where $\hat{\xi}_1(\epsilon)$ is the unique solution of the system $\alpha\xi_1 + \beta\xi_2 = c$, $h(\xi_1, \xi_2) = \epsilon$, which exists due to the transversality of these two curves. The union of the tori $T(\xi_1)$, $\xi_1^*(\epsilon) \leq \xi_1 \leq \hat{\xi}_1(\epsilon)$, forms a solid torus $V_\epsilon$. Since $\xi_1^*(0) = 0$, the axis of the solid torus shrinks to the point $p$ as $\epsilon \to 0$. Therefore, $V_0$ is a cone over the torus. The proposition is proved. □

Now we describe the foliation of the level sets of $V_\epsilon$ into the orbits of the induced action. The condition $\Delta \neq 0$ implies that in a sufficiently small neighborhood $V$, this foliation, which is defined by the system $H = \epsilon$, $K = \kappa$, is equivalent to the foliation defined by the conditions $\xi_1 = a(\epsilon, \kappa)$, $\xi_2 = b(\epsilon, \kappa)$. In the neighborhood of the point $(0, 0)$, the curves $h(\xi_1, \xi_2) = \epsilon$, $k(\xi_1, \xi_2) = \kappa$ form two families of transversal curves on the plane $(\xi_1, \xi_2)$. First let us consider Case A. In this case $V_\epsilon$ is a preimage of the piece of the curve $h(\xi_1, \xi_2) = \epsilon$ between the axes $\xi_1 = 0$ and $\xi_2 = 0$. The condition $\omega_1\omega_2 > 0$ implies that this piece is transversal to the axes (see Figure 4.1). It follows from the transversality of the piece $h(\xi_1, \xi_2) = \epsilon$ to the curves $k(\xi_1, \xi_2) = x$ that the function $k$ varies monotonically along this piece, and its minimum and maximum values are attained at the endpoints, i.e., at the points on the axes $\xi_1 = 0$ and $\xi_2 = 0$. Therefore, there is the following correspondence in $V_\epsilon$, which is a smooth three-dimensional sphere: the endpoints of the piece of the curve correspond to two smooth closed curves which are elliptic periodic trajectories of the IHVF belonging to $\Sigma_2 = \{x_1 = y_1 = 0\}$ and $\Sigma_1 = \{x_2 = y_2 = 0\}$, respectively. They are linked in $V_\epsilon$ with the linking coefficient 1. The other values of the function $k$ between the minimum and maximum values correspond to a unique two-dimensional Liouville torus.

In Case B, for $\epsilon \notin \{0, h_1^*, h_2^*\}$, $V_\epsilon$ is a solid torus which is a preimage of the piece of the curve $h(\xi_1, \xi_2) = \epsilon$ between the straight line $\alpha\xi_1 + \beta\xi_2 = c$ and one of the axes, $\xi_2 = 0$, for $\epsilon > 0$, and $\xi_1 = 0$, for $\epsilon < 0$ (see Figure 4.1). The endpoint of the piece of the curve $h(\xi_1, \xi_2) = \epsilon$ belonging to the axis corresponds to a closed elliptic periodic trajectory of the IHVF which belongs to $\Sigma_1$ or $\Sigma_2$. Other points of the piece of the curve correspond to two-dimensional Liouville tori. It follows from the transversality of the curves $h(\xi_1, \xi_2) = \epsilon$ and $k(\xi_1, \xi_2) = \kappa$ that the function $k$ varies monotonically on $V_\epsilon$ in the direction transversal to the tori. For $\epsilon = 0$, the periodic trajectory coincides with the point $p$. For $\epsilon \in \{h_1^*, h_2^*\}$, $V_\epsilon$ itself is a periodic elliptic trajectory of the IHVF. Therefore, we have proved the following result.

PROPOSITION 4.1.2. *The foliation of the level sets of $V_\epsilon$ into the orbits of the induced action has the following form:*

a) *In Case A, there are two compact one-dimensional elliptic orbits on a smooth three-dimensional sphere $V_\epsilon$ which are linked in $V_\epsilon$ with the linking coefficient* 1. *The other fibers are two-dimensional compact orbits (Liouville tori), and $V_0 = \{p\}$ for $\epsilon = 0$.*

b) *In Case B, there is one compact one-dimensional elliptic orbit in a solid torus $V_\epsilon$ ($\epsilon \notin \{0, h_1^*, h_2^*\}$), and this orbit is the axial line of a solid torus. The other fibers are two-dimensional compact orbits (tori surrounding the axial line). For $\epsilon = 0$, the foliation into tori is preserved, but the axial line of the solid torus has been contracted at a point. For $\epsilon \in \{h_1^*, h_2^*\}$, the set $V_\epsilon$ coincides with a compact one-dimensional elliptic orbit.*

REMARK 4.1. Since the change of variables $(\xi_1,\ \xi_2) \to (H,\ K)$ is a local diffeomorphism, the moment transformation $\mu : V \to (H,\ K)$ has the bifurcation diagram as in Figure 4.2.

Now we can prove the main theorem of this chapter. We consider two IHVFs $(X_H, K)$ and $(X_{H'}, K')$ on four-dimensional $C^\infty$-smooth symplectic manifolds $M$ and $M'$, respectively. We assume that they have simple elliptic singular points $p \in M$ (for $(X_H, K)$) and $p' \in M'$ (for $(X_{H'}, K')$). Let $V$, $V'$ be the corresponding neighborhoods constructed above. From now on, we will denote objects from the second system by the same letters we use for the first system, but with primes.

THEOREM 4.1.1. *The IHVF $(X_H, K)$ in $V$ is isoenergetically equivalent to the $(X_{H'}, K')$ in $V'$ if and only if the same case $A$ or $B$ occurs for both fields.*

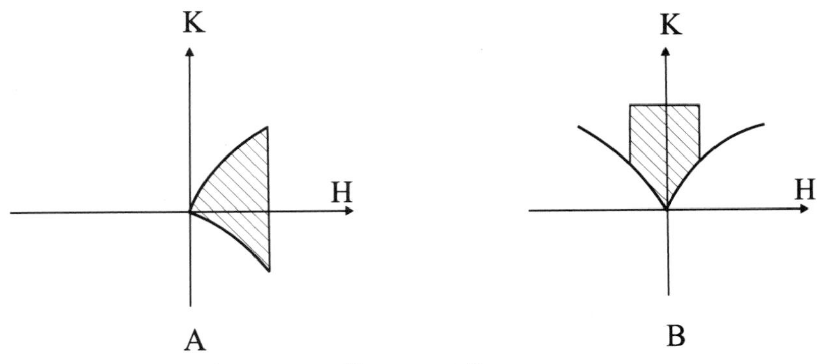


FIGURE 4.2

PROOF. *The "only if" part.* Let the IHVF $(X_H, K)$ in $V$ be isoenergetically equivalent to the IHVF $(X_{H'}, K')$ in $V'$. Then $V_\epsilon$ is homeomorphic to $V'_{\epsilon'}$ for some $\epsilon'$. However, for different cases, these sets are not homeomorphic (see Proposition 4.1.1).

*The "if" part.* Suppose that the same case occurs for both fields. Considering, if necessary, the field $(X_{-H}, K)$ instead of the field $(X_H, K)$, we can assume that $\omega_1 > 0$, $\omega'_1 > 0$ in (4.1). As before, by $\delta$ and $\delta'$ we denote curvilinear (in Case A) or linear (in Case B) triangles in the quadrant $\xi_1 \geq 0$, $\xi_2 \geq 0$ on the plane $(\xi_1, \xi_2)$ such that $V = \mu^{-1}(\sigma)$, $V' = {}'\mu^{-1}(\delta')$. Let us define a homeomorphism $q : \delta \to \delta'$, $q(0,0) = (0,0)$ preserving the foliation of these triangles into the curves $h(\xi_1, \xi_2) = \epsilon$ and $h'(\xi'_1, \xi'_2) = \epsilon'$ respectively, and such that the level set $h(\xi_1, \xi_2) = 0$ is mapped into the level set $h'(\xi'_1, \xi'_2) = 0$, and the curves with $\epsilon > 0$ are mapped into the curves with $\epsilon' > 0$. In Case A we will also require that the axis $\xi_1 = 0$ be mapped into the axis $\xi'_1 = 0$.

Let us define symplectic polar coordinates in $V$:

$$x_i = \sqrt{2\xi_i}\cos\varphi_i, \quad y_i = \sqrt{2\xi_i}\sin\varphi_i, \qquad i = 1, 2,$$

where $(x_1, x_2, y_1, y_2)$ are symplectic coordinates in which $H = h(\xi_1, \xi_2)$ and $K = k(\xi_1, \xi_2)$. Then for a fixed $\xi_1 = \xi_1^0$, $\xi_2 = \xi_2^0$, $\xi_1^0\xi_2^0 \neq 0$, $(\xi_1^0, \xi_2^0) \in \delta$, we have a torus in $V$ such that $(\varphi_1, \varphi_2)$ are the angle coordinates on it. For $\xi_1^0 = 0$ or $\xi_2^0 = 0$ we have the corresponding circle in $\Sigma_2$ or $\Sigma_1$, respectively, such that $\varphi_2$ or $\varphi_1$ respectively is the angle coordinate on the circle. Let us take two arbitrary homeomorphisms of circles $\varphi_i \to \varphi'_i \,(\mathrm{mod}\, 2\pi)$, $i = 1, 2$. Define a map $g : V \to V'$ as follows. Let $m \in V$. Denoting $\xi_1^0 = \xi_1(m)$, $\xi_2^0 = \xi_2(m)$, we obtain a point $z_0 = (\xi_1^0, \xi_2^0)$ in $\delta$. Therefore, we obtain a point $\mathring{z}' = (\mathring{\xi}_1{}', \mathring{\xi}_2{}')$ in $\delta'$. Let $\xi_1(m) \neq 0$, $\xi_2(m) = 0$. Then the Liouville torus $T(m)$ with coordinates $(\varphi_1, \varphi_2)$ passes through the point $m$, and the point $m$ corresponds to the point $(\mathring{\varphi}_1, \mathring{\varphi}_2)$. Due to the construction of $q$, the equality $\xi_1^0\xi_2^0 \neq 0$ holds at the point $\mathring{z}'$. Therefore, the point $\mathring{z}'$ defines a two-dimensional torus $T'$ with coordinates $(\varphi'_1, \varphi'_2)$ in $V'$. Let $g(m) = m' \in T'$ be such that $\varphi'_1(m') = \tau_1(\mathring{\varphi}_1)$, $\varphi'_2(m') = \tau_2(\mathring{\varphi}_2)$. Assume that, say, $\xi_1(m) = 0$. Then $m \in \Sigma_2$, and there is a one-dimensional closed trajectory of IHVF passing through $m$ such that $\varphi_2$ is the angle coordinate on it and the point $m$ corresponds to $\varphi_2 = \mathring{\varphi}_2$. By the construction of $q$, the point $\mathring{z}' = q(\mathring{z})$ belongs to the axis $\xi'_1 = 0$, i.e., to it there corresponds a circle on $\Sigma'_2$ with an angle coordinate $\varphi'_2$ in $V'$. Let $m' = g(m)$, where $\varphi'_2(m') = \tau_2(\mathring{\varphi}_2)$. Finally, let $g(p) = p'$. It is obvious that $g$ is one-to-one and continuous on $V\backslash(\Sigma_1 \cup \Sigma_2)$. Let us verify its continuity on $\Sigma_1 \cup \Sigma_2$. Suppose, for example, that $m \in \Sigma_2$ and $m' = g(m)$. In symplectic polar coordinates the point $m$ is defined by the conditions $\xi_1 = 0$, $\varphi_1$ is arbitrary, $\varphi_2 = \mathring{\varphi}_2$, $\xi_2 = \mathring{\xi}_2$. If $m_k \to m$ for a sequence of points $m_k \in V$, then for the symplectic polar coordinates of the points $m_k$ we have $\xi_{1,k} \to 0$, $\varphi_{2,k} \to \mathring{\varphi}_2$, $\xi_{2,k} \to \mathring{\xi}_2$. Note that the sequence $\varphi_{1,k}$ can be arbitrary. It follows from the construction of the map $g$ that for the points $m'_k = g(m_k)$ we have $(\xi'_{1,k}, \xi'_{2,k}) = q(\xi_{1,k}, \xi_{2,k})$, $\varphi'_{1,k} = \tau_1(\varphi_{1,k})$, $\varphi'_{2,k} = \tau_2(\varphi_{2,k})$. By the continuity of the maps $q$ and $\tau_i$, we have $\xi'_{1,k} \to 0$, $\varphi'_{2,k} \to \varphi'_2, \xi'_{2,k} \to 0$. Therefore, $m = g(m)$ is a limit point for the sequence $m'_k$. The theorem is proved. □

Now let $\phi$ and $\phi'$ be two PAs with simple singular points of the center-center type. The definition of a point of the center-center type implies that for every action in $\mathbb{R}^2$ one can choose two one-parameter subgroups so that the corresponding IHVF have an elliptic singular point for which, for example, Case A holds. One can choose neighborhoods for the corresponding IHVF as neighborhoods of singular points of the actions. Then the following theorem follows from Theorem 4.1.1.

THEOREM 4.1.2. *Two PA's $\phi$ and $\phi'$ with singular points $p$ and $p'$ of the center-center type are topologically equivalent at some neighborhoods of their singular points.*

CHAPTER 5

# Saddle-Center Type Singular Points

## 5.1. Extended neighborhood of a singular point, and its topology

Let $(X_H, K)$ be an IHVF and $p$ a singular point of the saddle-center type. Proposition 3.3.2 implies that the induced PA has four local one-dimensional $p$-orbits which lie on a smooth local two-dimensional symplectic disk foliated into one-dimensional orbits of the elliptic type. The point $p$ is a singular point of the saddle type for the restriction of the IHVF to this disk, and the local $p$-orbits are simply the local stable and unstable separatrices of the saddle $p$. Let us describe the garland of the point $p$. In order to do this we extend one of the local unstable separatrices along the trajectories of the field $X_H$. It is a noncompact one-dimensional orbit of the elliptic type. Hence, its limit set can contain only singular points of the saddle-center type (see Corollary 3.3.1). The point $p$ is one of these points. Since only one singular point belongs to the garland (condition 4 for a Hamilton function to be of the Morse type), the entire limit set of the orbit coincides with the point $p$, i.e., the orbit forms a loop $\gamma_1$. This means that upon continuation the separatrix returns to a neighborhood of the point $p$ and merges with one of the stable local separatrices of this point. The second loop $\gamma_2$ is obtained analogously. It is formed by continuation of the second local unstable separatrix, and it contains the second local stable separatrix. In a neighborhood of a point of the saddle-center type there are no two-dimensional orbits containing this point in their closure (Proposition 3.3.2). Therefore, a garland of the point $p$ is a "figure eight" consisting of the point $p$ and two loops. An extended neighborhood of the point $p$ will be chosen as an invariant neighborhood of the "figure eight".

We start with the construction of a local neighborhood $U$ of the point $p$ and the description of the structure of the level sets $U_\epsilon = U \cap \{H = \epsilon\}$. Let us choose a neighborhood $U_1$ such that in this neighborhood the functions $H$ and $K$ have the following form:

$$H = h(\xi, \eta) = \lambda\xi + \omega\eta + \cdots, \qquad K = k(\xi, \eta) = \mu\xi + \nu\eta + \cdots,$$

where $\xi = x_1 y_1$, $\eta = \frac{1}{2}(x_2^2 + y_2^2)$, $\Delta = \lambda\nu - \omega\mu \neq 0$. Without loss of generality we can assume that $\lambda \geq 0$, $\omega \geq 0$, $\Delta \geq 0$. Indeed, if $\omega \leq 0$, then we can use the Hamilton function $-H$. If $\lambda \leq 0$, we can perform a symplectic coordinate change $(x_1, y_1, x_2, y_2) \to (-y_1, x_1, x_2, y_2)$. Finally, for $\lambda \geq 0$, $\omega \geq 0$ the sign of $\Delta$ can be changed, if necessary, by replacing $K$ with $-K$. In all these cases we obtain isoenergetically equivalent IHVFs.

First we describe the bifurcation diagram of the restriction of the IHVF to the neighborhood $U_1$. Recall that in $U_1$ a set of one-dimensional and zero-dimensional orbits of the induced action (a degeneracy set) is located on the two two-dimensional symplectic disks $\Sigma_1 (x_2 = y_2 = 0)$ and $\Sigma_2 (x_1 = y_1 = 0)$. One of them, $\Sigma_1$, contains local $p$-orbits, and on the second, $\Sigma_2$, the point $p$ is a singular point of the center type with respect to the restriction of the field $X_H$. The bifurcation diagram in the plane of variables $(h, k)$ consists of the images of these disks under the moment map $\mu : x \to (H(x), K(x))$. Therefore, we have two smooth curves $l_1 = \mu(\Sigma_1)$ and $l_2 = \mu(\Sigma_2)$, which are the graphs of the functions $K = \psi_1(h) = \frac{\mu}{\lambda}h + O(h^2)$ and $K = \psi_2(h) = \frac{\nu}{\omega}h + O(h^2)$. Note that $\psi_1(h)$ is defined for $|h| \leq h^*$, and $\psi_2(h)$ is defined for $0 \leq h \leq h^*$ (the fact that $l_2$ belongs to the interval $0 \leq h \leq h^*$ follows from the agreement about the sign: $\omega \geq 0$.) The functions $\psi_1$, $\psi_2$ can be determined from the equations $h(\xi, 0) = h$ and $k(\xi, 0) = k$ (for $\Sigma_1$), and $h(0, \eta) = h$ and $k(0, \eta) = k$ (for $\Sigma_2$), if one applies the implicit function theorem to the first equation, which gives us the initial choice of the constant $h^*$. The conditions $\lambda\omega \geq 0$, $\Delta \geq 0$ imply that the curves are located as in Figure 5.1 and the region $\mu(U_1)$ is a half-neighborhood of zero, which lies above the curve $l_1$ and contains the curve $l_2$.

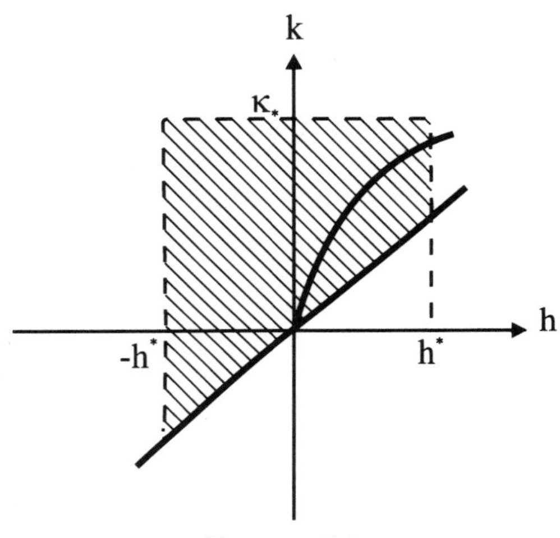


FIGURE 5.1

In the region $\mu(U_1)$ we choose a region $\sigma$ that will be used for the construction of the extended neighborhood $V$. In order to do this we choose a point $(0, \kappa_*)$ in the region $\mu(U_1)$ on the axis $k > 0$ such that the interval $[0, \kappa_*]$ of the axis $k$ belongs to $\mu(U_1)$. Then we choose sufficiently small $h^*$ so that the straight line $h = h^*$ intersects the curve $l_2$ at a certain point $(h_1, k_1)$, which satisfies the condition $\psi_1(h_1) < k_1 < \kappa_*$. Let us denote by $\sigma$ the set bounded by the lines $k = \kappa_*$, $h = \pm h^*$ and the curve $l_1$ (see Figure 5.1).

Let us introduce some functions which will be useful later. Let us denote by $u(\epsilon, \kappa)$, $v(\epsilon, \kappa)$ the solution of the system of the equations $h(\xi, x) = \epsilon$, $k(\xi, \eta) = x$. Since $\Delta = D(h, k)/D(\xi, \eta)\,|_{\xi=\eta=0} \neq 0$, such a solution exists, and it is unique for sufficiently small values of $|\xi|$, $|\eta|$, $|\epsilon|$, $|\kappa|$. Since $v_\kappa(0, 0) = \frac{\lambda}{\delta} > 0$, for a fixed $\epsilon$ the function $v(\epsilon, \kappa)$ increases with respect to $\kappa$, and since $u_\kappa(0, 0) = \frac{\omega}{\delta} < 0$, the function $u(\epsilon, \kappa)$ decreases. Let us also introduce two more functions $c(\eta, \epsilon)$ and $d(\xi, \epsilon)$, which are the solutions of the equation $h(\xi, \eta) = \epsilon$ with respect to $\xi, \eta$, respectively. Since $h(0, 0) = 0$, $h'_\xi(0, 0) = \lambda \neq 0$, $h'_\eta(0, 0) = \omega \neq 0$, these functions exist and are unique for sufficiently small values of $|\xi|$, $|\eta|$, $|\epsilon|$. Note that for all sufficiently small values of the parameter $|\epsilon| \leq 0$, the functions $c(\eta, \epsilon)$ and $d(\xi, \epsilon)$ are monotone in $\eta, \xi$, respectively. This follows from the fact that $c_\eta(0, 0) = -\frac{\omega}{\lambda} < 0$, $d_\xi(0, 0) = \frac{\lambda}{\omega} < 0$.

Let us choose an interval $h = \epsilon$ in $\sigma$. Along this interval, $k$ varies from $\psi_1(\epsilon)$ to $\kappa_*$. Meanwhile, the functions $\xi, \eta$ vary in the intervals $0 \le \eta \le \eta^* = v(\epsilon, \kappa_*)$, $U(\epsilon, x_*) \le \xi \le U(\epsilon, \psi_1(\epsilon))$. The same inequalities determine the range of the arguments and the values of the functions $c$ and $d$.

Now let us choose a neighborhood $U \subset U_1$. It is defined by the inequalities $H = \epsilon$, $|\epsilon| \le h^*$, $\psi_1(\epsilon) \le k \le \kappa_*$, $|x_1^2 - y_1^2| \le b$. It is clear that $\mu(U) = \sigma$. The set $U$ contains the point $p$, and if $|\epsilon| < h^*$, $\psi_1(\epsilon) \le k \le \kappa_*$, $|x_1^2 - y_1^2| < b$, the $U$ is open. As $h^*$, $\kappa_*$, $b$ approach zero, $U$ contracts to the point $p$. Indeed, for given $h^*$, $\kappa_*$ the variables $(x_1, y_1, x_2, y_2)$ vary within the closed region $|x_1^2 - y_1^2| \le b$, $|x_1 y_1| = |\xi| \le u_{\max}$, $0 \le \eta = \frac{1}{2}(x_2^2 + y_2^2) \le v_{\max}$, where $u_{\max} = \max\{\max_{|\epsilon|<h^*} |u(\epsilon, \kappa_*)|, \max_{|\epsilon|<h^*} |u(\epsilon, \psi_1(\epsilon))|\}$, $v_{\max} = \max \eta^*(\epsilon)$. Note that $u_{\max} \to 0$, $v_{\max} \to 0$ for $h^* \to 0$, $\kappa_* \to 0$. Therefore, as $h^*, \kappa_*, b \to 0$, the projection of $U$ to the plane $(x_1, y_1)$ contracts to the point $(0, 0)$. The same holds for the projection of $U$ to the plane $(x_2, y_2)$.

Now we describe the topology of the sets $U_\epsilon$.

LEMMA 5.1.1. *The set $U_\epsilon$ is homeomorphic to one of the following sets.*

1. *The disjoint union of two solid cylinders $U_\epsilon^1$, $U_\epsilon^2$, for $\epsilon < 0$.*
2. *The disjoint union of $U_\epsilon^1$, $U_\epsilon^2$, with two of their interior points identified, one from $U_\epsilon^1$, and the other from $U_\epsilon^2$ (union of $U_\epsilon^1$, $U_\epsilon^2$) for $\epsilon = 0$.*
3. *The connected sum $U_\epsilon^1 \# U_\epsilon^2$, for $\epsilon > 0$.*[1]

PROOF. First, find a projection of the set $U_\epsilon$ to the plane $(x_1, y_1)$. We use the notation $\pi : (x_1, y_1, x_2, y_2) \to (x_1, y_1)$. Then $\pi(U_\epsilon)$ is the set defined by the condition $\xi = a(\eta, \epsilon)$, where $\xi = x_1 y_1$, $\eta = \frac{1}{2}(x_2^2 + y_2^2)$. Here $\eta$ varies from 0 to $\eta^*(\epsilon)$, and $\xi$ varies from $u(\epsilon, \kappa_*)$ to $u(\epsilon, \psi_1(\epsilon)) = \frac{\epsilon}{\lambda} + O(\epsilon^2)$. The change of the variables $x_1$, $y_1$ is also determined by the inequalities $|x_1^2 - y_1^2| \le b$. Therefore, for $\epsilon < 0$ the set $\pi(U_\epsilon)$ on the plane $(x_1, y_1)$ is the union of two curvilinear rectangles $\Pi_1(\epsilon)$ and $\Pi_2(\epsilon)$ lying in the second and forth quadrants of the plane $(x_1, y_1)$. The boundaries of each rectangle are the curves $x_1^2 - y_1^2 = \pm b$, and pieces of the hyperbolas $x_1 y_1 = u(\epsilon, \kappa_*)$, $x_1 y_1 = u(\epsilon, \psi_1(\epsilon))$. Each rectangle $\Pi_i(\epsilon)$, $i = 1, 2$, is foliated by the curves $x_1 y_1 = \text{const} = c(\eta, \epsilon)$, i.e., we have one curve for any fixed $\eta = \eta_0$, $0 \le \eta_0 \le \eta^*(\epsilon)$. There is the cylinder $\frac{1}{2}(x_2^2 + y_2^2) = \eta_0$, $x_1 y_1 = c(\eta, \epsilon)$, $|x_1^2 - y_1^2| \le b$ above every such curve in $U_\epsilon$. For $\eta_0 = 0$ the cylinder degenerates into the curve $x_1 y_1 = c(0, \epsilon)$ on the plane $(x_1, y_1)$. The union of the cylinders and the curve is a solid cylinder $U_\epsilon^i$, which is projected on its rectangle $\Pi_i(\epsilon)$, $i = 1, 2$ (Figure 5.2 a) and Figure 5.5 a) in the next section).

As $\epsilon \to -0$, the two rectangles are glued at the point $x_1 = y_1 = 0$ forming two closed regions $\Pi_1(0)$ and $\Pi_2(0)$ glued at the point $(0, 0)$. The boundary of each region consists of the curves $|x_1^2 - y_1^2| = b$, $x_1 y_1 = u(0, \kappa_*) \le 0$, $x_1 y_1 = u(0, \psi_1(0)) = 0$. We denote by $U_0^1$ and $U_0^2$ the preimages of the regions $\Pi_i(0)$ under the map $\pi$. The region $U_0^1$ is foliated by the cylinders $\frac{1}{2}(x_2^2 + y_2^2) = \eta_0$, $0 \le \eta \le \eta^*(0)$, which for $\eta \to 0$ contract to a nonsmooth curve formed by two pieces of the coordinate axes $x_1 = 0$ and $y_1 = 0$ and belonging to the boundary of the corresponding rectangle.

[1]The connected sum of two oriented smooth manifolds (possibly, with boundaries) $N_1$ and $N_2$ is defined as a connected manifold $N = N_1 \# N_2$ obtained in the following way. First, cut out closed solid spheres $D_i$ with smooth boundaries $\partial D_i$ from $N_i \backslash \partial N_i$. Then define a homeomorphism $h : \partial D_1 \to \partial D_2$ that changes the orientation of the sphere $\partial D_2$. Identify points on the boundary spheres $\partial D_i$ in $(N_1 \setminus D_1) \cup (N_2 \setminus D_2)$ by means of $h$. The oriented manifold obtained is called the connected sum $N_1 \# N_2$ **[53]**.

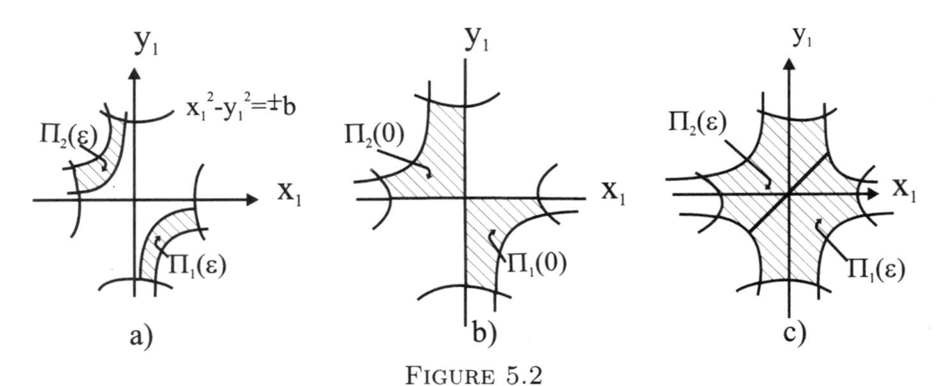


FIGURE 5.2

Thus, $U_0^i$ are also solid cylinders and $U_0$ is obtained by gluing $U_0^1$ and $U_0^2$ at the point $p$ (see Figure 5.2 b) and Figure 5.5 b) in the next section).

Suppose $\epsilon > 0$. In this case $\pi(U_\epsilon)$ is a connected set with the boundary consisting of the curves $x_1y_1 = u(\epsilon, x_*) \leq 0$, $x_1y_1 = u(\epsilon, \psi_1(\epsilon)) \geq 0$ and $|x_1^2 - y_1^2| = b$. We cut $\pi(U_\epsilon)$ into two regions by the straight line $x_1 = y_1$ and denote by $\Pi_i(\epsilon)$ the closures of the obtained regions. Note that, since $x_1y_1 = u(\epsilon, \psi_1(\epsilon)) = \frac{\epsilon}{\lambda} + O(\epsilon^2) > 0$, the boundary curves corresponding to $\eta = 0$ belong to the first and the third quadrants. Let us consider the set $\Pi_1(\epsilon)$ in the region $y_1 \geq x_1$. It consists of two parts $\Pi_1^+(\epsilon)$, $x_1y_1 \geq 0$, and $\Pi_1^-(\epsilon)$, $x_1y_1 \leq 0$. As above, it is easy to obtain that there is a set $U_1^-(\epsilon)$ homeomorphic to a product of the annulus $K$ and an interval, which lies above $\Pi_1^-(\epsilon)$. The difference with the case $\epsilon = 0$ is that, since $\eta = v(0, \epsilon) \geq 0$, a nonsmooth cylinder rather than a piecewise linear curve lies over the piecewise linear curve consisting of two pieces of the axes $x_1 = 0$, $y_1 \geq 0$, $x_1 < 0$, $y_1 = 0$. Before we turn to the description of the set $U_1^+(\epsilon)$ located over $\Pi_1^+(\epsilon)$, let us determine the topological type of the boundary of the figure obtained by gluing $\Pi_1(\epsilon)$ and $\Pi_2(\epsilon)$. This set is located over the interval $x_1y_1$, $0 \leq x_1y_1 \leq u(\epsilon, \psi_1(\epsilon))$. The points $x_2 = y_2 = 0$, $x_1 = y_1 = \pm\sqrt{u(\epsilon, \psi_1(\epsilon))}$ lie over the boundary points in $U_\epsilon$, and the circles $x_1 = y_1 = x_1^0$, $\frac{1}{2}(x_2^2 + y_2^2) = d((x_1^0)^2, \epsilon) > 0$ lie above the interior points of the interval. Therefore, a two-dimensional sphere $S$ lies over this interval. The set $\Pi_1^+(\epsilon)$ is foliated by the curves $x_1y_1 = \xi_0$, $0 \leq \xi_0 \leq c(0, \epsilon)$. Note that, for a fixed $\xi_0$, there are two curves. They lie in the first and the second quadrant, respectively. For $\xi \neq c(0, \epsilon)$, there is a cylinder $(x_2^2 + y_2^2)/2 = d(\xi_0, \epsilon)$, $x_1y_1 = \xi_0$ over each connected curve. This cylinder intersects the sphere $S$ along the circle that lies above the corresponding point of the interval $x_1 = y_1 = \pm\sqrt{\xi_0}$. For $\xi = c(0, \epsilon)$, the cylinder contracts to the interval intersecting $S$ at the boundary point of the interval $x_1 = y_1$. Thus, we have a solid cylinder with removed open ball with the boundary $S$. The set $U_\epsilon^1$ is obtained by gluing its parts $U_1^-(\epsilon)$ and $U_1^+(\epsilon)$ along the boundary cylinder corresponding to $\xi_0 = 0$. The obtained set is homeomorphic to the solid cylinder $D^2 \times I$ with removed open ball such that the closure of this ball belongs to the interior of the solid cylinder. The same is true for $U_\epsilon^2$. Therefore, $U_\epsilon$ is obtained by gluing $U_\epsilon^1$ and $U_\epsilon^2$ along the sphere $S$, i.e., we have the connected sum of two solid cylinders (see Figure 5.2 c) and Figure 5.5 c) in the next section). The lemma is proved. □

We note that the constructed neighborhood $U$ is canonical in the sense of Chapter 2. Therefore, its boundary $\partial U$ consists of smooth pieces $R^{||}$ and $R^{\perp}$.

Note that $R^{||}$ consists of pieces of the orbits of the action, $R^{\perp}$ is transversal to the orbits of the action, and $\pi(R^{\perp})$ consists of four disjoint smooth arcs defined by the equations $|x_1^2 - y_1^2| = b$, $u(\epsilon, \kappa_*) \leq \xi \leq u(\epsilon, \psi_1(\epsilon))$.

Let us start constructing an extended neighborhood $V$ of the point $p$. By construction, $R^{\perp} \subset \partial U$ consists of four smooth pieces which are diffeomorphic to $D^2 \times [-h^*, h^*]$. For a fixed $\epsilon \in [-h^*, h^*]$ the disk $D^2_\epsilon = D^2 \times \{\epsilon\}$ is foliated into the circles $K = \kappa$, $x \leq \kappa_*$, $H = \epsilon$. If $\kappa = \psi_1(\epsilon)$, then the circle degenerates to a point. Each loop $\gamma_i$ from the separatrix set intersects exactly two pieces from $R^{\perp}$, which we denote by $R_i^+$, $R_i^-$, $i = 1, 2$. The sign $+$ corresponds to local stable pieces of the loop $\gamma_i p$ and the sign $-$ corresponds to local unstable pieces of the loop $\gamma_i$. Let the trajectories of the field $X_H$ start at the points of $R_i^-$. Then they will arrive at $R_i^+$ in finite time, and the trajectories starting on the circle $K = \kappa$, $H = \epsilon$ on $R_i^-$ will arrive at the circle $K = \kappa$, $H = \epsilon$ on $R_i^+$. In particular, there is a smooth curve $l_i^-$ on $R_i^-$, which is the trace of $\Sigma_1$: $H = \epsilon$, $K = \kappa = \phi_1(\epsilon)$. Trajectories starting at the points of the curve $l_i^-$ arrive at the analogous curve $l_i^+$ on $R_i^+$. Pieces of the obtained trajectories between $R_i^-$ and $R_i^+$ form a set which is diffeomorphic to $D^2 \times [-h^*, h^*] \times I$. We denote it by $Z_i$ and call it a handle. The set $V = U \cup Z_1 \cup Z_2$ is called an extended neighborhood of the point $p$. It is clear from the construction that $\mu(V) = \mu(U) = \sigma$. Thus, the boundary $\partial V$ consists of two smooth pieces. One of them is projected into the piece $K = \kappa_*$ on the momentum plane $(h, k)$, and the other is projected into two pieces $h = \pm h^*$. The neighborhood $V$ contains a two-dimensional symplectic submanifold foliated into one-dimensional orbits of the action. This manifold is obtained from $\Sigma_1$ by gluing two strips formed by the pieces of the trajectories of the field $X_H$ connecting the curves $l_i^-$ and $l_i^+$, $i = 1, 2$. This manifold with the boundary consisting of three closed one-dimensional orbits is diffeomorphic to the disk $D^2$ with two disjoint disks removed. Again, we denote this manifold by $\Sigma_1$. It contains a separatrix "figure eight" (see Figure 5.3). The "figure eight" divides $\Sigma_1$ into two parts. In one of these parts $H > 0$, and in the other $H < 0$. All the trajectories of the field $X_H$ on $\Sigma_1$, except for the "figure eight", are closed. There is either one ("long") or two ("short") periodic trajectories on the level $H = \epsilon$ (see Figure 5.3). We recall that there is a second symplectic two-dimensional manifold $\Sigma_2$ in $U$, and therefore, in $V$. It is filled with closed one-dimensional orbits of the hyperbolic type (orbits on $\Sigma_1$ have the elliptic type; see Chapter 2). Closed trajectories on $\Sigma_1$ and $\Sigma_2$ are singular periodic trajectories (SPTs).

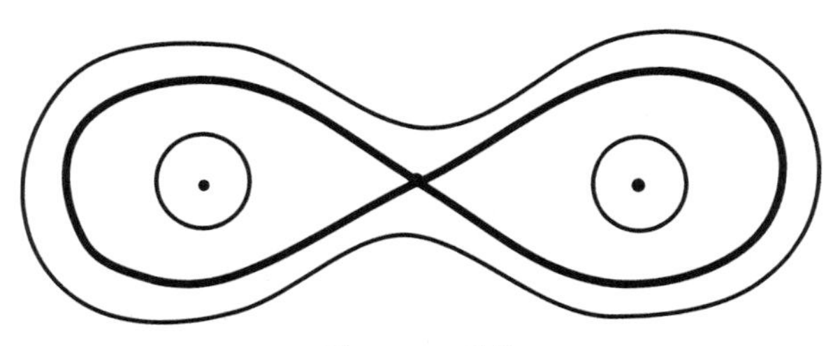

FIGURE 5.3

Note that due to our agreement about the sign of $\omega$ the manifold $\Sigma_2$ is situated in the region $H \geq 0$.

To describe the topology of the neighborhood $V$ and the level sets $V_\epsilon = V \cap \{H = \epsilon\}$, we note that there are only two possibilities when the loops are forming.

Namely, consider the set $U_0 = U \cap \{H = 0\}$. By Lemma 5.1.1, $U_0$ represents two solid cylinders $U_0^1$ and $U_0^2$ glued at the point $p$. When the loops $\gamma_i$ are forming, the local unstable separatrix belonging to $U_0^1$ merges (while continuing) either with the local stable separatrix, also belonging to $U_0^1$ (then the second loop is formed by continuing local separatrices from $U_0^2$), or with the local stable separatrix belonging to $U_0^2$ (then the second loop is also formed by separatrices from different $U_0^i$). Let us call the first possibility Case A, and the second Case B.

PROPOSITION 5.1.1. *In Case A the set $V_\epsilon$ is homeomorphic to:*

1. *a connected sum of two solid tori, for $\epsilon > 0$;*
2. *a union of two solid tori (glued at an interior point), for $\epsilon = 0$;*
3. *a union of two disjoint solid tori, for $\epsilon < 0$.*

*In Case B the set $V_\epsilon$ is homeomorphic to:*

1. *a connected sum of a solid torus and $S^2 \times S^1$, for $\epsilon > 0$;*
2. *a solid torus with two identified interior points, for $\epsilon = 0$;*
3. *a solid torus, for $\epsilon < 0$.*

*In both cases the boundary $\partial V$ of the extended neighborhood $V$ is homeomorphic to the connected sum of two copies of $S^2 \times S^1$.*

PROOF. Consider, for example, the statement in Case B with $\epsilon > 0$. By Lemma 5.1.1 the set $U_\epsilon$ is a connected sum of two solid cylinders $U_\epsilon^1$ and $U_\epsilon^2$. The neighborhood $V$ is obtained by attaching two handles $Z_1$ and $Z_2$, diffeomorphic to $D^2 \times [-h^*, h^*] \times I$, to $U$ along the sets $R_i^+$, $R_i^-$, $i = 1, 2$.

Case B implies that $R_1^+(\epsilon) = R_1^+ \cap \{H = \epsilon\}$ and $R_1^-(\epsilon) = R_1^- \cap \{H = \epsilon\}$ lie on different $V_\epsilon^1$ and $V_\epsilon^2$. The set $V_\epsilon$ is obtained from $U_\epsilon$ by gluing two solid cylinders, $Z_1(\epsilon) = Z_1 \cap \{H = \epsilon\}$ and $Z_2(\epsilon) = Z_2 \cap \{H = \epsilon\}$. The order of gluing is determined by the location of the sets $R_i^+(\epsilon)$ and $R_i^-(\epsilon)$ on $U_\epsilon$. Thus, $V_\epsilon$ is obtained by gluing two solid cylinders and two solid cylinders with removed solid balls. Due to the orientability of $V_\epsilon$, we obtain a solid torus with two disjoint solid balls with identified boundaries removed. It follows from [**54**] that the given manifold is homeomorphic to a connected sum of a solid torus and $S^2 \times S^1$. Analogous arguments can be applied to the other cases of the proposition.

Let us describe the topology of the boundary $\partial V$. It is a compact three-dimensional orientable topological manifold. Later in Theorem 5.2.1 we will prove that the extended neighborhoods in Cases A and B are homeomorphic. Therefore, we will only consider Case B. The boundary $\partial V$ can be viewed as a union of two topological manifolds: $W_1 = V_{h^*}$ and $W_2 = V_{-h^*} \cup P$, where

$$P = \{x \in M \mid K(x) = \kappa_*, \ -h_* \leq H(x) \leq h_*\} \cap V.$$

The topology of $W_1$ is known: it is a solid torus $\hat{W}_1$ with two solid balls with identified boundaries removed. The set $V_{-h^*}$ is homeomorphic to a solid torus (the statement of this proposition in Case B, $\epsilon < 0$), $\partial V_{-h^*}$ is a torus, and, for a fixed $K = \kappa_*$, we do not cross the bifurcation set on the bifurcation diagram when $\epsilon$ varies along the interval $K = \kappa_*$, $\epsilon < h_*$. Therefore, $P$ is diffeomorphic to $T^2 \times I$. Then it can be readily seen that $W_2$ is a solid torus. Thus, $\partial V$ is a manifold obtained by gluing a solid torus and the described manifold $W_1$ along the boundary torus $T_0$.

In constructing $W_1$ we removed from the solid torus the closed balls disjoint from the boundary torus $T_0$. Therefore, $\partial V$ can be viewed as a manifold obtained by gluing two solid tori $\hat{W}_1$ and $W_2$, and such that two balls are removed from $\hat{W}_1$

and the obtained spheres are glued. It allows us to observe how the solid tori are glued along $T_0$. Namely, we obtain a three-dimensional manifold with the Heegaard diagram [**53**] of genus 1. Its topological type depends on the gluing map. We will show that there is a simple closed curve nonhomotopic to zero on $T_0$ such that it is freely homotopic to zero in $\hat{W}_1$ and in $W_2$. Then from the classification of such manifolds it follows that the obtained manifold is homeomorphic to $S^2 \times S^1$ [**53**]. Let us consider a simple closed curve $\gamma$ nonhomotopic to zero on $T_0$, which under the projection $\pi$ is mapped into the point of intersection of two curves $x_1^2 - y_1^2 = -b$, $b > 0$, $y_1 > 0$ and $\xi = u(h^*, \kappa_*)$ on the plane $(x_1, y_1)$ (see Figure 5.2). In $\hat{W}_1$, $\gamma$ contracts to a point along the disk $x_1^2 - y_1^2 = -b$, $y_1 > 0$, $u(h^*, x_*) \leq \xi \leq u(h^*, \phi_1(h^*))$. The contraction in $W_2$ can be done in two steps: along $P$ and then along $V_{-h^*}$. The deformation along $P$ occurs in the annulus $x_1^2 - y_1^2 = -b$, $y_1 > 0$, $\xi = u(\epsilon, x_*)$, $-h^* \leq \epsilon \leq h^*$. At the end of this stage we obtain a circle $x_1^2 - y_1^2 = -b$, $y_1 > 0$, $\xi = u(-h^*, x_*)$. Finally, this circle is deformed along the annulus $x_1^2 - y_1^2 = -b$, $y_1 > 0$, $u(h^*, \kappa_*) \leq \xi \leq u(h^*, \phi_1(h^*))$, into a point in $V_{-h^*}$. Now we must remove two smooth closed disjoint balls from $S^2 \times S^1$ and, taking the closure of the obtained manifold, glue them along the boundary spheres. It follows from [**54**] that the obtained manifold is homeomorphic to a connected sum of two copies of $S^2 \times S^1$. The proposition is proved. □

REMARK 5.1. The topology of the neighborhood $V$ is clear from the construction. $V$ is obtained from a closed four-dimensional ball ($U$ in our case) with four disjoint closed three-dimensional spheres on its boundary, which are grouped by two (the sets $R_i^+$, $R_i^-$, $i = 1, 2$, in our case). A four-dimensional "handle" $D^3 \times I, I = [0, 1]$ is attached to each pair of balls at their bases $D^3 \times \{0\}$ and $D^3 \times \{1\}$. The obtained four-dimensional manifold can be called four-dimensional solid pretzel, by analogy with the standard three-dimensional manifold in $\mathbb{R}^3$ whose boundary is a pretzel (a sphere with two handles). The analogy is confirmed by the fact that the boundary of this three-dimensional pretzel can also be viewed as a connected sum of two tori $S^1 \times S^1$.

We can give another, more geometric, description of Cases A and B. Namely, in Case A there are two short (elliptic) SPTs from $\Sigma_1$ on the level set $V_\epsilon$, $\epsilon < 0$, and one long (elliptic) SPT from $\Sigma_1$ and one hyperbolic SPT from $\Sigma_2$, for $\epsilon > 0$. In Case B there is one long (elliptic) SPT from $\Sigma_1$ on the level set $V_\epsilon$, $\epsilon < 0$, and two short (elliptic) SPTs and one hyperbolic SPT from $\Sigma_2$, for $\epsilon > 0$. Notice that there are no other SPTs on $V_\epsilon$.

## 5.2. Isoenergetic equivalence of IHVFs

In order to prove the isoenergetic equivalence of two IHVFs in extended neighborhoods of singular points of the saddle-center type we need some foliations in $V$. One of these foliations ($H$-foliation) is a foliation into level sets of the function $H$. Fibers of this foliation in $V$ are the sets $V_\epsilon$. The second foliation ($B$-foliation) is defined in $U$ by the level surfaces of the function $b = x_1^2 - y_1^2$, and then it is extended to the whole $V$. The third foliation ($\Phi$-foliation) is also first defined in $U$, and then it is extended to $V$. Locally in $U$, it is defined by the polar coordinates $x_2 = \sqrt{2\eta}$, $\cos\varphi$, $y_2 = \sqrt{2\eta}$, $\sin\varphi$, and its fibers are the level sets $\varphi = \text{const} \pmod{2\pi}$.

First we describe the local (in $U$) structure of the $B$-foliation. Singular points of the function $b$ coincide with the part of the plane $x_1 = y_1 = 0$ that belongs to

$U$, i.e., with $\Sigma_2$. Therefore, for $b \neq 0$ the level sets are smooth, they form two three-dimensional disks, and the projections of these disks to the plane $(x_1, y_1)$ are two pieces of the hyperbola $x_1^2 - y_1^2 = b$. The fiber $b = 0$ consists of two pieces of three-dimensional planes $x_1 = y_1$ and $x_1 = -y_1$ transversally intersecting along $\Sigma_2$. Now let us consider the intersection of the fibers of the $B$-foliation with the level sets $V_\epsilon$. The corresponding foliation of $V_\epsilon$ is called $B_\epsilon$-foliation. Let us determine the fibers of these foliations. For $\epsilon < 0$, the set $U_\epsilon$ consists of two disjoint pieces $U_\epsilon^1$ and $U_\epsilon^2$ that do not intersect $\Sigma_2$. The intersection of the fiber $x_1^2 - y_1^2 = b$ and $U_\epsilon^i$, $\epsilon < 0$, is a smooth two-dimensional disk which is transversal to the axial curve of a solid cylinder $U_\epsilon^i$. This can be easily established by considering projections of the fibers of the $B$-foliation and the sets $U_\epsilon^i$ to the plane $(x_1, y_1)$ as in Lemma 5.1.1 (also, see Figure 5.2). For $\epsilon = 0$, the arguments are similar except for the fiber of the $B$-foliation for $b = 0$ passing through the point $p$. This fiber consists of two smooth disks such that each of them lies in its own $U_0^i$, $i = 1, 2$. These two disks are glued together at the point $p$.

Now let $\epsilon > 0$. In this case, as was proved in Lemma 5.1.1, $U_\epsilon$ consists of two closed sets, $U_\epsilon^1$ and $U_\epsilon^2$, each being a solid cylinder with a removed interior solid sphere. The boundary of this solid sphere is a sphere $S$ which is identified with the same sphere in the other copy. Here the projection of $S$ on the plane $(x_1, y_1)$ is an interval $x_1 = y_1$, $|x_1 y_1| \leq u(\epsilon, \psi_1(\epsilon))$. Therefore, we can see that any fiber of the $B_\epsilon$-foliation, except for the fiber $b = 0$ in $U_\epsilon^i$, is a smooth two-dimensional disk which is transversal either to the piece of SPT $x_2 = y_2 = 0$, $x_1 y_1 = u(\epsilon, \psi_1(\epsilon))$, $y_1 < 0$, or the piece of SPT $x_2 = y_2$, $x_1 y_1 = u(\epsilon, \psi_1(\epsilon))$, $y_1 > 0$. The fiber $b = 0$ is special. It is a sphere $S'$ with an annulus glued to it along SPT from $\Sigma_2$ such that it lies above the interval $u(\epsilon, \kappa_*) \leq \xi \leq 0$, $x_1 = -y_1$. Note that $y_2 > 0$ for one of $U_\epsilon^i$, and $y_2 < 0$ for the other (see Figure 5.4). Since $V \setminus U$ consists of two handles $Z_1, Z_2$ which are glued to $U$ along the fibers of the $B$-foliation (the sets $R_i^+, R_i^-$, $i = 1, 2$), the $B$-foliation can be easily extended from $U$ to $V$. The set $Z_i$ is diffeomorphic to the direct product $D^2 \times [-h^*, h^*] \times I$ and is smoothly foliated by the cylinders $H = \epsilon$, $K = \kappa$. Thus, there is no doubt about the possibility of extending the $B$-foliation with the condition of transversality of the fibers of the $B$-foliation to the $H$-foliation and $K$-foliation on $Z_i$.

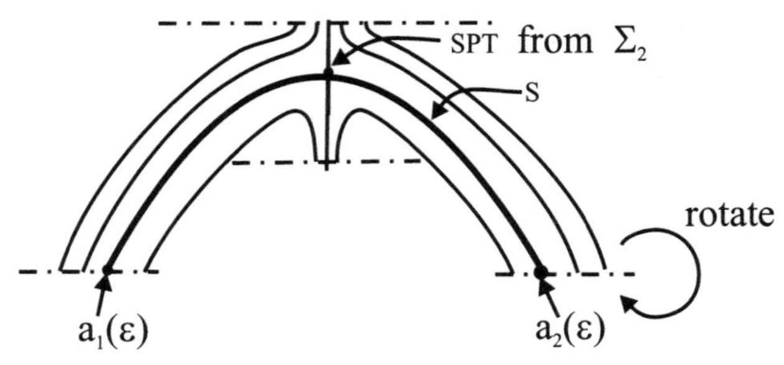


FIGURE 5.4

Let us turn to the structure of the $\Phi$-foliation. Its singularities in $U$ coincide with the plane $\Sigma_1 \cap U$, where $\eta = (x_1^2 + y_1^2)/2 = 0$ and $\phi$ is arbitrary. Let us describe the fibers of the $\Phi_\epsilon$-foliation obtained by the intersection of the fibers of the $\Phi$-foliation with $V_\epsilon$. In order to do this, we find the intersection of a fiber of the $B_\epsilon$-foliation and a fiber of the $\Phi_\epsilon$-foliation.

Let us fix $U_\epsilon^1$, where $y_1 \geq x_1$. If $\xi = \xi_0$ is given, then one can choose a point on the projection of a $B_\epsilon$-fiber on the plane $(x_1, y_1)$. Then one can define $\eta = d(\xi_0, \epsilon)$, and, for a fixed $\phi$, a point in $U_\epsilon^1$. Varying $\xi_0$ so that $u(\epsilon, \kappa_*) \leq \xi_0$, we obtain a curve that is the intersection of a $B_\epsilon$-fiber and a $\Phi_\epsilon$-fiber (see Figure 5.5). If $\epsilon \leq 0$, then the obtained curve tends to a unique boundary point at the piece of SPT as $\xi_0 \to u(\epsilon, \psi_1(\epsilon))$ (for $b = 0$, $\epsilon = +0$ it is the point $p$). Fixing $\phi_0$ and varying $b_0$ we obtain a curvilinear quadrangle. Thus, for $\epsilon < 0$, the fiber $\phi = \phi_0 \pmod{2\pi}$ lying in $U_\epsilon^i$ is a curvilinear rectangle. One of the sides of this rectangle is the axis of the solid cylinder $U_\epsilon^i$ (a piece of SPT), and the opposite side belongs to the boundary cylinder. When $\phi$ rotates by $2\pi$ the corresponding curvilinear rectangles foliate $U_\epsilon^i$ having the common piece of SPT (see Figure 5.5). The complete fiber $\phi = \phi_0$ in $U_\epsilon$ consists of two such rectangles, one in $U_\epsilon^1$ and the other in $U_\epsilon^2$. For $\epsilon = 0$, the structure of the foliation is the same with the only exception that the two axes have common point $p$.

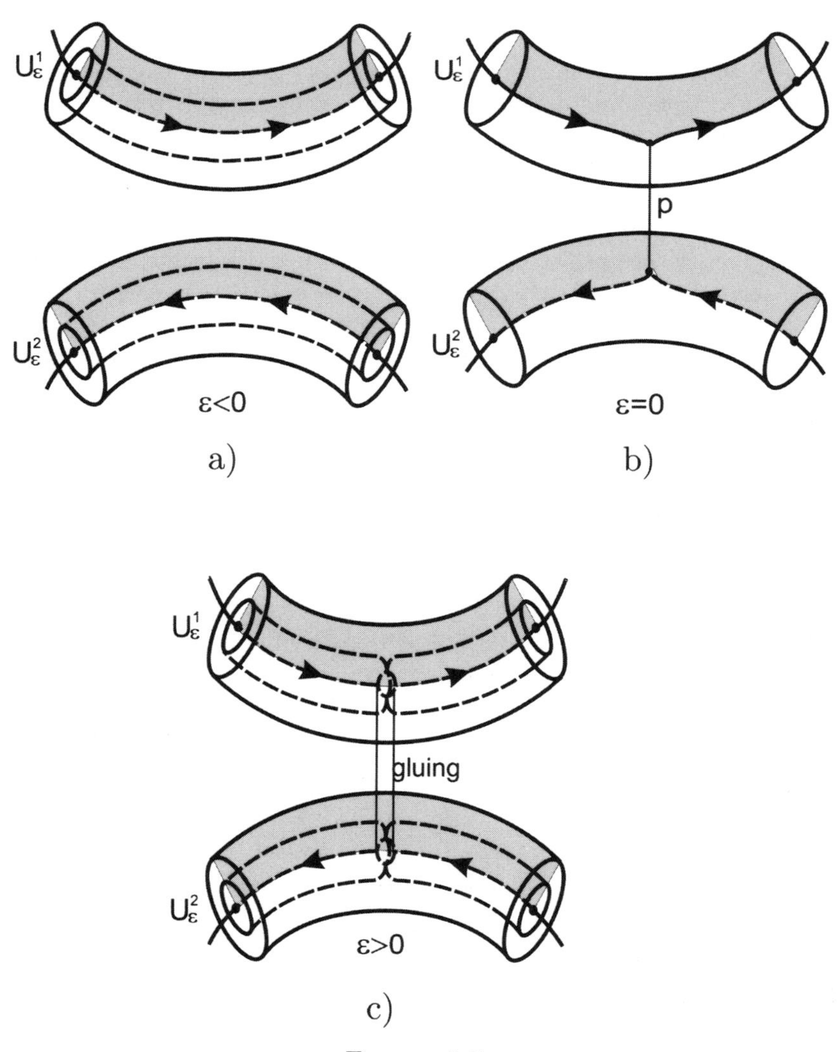


FIGURE 5.5

Let $\epsilon > 0$. Then for $b_0 \neq 0$ the intersection of the fiber $b = b_0$ of $B_\epsilon$-foliation and the fiber $\phi = \phi_0$ of $\Phi_\epsilon$-foliation is the same curve as above. As we mentioned, the fiber $b_0 = 0$ in $U^1_\epsilon$ is a sphere $S$ with an annulus glued to it along SPT from $\Sigma_2$. The projection of the fiber $b = b_0$ on the plane $(x_1, y_1)$ consists of two intervals $x_1 = y_1$ and $x_1 = -y_1$, which intersect transversally at $(0,0)$. Moreover, half of the interval $x_1 = -y_1$ belongs to $U^1_\epsilon$. As above, $\xi$ is monotone on the interval $x_1 = -y_1$ from $u(\epsilon, \kappa_*)$ to 0, and $\eta = d(\xi, \epsilon)$ decreases from $\eta^*(\epsilon) = v(\epsilon, x_*)$ to $v(\epsilon, \psi_2(\epsilon)) > 0$. Thus, fixing $\phi_0$ and varying $\xi$ we obtain an interval $l_i$ in $U^i_\epsilon$ such that it goes into the point $q(\phi)$ of the boundary sphere $S$. The sphere $S$ lies over the interval $x_1 = y_1$, $|\xi| \leq u(\epsilon, \psi_1(\epsilon))$. At each half of this interval the function $\xi$ varies monotonically from 0 to $u(\epsilon, \psi_1(\epsilon))$ and the function $\eta$ decreases from $\eta^*(\epsilon)$ to 0. Fixing $\phi_0$ we obtain an arc $\tau$ on $S$ with ends at the points $a_1$ and $a_2$ belonging to SPT from $\Sigma_1$. Note that the obtained arc on $S$ contains the point $q(\phi)$. Therefore, the intersection of the fiber $b = b_0$ and $\phi = \phi_0$ is a union of $\tau \subset S$ and the set $\{b = b_0\} \cap \{\phi = \phi_0\}$. Taking the union of the intervals $l_i$ for all $b_0$ and a fixed $\phi_0$ we obtain the complete fiber $\phi = \phi_0$ of the $\Phi_\epsilon$-foliation in $U^i_\epsilon$ (see Figure 5.5). The two pieces of the fiber in $U^1_\epsilon$, $U^2_\epsilon$ are glued into one fiber $\phi = \phi_0$ along the arc $\tau$ of the sphere $S$ which is common for these two fibers. In particular, the intersection of the fibers $b = b_0$ and $\phi = \phi_0$ is a "cross" formed by the arcs $\tau, l_1, l_2$.

Let us extend $\Phi$-foliation to the handles. The $\Phi$-foliation constructed in $U$ induces a foliation on $R^+_i, R^-_i$, $i = 1, 2$. It is easy to see from the construction of the $\Phi$-foliation that on $R^\pm_i$ this foliation is diffeomorphic to the foliation of the standard cylinder $\rho \leq 1$, $0 \leq \phi \leq 2\pi$, $|z| \leq 1$ by the level sets $\phi = \text{const}$. Here a fiber is a rectangle and all the rectangles have common axis $\rho = 0$, which corresponds to the trace of $\Sigma_1$ on $R^\pm_1$. To construct the extension we note that in each $R^+_i, R^-_i$ there is one fiber corresponding to one value of $\phi = \phi_0$. Also, the handle $Z_i$ connecting $R^+_i$ and $R^-_i$ is diffeomorphic to $R^-_i \times I$ and is smoothly foliated by the level sets $H = \epsilon$ and $K = \kappa$. Therefore, the $\Phi$-foliation can be easily extended to $Z_i$ in such a way that $\phi_0$-fiber in $\Sigma_i$ connects the traces of the two $\phi_0$-fibers in $R^+_i, R^-_i$, $i = 1, 2$, transversally to the fibers $H = \epsilon$ and $K = x$. We note the following properties of the $\Phi$-foliation.

1. By construction, the fibers of the $\Phi$-foliation are in one-to-one correspondence with the points of the circle $[0, 2\pi] \pmod{2\pi}$. Such correspondence can be defined, for example, by the coordinate $\phi$ in $U$. Therefore, from now on, we will talk about the fiber $\phi = \phi_0$.

2. Each fiber of the $\Phi$-foliation intersects each fiber of $B$-foliation. In particular, on each $V_\epsilon$ a fiber of the $B_\epsilon$-foliation and a fiber of the $\Phi_\epsilon$-foliation intersect transversally either along a curve (simple $B\Phi(\epsilon)$-fiber) if the fiber of the $B_\epsilon$-foliation is a disk, or along a "cross" $\tau \cup l_1 \cup l_2$ in $U$ (special $B\Phi(\epsilon)$-fiber).

LEMMA 5.2.1. *On every level $V_\epsilon$ the function $K$ varies monotonically along any simple $B\Phi(\epsilon)$-fiber on the disk of $B_\epsilon$-foliation. The function $K$ varies monotonically on a special $B\Phi(\epsilon)$-fiber if it is restricted to the broken line $L$ which is formed by the interval $l_i$ and one of the two halves of the arc $\tau$ divided by the point $q(\phi)$.*

PROOF. The $B\Phi(\epsilon)$-fibers are simple on the handles $Z_1$, $Z_2$, and the monotonicity of $K$ follows from the transversality of the $B$- and $\Phi$-foliations and the cylinders $H = \epsilon$, $K = \kappa$ (by construction). In $U$ the fiber $K = \kappa$ is defined by the condition $\eta = v(\epsilon, \kappa)$ for a fixed $\epsilon$, and $\xi = c(\eta, \epsilon)$. A simple $B\Phi(\epsilon)$-fiber is located on the disk $b = b_0$ of $B_\epsilon$-foliation which is foliated into circles $\eta = \text{const}$, and there is a correspondence between simple $B\Phi(\epsilon)$-fibers and radii $\phi = \phi_0$ on this disk. Since

$$\frac{d}{d\eta}K(c(\eta,\epsilon),\eta) = \frac{\Delta}{\lambda} + O(\epsilon,\kappa) \neq 0,$$

the fact that $K$ is monotone along the simple $B\Phi(\epsilon)$-fiber means that the condition $\frac{d}{d\eta}K(c(\eta,\epsilon),\eta) \neq 0$ is satisfied. It follows from the description of the $B\Phi(\epsilon)$-fiber that under the projection $\pi$ a special $B\Phi(\epsilon)$-fiber is homeomorphically mapped to a cross formed by two transversal intervals $x_1 = y_1$ and $x_1 = -y_1$ in such a way that the image of the broken line $L$ is a broken line, formed by a half of the interval $x_1 = -y_1$ (for example, $y_1 \geq 0$) and a half of the interval $x_1 = y_1$ (for example, $y_1 \geq 0$). For a fixed $\epsilon$, $\xi = u(\epsilon, x)$ and $\eta = v(\epsilon, x)$ vary monotonically along this broken line, so that $K = k(c(\eta,\epsilon),\eta)$ varies monotonically, as in the case of a simple $B\Phi(\epsilon)$-fiber. The lemma is proved. □

Now, let us turn to the main result of this chapter: the isoenergetic equivalence theorem. Suppose that $(X_H, K)$ and $(X_{H'}, K')$ are two IHVFs on symplectic manifolds $M$, $M'$ with singular points $p$ and $p'$ of the saddle-center type, and $V \subset M$ and $V' \subset M'$ are their extended neighborhoods. Without loss of generality we may assume that all the agreements of Section 5.1 about the signs of $\lambda, \omega, \Delta$ and, respectively, $\lambda', \omega', \Delta'$ are satisfied.

THEOREM 5.2.1. *An IHVF $(X_H, K)$ on $V$ is isoenergetically equivalent to an IHVF $(X_{H'}, K')$ on $V'$ if and only if the same case A or B holds for both IHVFs.*

PROOF. The "only if" part of the theorem is obvious since, by Proposition 5.1.1, the topology of the level sets $V_\epsilon$ is different in Cases A and B. Let us now prove the "if" part.

In order to construct the desired homeomorphism $g : V \to V'$ we will first define a homeomorphism $\rho : \Sigma_1 \to \Sigma_1'$ that maps:

a) the set $H = \epsilon$ to the set $H' = \epsilon'$;

b) the induced $B$-foliation on $\Sigma_1$ to the corresponding $B'$-foliation on $\Sigma_1'$.

In particular, this homeomorphism $\rho$ induces a homeomorphism $q_1 : [-h_*, h_*] \to [-h_*', h_*'], q(0) = 0$. Since we consider the same cases, $q_1(h_*) = h_*'$. We will also require that the fibers of the $B$-foliation corresponding to the sets $R_i^{\pm}$ be mapped to the same fibers of the $B'$-foliation.

Let us define a homeomorphism $q : \sigma_1 \to \sigma_1'$, where, as above, $\sigma = \mu(V), \sigma' = \mu'(V')$, and $\mu$, $\mu'$ are the moment maps. The homeomorphism $q$ must have the following properties.

1. The intervals $h = \epsilon$ are mapped onto the intervals $h' = q_1(\epsilon)$.
2. The curve $l_2 : k = \phi_2(h)$ is mapped onto the curve $l_2'$.

Coordinatewise $q$ has a triangular form: $h' = q_1(h)$, $k' = q_2(h,k)$. Finally, we define an arbitrary homeomorphism $\Phi : [0, 2\pi] \to [0, 2\pi] \,(\mathrm{mod}\, 2\pi)$, which will determine the correspondence between the fibers of the $\Phi$-foliations.

Let us represent $V$ as a union $V = V^+ \cup V^-$, where $V^+ = \{x \in V \mid H(x) \geq 0\}$, $V^- = \{x \in V \mid H(x) \leq 0\}$, $V^+ \cap V^- = V_0$. The homeomorphism $g$ must preserve the $H$-foliation so that $g$ can be constructed separately in $V^+, V^-$ provided that the compatibility condition on $V_0$ is satisfied.

We start with $V^-$. It follows from the description of the $B$-foliation that everywhere, except for the point $p$, $V$ is a bundle in $\Sigma_1$ defined by the inequality $H \leq 0$, whose fiber is the corresponding disk of $B$-foliation. There are two disks above the point $p$ in $V^-$ glued at this point. Consider an arbitrary point $m \in V, \epsilon = H(m)$ different from $p$. There is a unique disk $D$ in $V_\epsilon^-$ that passes through the point $m$ and contains a point $s \in \Sigma_1$. There is also a unique fiber of the $\Phi_\epsilon$-foliation passing through the point $m$. At first we assume that $s \neq p$. The point $s' = \rho(s)$ and the disk $D'$, which lies on the level $H' = \epsilon' = q_1(\epsilon)$ and passes through $s'$, are defined on $\Sigma_1'$ for the IHVF $(X_{H'}, K')$. If $s = p$, then $\epsilon = 0$, $s' = p'$, and there are two disks passing through $p'$. As was mentioned above while describing the $B$-foliation, for $\epsilon = 0$ the fiber $b = 0$ (lying in $U$) consists of two disks glued at the point $p$, and each disk belongs to its component $U_0^i$, $i = 1, 2$. Each $U_0^i$ is characterized by the property that it contains a pair of the local separatrices of the point $p$. In Case A this pair belongs to one loop of a "figure eight." In Case B each element of the pair belongs to its own loop, and one of them lies on the local stable manifold, and the other on the local unstable manifold of the restriction of the field $X_H$ to $\Sigma_1$. Analogous properties are true for $X_{H'}$ on $\Sigma_1'$. It is clear from the topologies of $\Sigma_1$ and $\Sigma_1'$ that the map $\rho$ preserves decomposition of the local separatrices into pairs. Thus, $\rho$ induces the correspondence between $U_0^i$ and $'U_0^j$, and this means that the disk $D'$ is uniquely determined by the disk $D$.

The fiber $\phi = \phi_m$ uniquely defines $\phi' = \Phi(\phi_m)$, hence determines the fiber of the $\Phi$-foliation intersecting $D'$ on the level $H' = \epsilon'$ along the unique simple $B\Phi(\epsilon)$-fiber. On a given trajectory the point $m' = g(m)$ is uniquely chosen by the condition $K'(m') = q_2(\epsilon, K(m))$ in view of Lemma 5.2.1. It is clear that the constructed map is a homeomorphism which can be extended by continuity to the point $p$ using the equality $p' = g(p)$.

Let us pass to $V^+$. We represent $V^+$ in the form $V^+ = Z_1^+ \cup Z_2^+ \cup U^+$, where $Z_i^+ = V^+ \cap Z_i$, $U^+ = U \cap V^+$. In turn, $U^+$ is divided by the fiber $b = 0$ into two parts $U_+^i = \bigcup_{\epsilon \geq 0} U_\epsilon^i$, $i = 1, 2$. For $'V^+$ there is an analogous representation. The homeomorphism is constructed separately for each of the obtained sets. It is only necessary to check the compatibility condition on the common boundaries.

The correspondence between the handles $Z_i^+$ and $'Z_i^+$ is established by the homeomorphism $\rho$ using the correspondence between the loops. Moreover, $Z_i^+$ and $'Z_i^+$ are foliated by disks which are fibers of the $B_\epsilon$-foliation. Each disk contains a unique point of $\Sigma_1$ $(\Sigma_1')$. The homeomorphism $\rho$ defines the correspondence between the disks in $Z_i^+$ and $'Z_i^+$. The homeomorphism on the disks is constructed in the same way as for $V^-$.

Now let us construct a homeomorphism on $U^+$. Notice that the homeomorphism $\rho$ uniquely defines the correspondence of the parts $U_\epsilon^i$ and $'U_\epsilon^i$. Continuous coordinates $(h, k, b, \phi)$ are defined on each $U_+^i$. By Lemma 5.2.1, the function $K$ decreases on the curve $H = \epsilon$, $\phi = \phi_0$, $b = b_0$, from $K = \kappa_*$ to $K = \psi_1(\epsilon)$ (recall that the curve $K = \psi_1(\epsilon)$ is $\mu(\Sigma_1)$). Thus, the homeomorphism $g$ in $U_+^i$ is defined by the formula $(h, k, b, \phi) \to (q_1(h), q_2(h, k), b', \Phi(\phi))$, where $b'$ is defined by the homeomorphism $\rho$. The agreement of the obtained homeomorphisms on the boundaries follows immediately from the construction. Theorem 5.2.1 is proved. □

Now, let us consider two Poisson actions $\Phi, \Phi'$ of the Morse type, with singular points $p, p'$ of the saddle-center type. Due to the simplicity condition for a singular point, one can choose one-parameter subgroups of the action $\Phi$ in such a way that we obtain an IHVF which realizes, for instance, Case A of Theorem 5.2.1. Indeed, it follows from the definition of a point of the saddle-center type that one can choose one-parameter subgroups and local symplectic coordinates in such a way that the corresponding Hamilton functions $H$ and $K$ have the forms as in Theorem 2.2.1. We may assume here that $\nu_1/\lambda_1 < 0$, $\lambda_2 > 0$, $\nu_2/\lambda_2 > 0$ (passing, if necessary, to linear combination of $H$ and $K$). Now, suppose that for $(X_H, K)$ we have Case B. Then for $(X_K, H)$ we have Case A. One-parameter subgroups for $\Phi'$ can be chosen analogously. Then we can take the neighborhoods constructed above for the IHVF as extended neighborhoods $V$ and $V'$ of the points $p$ and $p'$. Therefore, from Theorem 5.2.1 we obtain the following result.

THEOREM 5.2.2. *Actions $\Phi$ and $\Phi'$ with singular points $p$ and $p'$ of the saddle-center type are topologically equivalent in the extended neighborhoods $V$ and $V'$.*

CHAPTER 6

# Saddle Type Singular Points

## 6.1. The degeneracy set

Let $(X_H, K)$ be an IHVF and $p$ a singular point of the saddle type. Let $U$ be a local neighborhood of the point $p$ such that Theorem 2.2.1 is valid in this neighborhood. Let $(x_1, x_2, y_1, y_2)$ be local coordinates in $U$ in which the functions $H$ and $K$ have the form $H = h(\xi_1, \xi_2) = \lambda_1\xi_1 + \lambda_2\xi_2 + \cdots$, $K = k(\xi_1, \xi_2) = \mu_1\xi_1 + \mu_2\xi_2 + \cdots$, where $\xi_1 = x_1y_1$, $\xi_2 = x_2y_2$, $\Delta = \lambda_1\mu_2 - \lambda_2\mu_1 \neq 0$.

Without loss of generality we may assume everywhere in this chapter that $\lambda_1, \lambda_2 > 0$, $\Delta > 0$. Indeed, if $\lambda_1 < 0$, then after the symplectic change of variables $(x_1', x_2', y_1', y_2') = (-y_1, x_2, x_1, y_2)$ the monomial $x_1y_1$ becomes $-x_1'y_1'$, i.e., $\lambda_1$ changes sign. If $\Delta < 0$, then the symplectic change of variables $(x_1', x_2', y_1', y_2') = (x_2, x_1, y_2, y_1)$ changes the sign of $\Delta$ but preserves the signs of $\lambda_1, \lambda_2$. Additionally, we will assume that $H(p) = K(p) = 0$, which is also not a restriction.

We recall the structure of the trajectories of the field $X_H$ on the degeneracy set in $U$. (A set in $U$ is called a degeneracy set if orbits of the induced action of dimension less than two pass through its points.) This set consists of two disks, $\Sigma^1_{\text{loc}} : x_2 = y_2 = 0$ and $\Sigma^2_{\text{loc}} : x_1 = y_1 = 0$, which intersect transversally at the point $p$. Consider $\Sigma^i_{\text{loc}}$, $i = 1, 2$. The restriction of $X_H$ to this disk is a Hamiltonian vector field with one degree of freedom and a saddle singular point at $p$. Its local stable and unstable manifolds are the intervals $x_i = 0$ and $y_i = 0$. Notice that all trajectories of the field $X_H$ on $\Sigma^i_{\text{loc}}$ (except for the point $p$) are one-dimensional (hyperbolic) orbits of the action.

The stable (unstable) manifold of the point $p$ on $\Sigma^i_{\text{loc}}$ consists of the point $p$ and two semitrajectories of the field $X_H$, which will be called stable (unstable) separatrices. These separatrices are local $p$-orbits of the action $\Phi$ (see Proposition 3.3.2).

Let us describe the garland of the point $p$ and its separatrix set $W$. Due to Proposition 3.3.5, each two-dimensional orbit containing the point $p$ in its closure is homeomorphic to a plane. Therefore, there are no closed one-dimensional orbits in their limit sets, i.e., the garland consists of the point $p$. This fact and Proposition 3.3.5 imply that noncompact one-dimensional orbits which contain the point $p$ in their closures form loops containing local one-dimensional $p$-orbits. Thus, there are four such loops. For the field $X_H$ these loops are obtained when the four local unstable separatrices (two from $\Sigma^1_{\text{loc}}$ and two from $\Sigma^2_{\text{loc}}$) are continued with time. Then these separatrices coincide with the local stable separatrices of the point $p$ on $\Sigma^1_{\text{loc}}$ and $\Sigma^2_{\text{loc}}$. It turns out that a pair of local stable and unstable separatrices belonging to one loop is not arbitrary.

PROPOSITION 6.1.1. *Local stable and unstable separatrices belonging to one loop belong to the same disk* $\Sigma^i_{\text{loc}}$.

PROOF. Assume the contrary, namely that, for example, the loop $\gamma$ leaves $\Sigma^1_{\rm loc}$ and returns to $\Sigma^2_{\rm loc}$. Notice that the restriction of $X_H$ to $U$ has the form

$$\dot{x}_i = -\lambda_i x_i + \cdots, \qquad \dot{y}_i = \lambda_i y_i + \cdots, \qquad i = 1, 2,$$

and $\lambda_1 \neq \lambda_2$, $\lambda_1 > 0$, $\lambda_2 > 0$. This means that the local (unstable) separatrix from $\Sigma^1_{\rm loc}$ satisfies the equation

$$\dot{y}_1 = \lambda_1 y_1 + O(y_1^2), \qquad x_1 = x_2 = y_2 = 0,$$

and the local stable separatrix from $\Sigma^2_{\rm loc}$ satisfies the equation

$$\dot{x}_2 = -\lambda_2 x_2 + O(x_2^2), \qquad x_1 = y_1 = y_2 = 0.$$

Consider the Hamiltonian vector field $X_Q$ in $U$ with the Hamilton function

$$Q = \alpha H + \beta K = (\alpha\lambda_1 + \beta\mu_1)x_1 y_1 + (\alpha\lambda_2 + \beta\mu_2)x_2 y_2 + \cdots$$

(it corresponds to another choice of a one-parameter subgroup for the action $\Phi$).

Since $\Delta = \lambda_1\mu_2 - \lambda_2\mu_1 \neq 0$, it is possible to choose $\alpha$ and $\beta$ so that $\nu_1 = \alpha\lambda_1 + \beta\mu_1 > 0$ and $\nu_2 = \alpha\lambda_2 + \beta\mu_2 < 0$. Since the loop $\gamma$ is a one-dimensional orbit of the action $\Phi$, it is also a trajectory of the vector field $X_Q$. Since locally in $U$ the vector field $X_Q$ has the same form as $X_H$, with $\lambda_i$ changed to $\nu_i$, it follows from the choice of the signs of $\nu_1$ and $\nu_2$ that both local separatrices of the vector field $X_Q$ that belong to $\gamma$ are stable. This implies that there is a singular point of the vector field $X_Q$ on $\gamma$. However, there are no singular points of $X_H$ on $\gamma$. Since $X_H$ and $X_Q$ commute, $\gamma$ consists of the singular points of the vector field $X_Q$, which contradicts its local representation. The proposition is proved. □

Clearly, the separatrix set $W$ of the garland contains, along with the four loops $\gamma_1, \dots, \gamma_4$, two-dimensional orbits that contain the point $p$ (hence also $\gamma_i$) in their closure. The topology of the set $W$ will be described in the next section.

The proposition proved above enables us to extend the local symplectic disks $\Sigma^i_{\rm loc}$, $i = 1, 2$, along the trajectories of the vector field $X_H$. In order to do that, let us consider the local three-dimensional cross-sections $N_1, \dots, N_4$ to the local separatrices on $\Sigma^1_{\rm loc}$ such that they do not intersect $\Sigma^2_{\rm loc}$. Assume, for example, that the loops $\gamma_1$ and $\gamma_2$ intersect $\Sigma^1_{\rm loc}$ and that the loop $\gamma_1$ intersects the sections $N_1$ and $N_2$, whereas the loop $\gamma_2$ intersects the sections $N_3$ and $N_4$. By the theorem on the continuous dependence of solutions of differential equations on the initial conditions, all trajectories of the vector field $X_H$ that start on $N_1$ close enough to the trace of the loop $\gamma_1$ remain close to the global piece of $\gamma_1$ and intersect $N_2$. The similar statement holds for the loop $\gamma_2$ and the sections $N_3$ and $N_4$. Since all the trajectories of the vector field $X_H$ on $\Sigma^1_{\rm loc}$ are also one-dimensional orbits of the action, the trajectories that start at $N_1 \cap \Sigma^1_{\rm loc}$ come back to $N_2 \cap \Sigma^1_{\rm loc}$. Moreover, since the restriction of the vector field $X_H$ to $\Sigma^1_{\rm loc}$ does not have singular points on $\gamma_1$, the function $H$ varies monotonically along the segment transversal to the loop on $\Sigma^1_{\rm loc}$.

Since the values of the integral of $H$ over the trajectories are preserved, there are only two possibilities:

a) a trajectory different from $\gamma_i$, but close to it becomes a closed trajectory immediately after going around the loop (a short singular periodic trajectory);

b) a trajectory becomes closed after going around $\gamma_1$ first and then around $\gamma_2$ (a long singular periodic trajectory).

Thus, in a neighborhood of the global parts of the loops $\gamma_1$ and $\gamma_2$, $\Sigma^1_{\rm loc}$ may be extended along the periodic trajectories. As a result of the extension we obtain a two-dimensional manifold $\Sigma_1$ with the boundary consisting of one long SPT and two short ones (Figure 5.3). Without loss of generality we may assume that the boundaries are determined by the equations $H = \pm h^*$. Notice that $\Sigma_1$, being an extension of the symplectic disk $\Sigma^1_{\rm loc}$ along the trajectories of the Hamiltonian vector field $X_H$, is itself symplectic. The structure of the symplectic submanifold $\Sigma_2$ that contains the loops $\gamma_3$, $\gamma_4$ is the same.

## 6.2. The types of loops and SPTs

The purpose of this section is to study the possible types of SPTs from the point of view of orientability of the stable and unstable manifolds. It turns out that these properties are completely determined by the corresponding properties of the loops.

It follows from Proposition 3.3.2 and the local representation of the vector fields $X_H$, $X_K$ in $U$ that in the neighborhood of $U$ there are four $X_H$-invariant Lagrange disks passing through the point $p$. Two of them, $W^s_{\rm loc} : y_1 = y_2 = 0$ and $W^u_{\rm loc} : x_1 = x_2 = 0$, are the local stable and unstable manifolds of the point $p$. The restrictions of the vector field $X_H$ to the other two, $W^c_1 : x_1 = y_2 = 0$ and $W^c_2 : x_2 = y_1 = 0$, have a saddle point at $p$. The union of these four disks is the complete set of solutions of the system $H = K = 0$ in $U$. Thus, one can see that $W^u_{\rm loc}$ intersects $W^c_1$ in the segment $x_1 = x_2 = y_2 = 0$, and $W^c_2$ in the segment $x_1 = x_2 = y_1 = 0$.

Similarly, $W^s_{\rm loc}$ intersects $W^c_1$ in the segment $x_1 = y_1 = y_2 = 0$, and $W^c_2$ in the segment $x_2 = y_1 = y_2 = 0$.

Consider a loop $\gamma$ that belongs, for example, to $\Sigma_1$ and assume, for definiteness, that its intersection with $W^u_{\rm loc}$ is the segment $x_1 = x_2 = y_2 = 0$, $y_1 > 0$, and its intersection with $W^s_{\rm loc}$ is the segment $y_1 = y_2 = x_2 = 0$, $x_1 > 0$.

Then the segment of the loop $\gamma$ that starts at $p$ is the intersection of $W^u_{\rm loc}$ and $W^c_1$, and the segment that arrives at $p$ is the intersection of $W^s_{\rm loc}$ and $W^c_2$. As we know (Proposition 3.3.4), $\gamma$ is a one-dimensional hyperbolic orbit of the action $\Phi$. Therefore, locally near each of its points it is the intersection of two smooth rectangles, each rectangle consisting of two pieces of two-dimensional orbits adjacent to $\gamma$, and $\gamma$ itself. Locally in $U$, these rectangles are $W^u_{\rm loc}$ and $W^c_2$ for the segment that starts at $p$, and $W^s_{\rm loc}$ and $W^c_2$ for the segment that ends at $p$. Therefore, after going around $\gamma$ the extension of $W^s_{\rm loc}$ along the trajectories of the vector field $X_H$ must coincide with either $W^u_{\rm loc}$ or $W^c_2$. Then, after the extension, $W^c_1$ coincides with either $W^c_2$ or $W^s_{\rm loc}$, respectively. In the first case the extensions $W^s$ and $W^u$ of the local manifolds $W^s_{\rm loc}$ and $W^u_{\rm loc}$ that belong to the level set $H = 0$ merge along $\gamma$. In the second case they are transversal along $\gamma$. Let us show that, in fact, the first case is impossible.

PROPOSITION 6.2.1. *The manifolds $W^s$ and $W^u$ are transversal along $\gamma$ in the level set $H = 0$.*

PROOF. Without loss of generality we may assume that the set $H > 0$ is disconnected on $\Sigma_1$, i.e., in a neighborhood of the loop $\gamma$ short SPTs belong to the domain $H > 0$. Let us choose in $U$ a section $N^u = \{y_1 = \rho_1 > 0\}$ transversal to the segment $l^u$ of the loop $\gamma$ that starts at $p$, and a section $N^s = \{x_1 = \rho_2 > 0\}$ transversal to the segment $l^s$ that ends at $p$. The trace of $\Sigma_1$ on each of the manifolds

$N^u$ and $N^s$ is determined by the equation $x_2 = y_2 = 0$. The traces of the manifolds $W^u_{\text{loc}}$ and $W^c_1$ on $N^u$ are the segments $x_1 = x_2 = 0$ and $x_1 = y_2 = 0$, and the traces of the manifolds $W^s_{\text{loc}}$ and $W^c_2$ on $N^s$ are the segments $y_1 = y_2 = 0$ and $x_2 = y_1 = 0$, respectively. Assume now that the continuation of the manifold $W^u_{\text{loc}}$ along the trajectories of the vector field $X_H$ near the global piece of $\Gamma_2$ merges with $W^s_{\text{loc}}$. That means that the segment $l^2_u(0) = \{x_1 = x_2 = 0,\ y_1 = \rho_1\}$ is mapped by the global map $\varphi : N^u \to N^s$ into the segment $l^2_s(0) = \{y_1 = y_2 = 0,\ x_1 = \rho_2\}$. Then, as was pointed out above, the trace of $W^c_1$ on $N^u$ is mapped into the trace of $W^c_2$ on $N^s$, i.e., the segment $l^1_s(0) = \{x_1 = y_2 = 0,\ y_1 = \rho_1\}$ is mapped into the segment $l^2_s(0) = \{x_2 = y_1 = 0,\ x_1 = \rho_2\}$. For every sufficiently small $\epsilon$ on $H = \epsilon$ there is a single hyperbolic SPT $\gamma(\epsilon)$ on $\Sigma_1$ close to $\gamma$ and such that there are two smooth local manifolds that belong to the same level set $H = \epsilon$, both are homeomorphic to either a cylinder or Möbius band, and intersect transversally along $\gamma(\epsilon)$ (Section 2.3); see Figure 6.1. These (and only these) manifolds contain two-dimensional orbits of the action adjacent to $\gamma(\epsilon)$. The piece of the given SPT $\gamma(\epsilon)$ that belongs to $U$ is determined by the equations $x_2 = y_2 = 0$, $x_1 y_1 = \xi_1(\epsilon)$, where $\xi_1(\epsilon)$ is the solution of the equation $h(\xi_1, 0) = \epsilon$. Also, the two-dimensional orbits passing through $\gamma(\epsilon)$ can be found from the system $h(\xi_1, \xi_2) = \epsilon$, $k(\xi_1, \xi_2) = \kappa(\epsilon)$, where $\kappa(\epsilon) = k(\xi_1(\epsilon), 0)$. It is easy to see that this solution satisfies the equations $\xi_1 = \xi_1(\epsilon)$, $\xi_2 = 0$. The intersection of the two-dimensional orbits on $N^u$ is the cross $y_1 = \rho_1$, $\xi_2 = 0$, $x_1 = \xi_1(\epsilon)/\rho_1$. There is a similar cross $x_1 = \rho_2$, $\xi_2 = 0$, $y_1 = \xi_1(\epsilon)/\rho_2$ on $N^s$. Let us choose the segment $l^1_s(\epsilon) = \{x_1 = \rho_2,\ y_2 = 0,\ y_1 = \xi_1(\epsilon)/\rho_2\}$ of the cross on $N^s$ and find its image under the local map $\psi : N^s \to N^u$ along the trajectories of the vector field $X_H$. It follows from the form of the Hamilton function $H = h(\xi_1, \xi_2)$ that the disk $y_2 = 0$ is invariant with respect to the vector field $X_H$ (and so are all three disks $x_2 = 0$; $x_1 = 0$; $y_1 = 0$). Therefore, the image of the segment $l^1_s(\epsilon)$ under the map $\phi$ belongs to the segment $l^1_u(\epsilon) = \{y_2 = 0,\ y_1 = \rho_1,\ x_1 = \xi_1(\epsilon)/\rho_1\}$ of the cross. Similarly, the segment $l^2_s(\epsilon) = \{x_2 = 0,\ x_1 = \rho_2,\ y_1 = \xi_1(\epsilon)/\rho_1\}$ is mapped to the segment $l^2_u(\epsilon) = \{x_2 = 0,\ y_1 = \rho_1,\ x_1 = \xi_1(\epsilon)/\rho_1\}$. The global map transforms the traces of the two-dimensional orbits on $N^u$ into the traces of the two-dimensional orbits on $N^s$. Therefore, the cross on $N^u \cap \{H = \epsilon\}$ is mapped to the cross $N^s \cap \{H = \epsilon\}$. Moreover, due to the smoothness of $\varphi$, both smooth arcs, $l^1_u(\epsilon)$ and $l^2_u(\epsilon)$, that form the cross on $N^u$ are mapped to the smooth arcs $l^1_s(\epsilon)$ and $l^2_s(\epsilon)$ that form the cross on $N^s$. It follows from the fact that $W^u_{\text{loc}}$ merges with $W^s_{\text{loc}}$ that the image of the segment $l^2_u(\epsilon)$ is the segment $l^1_s(\epsilon)$, and the image of $l^1_u(\epsilon)$ is the segment $l^2_s(\epsilon)$. This leads to a contradiction since the set that we have obtained is not locally homeomorphic to a transversal intersection of two smooth cylinders or two Möbius bands. The proposition is proved. $\square$

Now let us pass to the determination of the orientability types of loops. Denote by $d$ the segment $W^u_{\text{loc}} \cap W^c_2$. It divides each manifold $W^u_{\text{loc}}$ and $W^c_2$ into two half-disks. Among these four half-disks, let us choose the pair that intersects $\gamma$. This pair of half-disks forms a topological disk $D^1_0$ that is glued along $d$ and contains the first and the last segments of the loop. Let us glue $D^1_0$ to a strip $B^1_0$ formed by segments of trajectories of the vector field $X_H$ that are close to the loop, start on the half-disk $D^1_0 \cap W^u_{\text{loc}}$, and arrive at the half-disks $D^1_0 \cap W^c_2$ (Figure 6.2). We obtain a two-dimensional topological manifold $C^1_0 = D^1_0 \cup B^1_0$ which is homeomorphic to either a cylinder or a Möbius band.

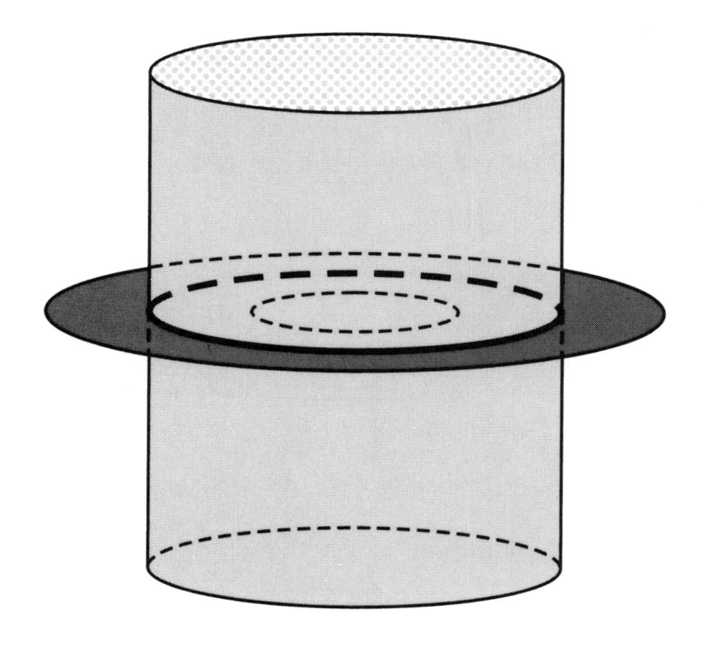

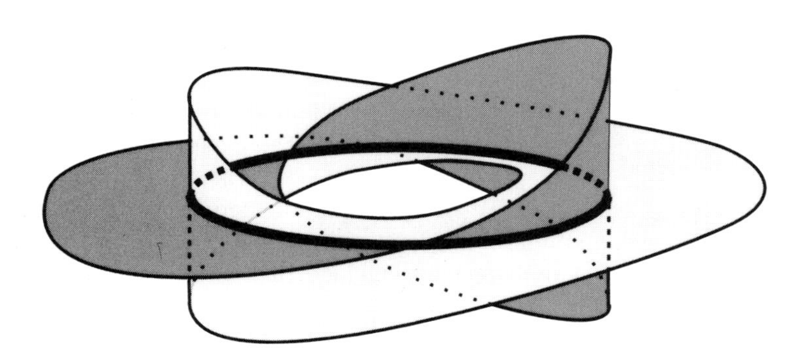

FIGURE 6.1

DEFINITION 6.1. The loop $\gamma$ is called orientable if $C_0^1$ is homeomorphic to a cylinder, and nonorientable if $C_0^1$ is homeomorphic to a Möbius band.

REMARK 6.1. A similar construction for the loop $\gamma$ may be realized using $W^s_{\text{loc}}$ and $W^c_1$. It is easy to show that the orientability type of the loop defined in such a way coincides with the one introduced in Definition 6.1.

REMARK 6.2. For the second loop on $\Sigma_1$ we construct the manifold $C_0^2 = D_0^2 \cup B_2^1$, where $D_0^2$ is formed by the pair of half-disks from $W^u_{\text{loc}}$, $W^c_2$, and $B_0^2$ is a strip that contains the segment of the second loop between $W^u_{\text{loc}}$ and $W^c_2$.

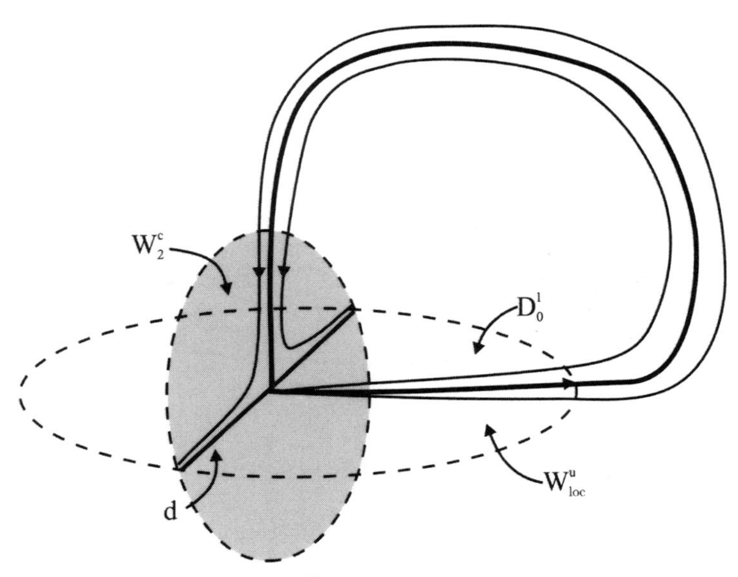


FIGURE 6.2

Our next goal is to describe the topology of the separatrix set $W$ for the garland of the point $p$. Renumber the loops $\gamma_1, \gamma_2, \gamma_3, \gamma_4$ according to the cyclic order of the local segments of these loops on $W^u_{\text{loc}}$. Moreover, in this order $\gamma_1$ follows $\gamma_4$. Assume that $\gamma_1$ is a loop on $\Sigma_1$ that starts in the direction $x_1 = x_2 = y_2 = 0$, $y_1 > 0$, $\gamma_3$ is the second loop on $\Sigma_1$, $\gamma_2$ is a loop on $\Sigma_2$ that starts in the direction $x_1 = x_2 = y_1 = 0$, $y_2 > 0$, and $\gamma_4$ is the second loop on $\Sigma_2$. On each of the manifolds $W^s_{\text{loc}}$, $W^c_1$, $W^c_2$ the segments of the loops that belong to the same symplectic manifold are not adjacent. Therefore, on each of the manifolds the cyclic order of the loops is preserved.

Each of the manifolds $W^u_{\text{loc}}$, $W^s_{\text{loc}}$, $W^c_1$, $W^c_2$ is divided by the segments of the loops that belong to this manifold into four quadrants. A quadrant is called stable, unstable, or saddle according to the type of the local manifold to which the quadrant belongs.

Now let us consider a two-dimensional orbit $L$ that contains the point $p$ in its closure. The local two-dimensional $p$-orbits belong to the four Lagrangian disks $W^u_{\text{loc}}$, $W^s_{\text{loc}}$, $W^c_1$, $W^c_2$ (Proposition 3.3.2), and coincide with the quadrants described above. It was proved in Chapter 3 that the boundary $\partial L$ of the two-dimensional orbit is a connected set and consists of the orbits of dimension 1 and 0. In our case, $\partial L$ consists of the point $p$ and several loops. Let us add to $L$ the set of points from $\partial L$ which are accessible points of the boundary. We recall this notion [**55**].

Let $l : [0,1] \to \bar{L}$ be a simple arc (here $\bar{L}$ is the closure of $L$) and $z = l(1) \in \partial L$, $l([0,1)) \subset L$. Two simple arcs $l_1$ and $l_2$ with the common point $z$ are called equivalent if there exists a neighborhood $v$ of the point $z$ in $M$ such that any point $a \in l_1 \cap v \cap L$ can be connected with any point $b \in l_2 \cap v \cap L$ by a continuous path in $L \cap v$. Clearly, this is an equivalence relation on the set of arcs in $\bar{L}$ with the common boundary point $z \in \partial L$. An equivalence class of such arcs determines an accessible point. The set of all points of $L$ together with the points accessible from $L$ is called the Carathéodory closure of $L$, and is denoted by $[L]$. Denote by $\pi : [L] \to \bar{L}$ the natural projection. It follows from the description of the local structure of the action in a neighborhood of a hyperbolic orbit (Proposition 3.3.4) and in a neighborhood of a point of the saddle-saddle type (Proposition 3.3.2) that every point of the set $\partial L$ determines at least one accessible point. Let us describe the structure of the set $[L]$ near the points of $[L] \setminus L$. Let $m \in [L] \setminus L$, and $\tilde{m} = \pi(m) \in \partial L$. Assume, for example, that $\tilde{m}$ belongs to the loop $\gamma_1$. Near the point $\tilde{m}$ consider the closure of the points that belong to the two-dimensional orbits. On the level set $H$ that contains the loop this set forms two disks that intersect transversally along the piece of the loop $\gamma_1$. Therefore, locally there are four pieces of two-dimensional orbits (half-disks) that are adjacent to $\tilde{m}$. Among them there is only one that corresponds to the point $m$, i.e., contains the equivalence class of the arcs in $L$ which have $\tilde{m}$ as their boundary point. Obviously, all the points accessible from this half-disk form a segment in $[L]$ which is the preimage of a segment from $\gamma_1$. Since the half-disk belongs to $L$, the restriction of $\pi^{-1}$ to it is a homeomorphism. If we add this segment in $[L] \setminus L$ to the preimage of the half-disk we obtain a neighborhood of the point $m$ in $[L]$. We do the same thing in the case $\tilde{m} = p$, choosing the corresponding quadrant on one of the Lagrangian disks. Now, we can describe the entire set $[L]$.

LEMMA 6.2.1. *Let $L$ be a two-dimensional orbit and $[L]$ its Carathéodory closure. The set $[L]$ is a two-dimensional cell complex* [**71**] *which has four vertices $p_1, \dots, p_4$, four one-dimensional cells $l_1, \dots, l_4$, and one two-dimensional cell $\hat{L}$. Moreover, $\pi(p_i) = p$, $i = 1, 2, 3, 4$, $\pi(l_i)$ is a one-dimensional noncompact orbit, and $\pi(\hat{L}) = L$. The image of a neighborhood of the vertex $p_i$ ("angle") is a quadrant. Moreover, two of those quadrants are saddle quadrants on different $W_1^c$ and $W_2^c$, one is stable and the other is unstable.*

PROOF. Since $p \in \partial L$, $L$ contains at least one quadrant $Q_1$. According to the construction there is a corresponding "angle" $A_1$ in $[L]$. Assume, for example, that its vertex is $p_1$. There are two one-dimensional cells adjacent to it that start at the point $p_1$. Assume that they are $l_1$ and $l_2$, and that $\pi(A_1) = Q_1$. The boundary of the quadrant $Q_1$ contains two local $p$-orbits that belong to the loops $\gamma_i$ and $\gamma_j$, $i \neq j$. If $Q_1$ is the unstable sector, i.e., $Q_1 \subset W^u_{\rm loc}$, then the set $W^u_{\rm loc}$, continued along the trajectories of the vector field $X_H$, merges with $W_1^c$ (along one of the loops, for example, $\gamma_i$), and $W^u_{\rm loc}$ merges with $W_2^c$ (along the other loop $\gamma_j$) (Proposition 6.2.1). Therefore, we have determined two different saddle quadrants that belong to $L$. In $[L]$ we have determined two angles $A_2$ and $A_4$ adjacent to $A_1$. In turn, every saddle quadrant from $W_i^c$, $i = 1, 2$, determines another local one-dimensional $p$-orbit that belongs to the boundary of that quadrant. These local $p$-orbits belong to some loops (may be the same ones) along which $W_i^c$ merges with $W^s_{\rm loc}$. In general, this determines two stable quadrants (see Figure 6.3). Let us show that it is, in fact, the same quadrant. Indeed, consider a smooth arc $s$ in $Q_1 \subset W^u_{\rm loc}$ which is transversal to the trajectories of the vector field $X_H$, including the segments of the local one-dimensional $p$-orbits. Since there are no singular points of the vector field $X_H$ in $L$, all the trajectories passing through the points of $s \cap L$ tend to the point $p$. Obviously, the set of points in $s \cap L$ whose trajectories approach a given quadrant in $W^s_{\rm loc}$, is open. If we assume that there are trajectories that pass through $s \cap L$ and run into different quadrants on $W^s_{\rm loc}$, then there is a point on $s$ which belongs to a trajectory that corresponds to a one-dimensional orbit of the action $\Phi$. But $s \cap L$ contains only points from $L$. Thus, there is only one stable quadrant in $L$, and there is a single angle $A_3$ in $[L]$ that corresponds to it. If the quadrant is stable, then the same argument works if we reverse the time. If $Q_1$ is a saddle quadrant, then the adjacent one will be either stable or unstable, and then we repeat the above argument. The lemma is proved. □

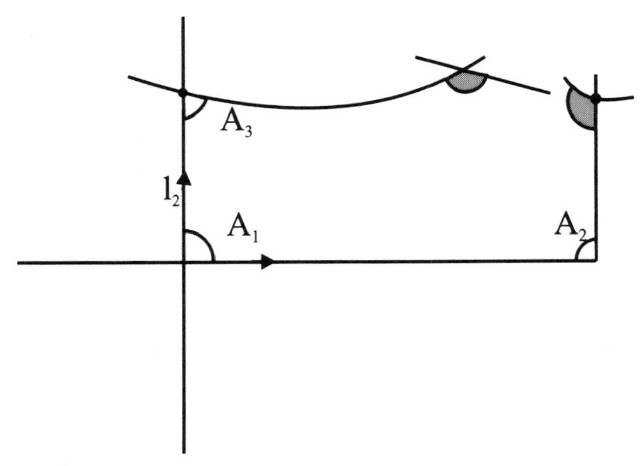


FIGURE 6.3

COROLLARY 6.2.1. *The separatrix set $W$ is a cell complex that consists of a single vertex (point $p$), four one-dimensional cells (the loops $\gamma_1, \dots, \gamma_4$) and four two-dimensional cells (two-dimensional orbits that contain $p$ in their closures).*

PROOF. By the previous lemma every two-dimensional orbit contains exactly four quadrants from different Lagrangian disks. Therefore, there are only four orbits with the structure described above. The cell map is given by the map $\pi_i([L_i]) \to \bar{L}_i$. □

Let us give a canonical description of the set $W$. For this purpose let us cut $W$ along the loops $\gamma_1, \dots, \gamma_4$, i.e., let us consider the connected components of the set $W \backslash \{\gamma_1, \dots, \gamma_4\}$. Each of the four components $L_i$ obtained in this way coincides with a two-dimensional orbit of the action. Now, consider the Carathéodory closure of each $L_i$. If we identify all the vertices of the rectangles $[L_i]$ into a point and glue the boundaries of $[L_i]$ along the segments of the corresponding loops with orientations induced by the vector field $X_H$, then we obtain the canonical representation of $W$.

Such a representation allows a visual interpretation of the orientability property for loops. Consider, for example, the loop $\gamma_1$, and the two $[L_i]$ and $[L_j]$ that contain the two quadrants in $W^u_{\text{loc}}$ adjacent to $\gamma_i$. Glue these components together along $\gamma_1$ according to its orientation. It follows from the transversality condition that if we start on $W^u_{\text{loc}}$ and move close to $\gamma_1$, we reach $W^c_2$ on which there are two quadrants adjacent to $\gamma_1$ (see Figure 6.4). Therefore, $D^1_0 \cup B^1_0$ is the result of gluing of the shaded area shown in Figure 6.4 along the segment $d$. The loops $\gamma_2$ and $\gamma_4$ are adjacent to $\gamma_1$ on $W^c_2$ and $W^u_{\text{loc}}$. The loop is nonorientable if the gluing is made as shown in Figure 6.4 a), and orientable if the gluing is made as shown in Figure 6.4 b).

It turns out that not all the combinations of the orientability types of the loops $\gamma_1, \dots, \gamma_4$ are possible for IHVF.

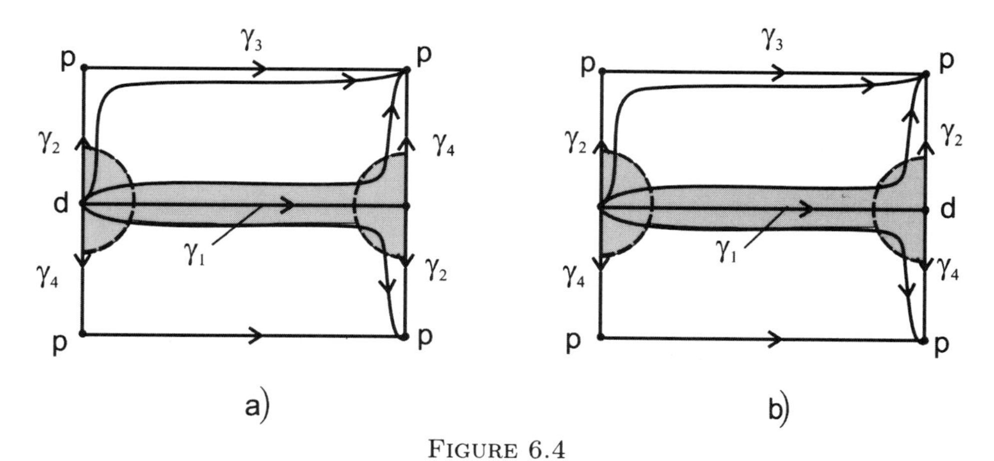


FIGURE 6.4

LEMMA 6.2.2. *If two loops adjacent with respect to the numbering are nonorientable, then all the loops on $W$ are nonorientable.*

PROOF. Let $\gamma_1$ and $\gamma_2$ be the nonorientable loops and $\gamma_3$ the orientable one. Choose some $L_i \subset W$ such that $\gamma_1$ and $\gamma_2$ belong to the boundary of the unstable quadrant. Consider a trajectory of the vector field $X_H$ in $L_i$ which is close to $\gamma_1$. This trajectory starts in the unstable quadrant and goes into the saddle quadrant whose boundary contains, due to the nonorientability of $\gamma_1$, the loops $\gamma_1$ and $\gamma_4$ (see Figure 6.5). Then this trajectory goes into the stable quadrant whose boundary contains $\gamma_4$. Now, consider another trajectory in $L_i$ close to $\gamma_2$ (see Figure 6.5). This trajectory passes the saddle quadrant bounded by $\gamma_2$ and $\gamma_3$ (due to the nonorientability of $\gamma_2$), and goes into the stable quadrant whose boundary contains $\gamma_2$ and $\gamma_3$ (due to the orientability of $\gamma_3$). This contradicts the fact that there is only one stable quadrant in $[L_j]$. The lemma is proved. □

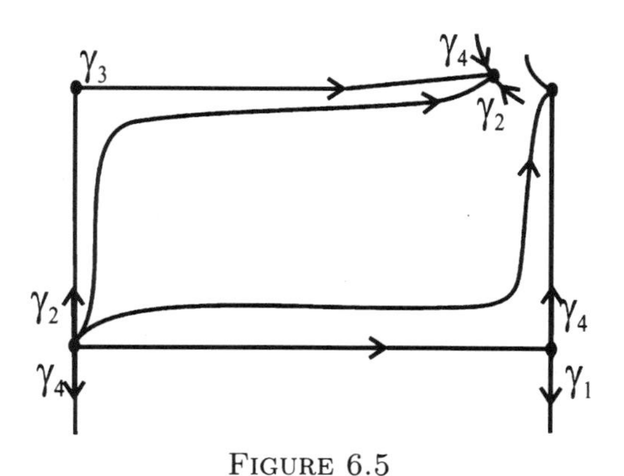


FIGURE 6.5

Recall that the loops adjacent with respect to the numbering belong to different symplectic manifolds $\Sigma_1$, $\Sigma_2$. The lemma immediately implies the following result.

COROLLARY 6.2.2. *There are only four possible combinations of the orientability types of the loops in $W$.*

1. *All the loops are orientable.*
2. *All the loops are nonorientable.*

3. *Three loops are nonorientable and one is orientable.*
4. *On one of the symplectic manifolds both loops are orientable, and on the other both loops are nonorientable.*

Thus, it follows from Lemma 6.2.2 and Corollary 6.2.2 that for the set $W$ there are four canonical representatives, which are shown in Figure 6.6.

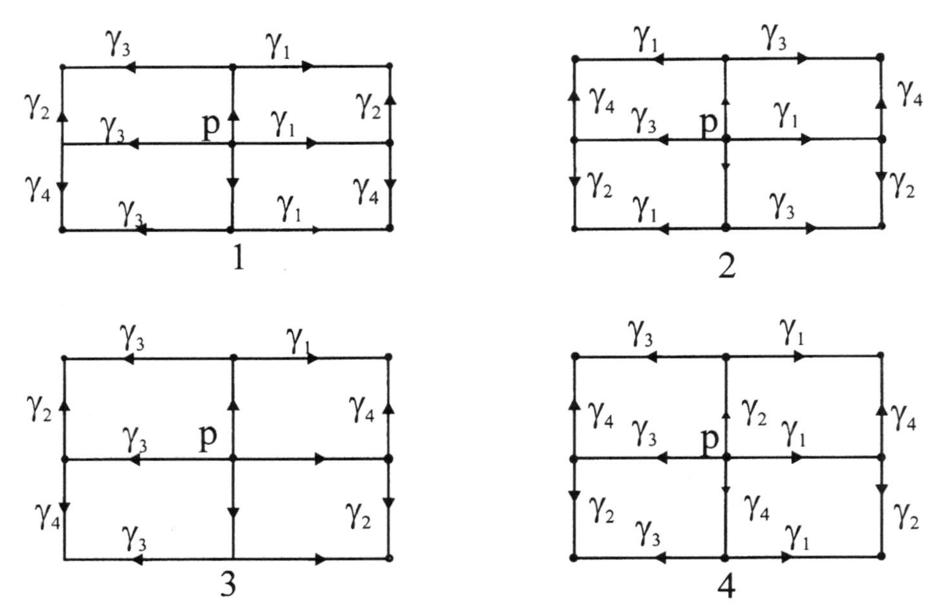


FIGURE 6.6

Let us now prove the following statement.

PROPOSITION 6.2.2. *Let $(X_H, K)$ and $(X_{H'}, K')$ be two IHVFs with Morse Hamilton functions such that they both have a singular point of the saddle type. Then their separatrix sets $W$ and $W'$ are homeomorphic if and only if they correspond to the same case in Corollary 6.2.2.*

PROOF. If the cases are the same, then, obviously, the fact that the separatrix sets $W$ and $W'$ are homeomorphic follows from the canonical representation. Conversely, let $h : W \to W'$ be the homeomorphism. Then $h$ is a cell homeomorphism, i.e., the restriction of $h$ to an open cell in $W$ homeomorphically maps this cell to a cell in $W'$. Indeed, the neighborhood of the point $p$ in $W$ is a union of four disks. The neighborhood of $a \in \gamma_i$ is the union of two disks intersecting each other along $\gamma$. Finally, the neighborhood of $b \in L_j$ is a disk. Thus, $h$ maps a given cell into a cell of the same dimension, and the closure of any two-dimensional cell into the closure of its image. It follows from the canonical representation that, in Case 1, the closure of each of the four cells is a two-dimensional torus, in Case 2, the closure of each cell is a square, with all the vertices identified, in Case 3, the closures of two cells are tori and the closures of the other two are cylinders in which one point of one boundary circle is identified with a point of another boundary circle, and in Case 4 the closure of each of the four cells is a cylinder with identified points as in Case 3. This shows that in all cases $W$ and $W'$ are homeomorphic. The proposition is proved. □

Now we establish the connection between the orientability type of the loops and the orientability type of the periodic trajectories of the vector field $X_H$ on $\Sigma_i$.

LEMMA 6.2.3. *Let $\gamma_1$ and $\gamma_3$ be loops on $\Sigma_1$.*

1. *If both $\gamma_1$ and $\gamma_3$ are orientable, then both short and long SPTs on $\Sigma_1$ are orientable.*
2. *If both $\gamma_1$ and $\gamma_3$ are nonorientable, then the short SPTs are nonorientable and the long SPTs are orientable.*
3. *If $\gamma_i$ is orientable and $\gamma_j$ is nonorientable, $i \neq j$, $i, j \in \{1, 3\}$, then the long SPTs are nonorientable, the short SPTs that are freely homotopic to $\gamma_i$ are orientable, and the short SPTs that are freely homotopic to $\gamma_j$ are nonorientable.*

*The similar statement holds for $\gamma_2$ and $\gamma_4$ on $\Sigma_2$.*

PROOF. We assume that the segment of $\gamma_1$ that ends at $p$ is defined by the relations $x_2 = y_2 = y_1 = 0$, $x_1 > 0$. In these coordinates the segment $d = W^u_{\text{loc}} \cap W^c_2$ is determined by the equations $x_1 = x_2 = y_1 = 0$, $|y_2| \leq \rho_1$, and the disk $D^1_0$ is the union of the half-disk $x_1 = x_2 = 0$, $y_1^2 + y_2^2 \leq \rho_2$, $y_1 \geq 0$, on $W^u_{\text{loc}}$ and the half-disk $x_2 = y_1 = 0$, $x_1^2 + y_2^2 \leq \rho_2$, $x_1 \geq 0$, on $W^c_2$.

Similarly, the disk $D^2_0$ that contains $\gamma_3$ is the union of the half-disk $x_1 = x_2 = 0$, $y_1^2 + y_2^2 \leq \rho^2$, $y_1 \leq 0$ and the half-disk $x_2 = y_1 = 0$, $x_1^2 + y_2^2 \leq \rho^2$, $x_1 \leq 0$. Notice that the given properties of $\gamma_1$ and $\gamma_3$ imply that the short SPTs correspond to $H > 0$ and the long ones to $H < 0$. If $\epsilon > 0$, then the set $\Sigma_1 \cap U$ in the level set $H = \epsilon$ contains a piece of the short SPT $\gamma_1(\epsilon)$ which is close to $\gamma_1$ and is determined by the equations $x_2 = y_2 = 0$, $H = \epsilon$. Therefore, $x_1 y_1 = \xi_1(\epsilon) = \epsilon/\lambda_1 + o(\epsilon^2)$, where $\xi_1(\epsilon)$ is the solution of the equation $h(\xi_1, 0) = \epsilon$. Thus, $K = x(\epsilon) = k(\xi_1(\epsilon), 0) = \nu_1 \epsilon/\lambda_1 + o(\epsilon^2)$. Now let us find the solutions of the system of equations $H = \epsilon$, $K = \kappa(\epsilon)$ that determine the stable and unstable manifolds of the SPT $\gamma_1(\epsilon)$. Locally in $U$, the solution is given by the equations $x_1 y_1 = \xi_1(\epsilon)$, $x_2 y_2 = 0$. Indeed, it follows from the condition $\Delta \neq 0$ that the system of equations $H = \epsilon$, $K = \kappa(\epsilon)$ in $U$ is equivalent to the system of equations $\xi_1 = u(\epsilon, \kappa)$, $\zeta_2 = v(\epsilon, \kappa)$. Since for $x = x(\epsilon)$ the points of the curve $\gamma_1(\epsilon)$ are the solutions of this system, we have $v(\epsilon, \kappa(\epsilon)) = 0$. This shows that in the domain $x_1 > 0$, $y_1 > 0$, this solution is diffeomorphic to the product of the "cross" $x_2 y_2 = 0$ and the interval $x_1 y_1 = \xi(\epsilon)$. Thus, the obtained set is a union of two rectangles, which are the intersections of $U$ with the stable and unstable manifolds of the SPT $\gamma_1(\epsilon)$, respectively. The topological limit, as $\epsilon \to 0$, of one of the rectangles $D^1_\epsilon : x_2 = 0, x_1 y_1 = \xi_1(\epsilon)$, is the disk $D^1_0$. We extend $D^1_\epsilon$ following the trajectories of the vector field $X_H$ along the global piece of the trajectory $\gamma_1(\epsilon)$ and obtain the strip $B^1_\epsilon$. Since the strips $B^1_\epsilon$ converge smoothly to $B^1_0$, it follows from our construction that $D^1_\epsilon \cup B^1_\epsilon$ is homeomorphic to $D^1_0 \cup B^1_0$. Thus, all the statements of the lemma concerning the short SPT are proved.

Now let $\epsilon < 0$. The level set $H = \epsilon$ contains a long SPT $\gamma(\epsilon)$, and $U \cap \Sigma_1$ contains two pieces of this trajectory, which are determined by the equations $x_2 = y_2 = 0$, $H = \epsilon$, i.e., $x_1 y_1 = \zeta(\epsilon)$. Moreover, $x_1 > 0$, $y_1 < 0$ for one of the pieces, $x_1 < 0$, $y_1 > 0$ for the other, and $K = x(\epsilon)$ on both pieces. As before, consider two disks $D^3_0$ and $D^4_0$. The disk $D^3_0$ consists of the half-disks $x_1 = x_2 = 0$, $y_1^2 + y_2^2 \leq \rho^2$, $y_1 \geq 0$ and $x_2 = y_1 = 0$, $x_1^2 + y_2^2 \leq \rho^2$, $x_1 \leq 0$, and the disk $D^4_0$ consists of the half-disks $x_1 = x_2 = 0$, $y_1^2 + y_2^2 \leq \rho^2$, $y_1 \leq 0$ and $x_2 = y_1 = 0$, $x_1^2 + y_2^2 \leq \rho^2$, $x_1 \geq 0$.

Gluing the strips $B_0^1$ and $B_0^2$ to $D_0^3 \cup D_0^4$ we obtain a manifold with boundary that self-intersects along the segment $d$. We obtain a new topological manifold without boundary by cutting the constructed manifold along $d$ and gluing the obtained four half-disks pairwise, i.e., gluing together the pairs that belong to $D_0^3$ and $D_0^4$. It is easy to see that the manifold $S_0$ without boundary obtained in this way is homeomorphic to a cylinder if both $\gamma_1$ and $\gamma_3$ are orientable or both $\gamma_1$ and $\gamma_3$ are nonorientable, and to a Möbius band if one of them is orientable and the other is not. As before, the solutions of the system $H = \epsilon$, $K = \kappa(\epsilon)$ in $U$ contain two disks $D_\epsilon^3$ and $D_\epsilon^4$ that are $C^0$-close to $D_0^3$ and $D_0^4$, respectively. There are strips $B_\epsilon^1$ and $B_\epsilon^2$ that are $C^0$-close to $B_0^1$ and $B_0^2$, glued to $D_\epsilon^3$ and $D_\epsilon^4$. Clearly, the manifold obtained in this way is homeomorphic to $S_0$. The lemma is proved. □

## 6.3. Construction of the extended neighborhood

In this section we will construct an extended neighborhood of the singular point $P$. In fact, it will be a certain special neighborhood $V$ of the separatrix set $W$ such that in this neighborhood the degeneracy set coincides with $\Sigma_1 \cup \Sigma_2$. First of all we specify a choice of a local neighborhood $U$ of the point $P$. We start from the description of the bifurcation diagram, i.e., $\mu(\Sigma_1 \cup \Sigma_2)$, where $\mu$ is the moment map.

Notice that all the trajectories of the field $X_{H_i}$ lying on $\Sigma_i$, $i = 1, 2$, intersect the set $U \cap \Sigma_i = \Sigma_{\text{loc}}^i$, and, therefore, $\mu(\Sigma_i) = \mu(\Sigma_{\text{loc}}^i)$. The degeneracy set in $U$ consists of the two disks $\Sigma_{\text{loc}}^i$, $i = 1, 2$, defined by the equations $x_2 = y_2 = 0$ for $\Sigma_{\text{loc}}^1$ and $x_1 = y_1 = 0$ for $\Sigma_{\text{loc}}^2$. Thus, it is clear that $\mu(\Sigma_{\text{loc}}^1)$ is defined as a parametrized curve $H = h(\xi_1, 0)$, $K = k(\xi_1, 0)$ or $K = \phi_1(H) = \frac{\mu_1}{\lambda_1} H + \cdots$.

Analogously, for $\Sigma_{\text{loc}}^2$ we have the expression $K = \phi_2(H) = \frac{\mu_2}{\lambda_2} H + \cdots$.

Therefore, we have two smooth curves passing through the point $(0, 0)$. Moreover, it follows from the condition $\Delta \neq 0$ that $\varphi_1'(0) \neq \varphi_2'(0)$, which means that their graphs intersect transversally at the point $(0, 0)$ (Figure 6.7). We can readily see that $\mu(U)$ is a neighborhood of zero on the plane $(H, K)$. Let us choose a rectangle $\delta : |H| \leq h_*$, $|K| \leq k_*$ in $\mu(U)$ such that the intervals $K = \pm k_*$ do not intersect the bifurcation diagram curves, and the intervals $H = \pm h_*$ intersect these curves transversally (see Figure 6.7). Next choose a neighborhood $U_1$ of the point $p$ inside $U$ as the set of points in $U$ that satisfy the inequalities

$$|H(x_1, x_2, y_1, y_2)| \leq h_*, \quad |K(x_1, x_2, y_1, y_2)| \leq k_*,$$
$$|x_1^2 - y_1^2| \leq 2b_1^*, \quad |x_2^2 - y_1^2| \leq 2b_2^*.$$

It follows from the representations $H = h(\xi_1, \xi_2)$, $K = k(\xi_1, \xi_2)$, where $\xi_1 = x_1 y_1$, $\xi_2 = x_2 y_2$ in $U$, that for $h_*, k_*$ small enough we have

$$|x_1 y_1| \leq \xi_1^*, \quad |x_2 y_2| \leq \xi_2^*.$$

These inequalities together with the inequalities

$$|x_1^2 - y_1^2| \leq 2b_1^*, \quad |x_2^2 - y_1^2| \leq 2b_2^*$$

determine a neighborhood of the point $p$ in $U$. When the constants $h_*$, $k_*$, $b_1^*$, $b_2^*$ tend to zero, this neighborhood contracts to the point $p$. First, we choose as the neighborhood $V_0$ the union of all orbits of the induced action $\Phi$ intersecting the neighborhood $U_1$. Clearly, for this choice we have $\mu(V_0) = \mu(U_1.)$ For further

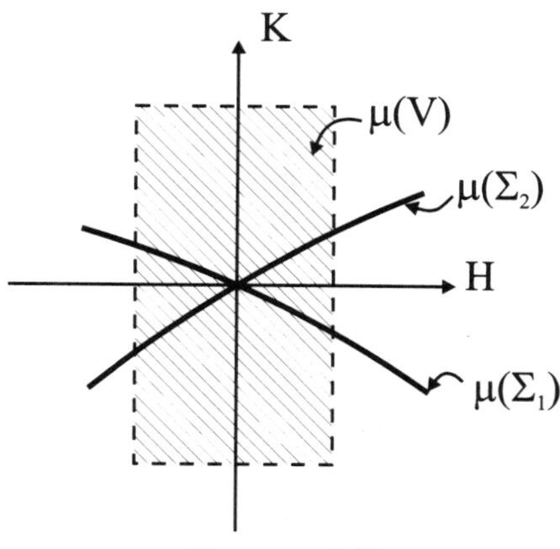


FIGURE 6.7

discussion it is useful to describe the geometry of the neighborhood $V_0$. In order to do this, notice that the sets in $U$ defined by the relations

$$x_1^2 - y_1^2 = \pm 2b_1^*, \quad |x_2^2 - y_2^2| \leq 2b_2^*, \quad |H| \leq h_*, \quad |K| \leq k_*$$

(consisting of four connected components $N_1^j$, $j = 1, \dots, 4$) and

$$x_2^2 - y_2^2 = \pm 2b_2^*, \quad |x_1^2 - y_1^2| \leq 2b_1^*, \quad |H| \leq h_*, \quad |K| \leq k_*$$

(the sets $N_2^j$, $j = 1, \dots, 4$) determine eight three-dimensional manifolds transversal to the trajectories of the field $X_H$, and hence to the orbits of the induced action $\Phi$. Each $N_i^j$ is foliated into two-dimensional disks by the level sets $H = \epsilon$, $|\epsilon| \leq h_*$. Moreover, each disk $N_i^j(\epsilon) = \{H = \epsilon\} \cap N_i^j$ is foliated by the level sets $K = \kappa$ (traces of the orbits of the action $\Phi$), and each of the disks $N_i^j(\epsilon)$ contains a saddle critical point of the restriction of the function $K$ to this disk. On $N_i^j$ the boundary of the disk consists of four pieces of the level curves $K = \pm k_*$, and of four pieces of the level curves of the function $x_k^2 - y_k^2 = \pm 2b_k^*$, $k \neq i$, $i = 1, 2$, $k = 1, 2$ (see Figure 6.8).

The union of all critical points over $\epsilon$ is a smooth curve, namely, the trace of the submanifold $\Sigma_i$ on $N_i^j$. There is a loop on $\Sigma_i$ passing through the critical point on $N_i^j(0)$. It also passes through one more disk $N_i^k$ , $k \neq j$. Thus, the disks, hence also the corresponding sets $N_i^j$, split into linked pairs. Let $N_i^j$ and $N_i^k$ be such a pair. The neighborhoods of the critical points on $N_i^j(\epsilon)$ and $N_i^k(\epsilon)$ can be chosen (uniformly with respect to $\epsilon$) so that all the trajectories intersecting one of these neighborhoods intersect the other. The obtained set of the intervals of the trajectories of the field $X_H$ between $N_i^j$ and $N_i^k$ will be called a handle. Thus, we have four handles. It is clear from the construction of the handles that the union $Q$ of the neighborhood $U_1$ and the four handles is a neighborhood of the set $\Sigma_1 \cup \Sigma_2$. Topologically, it is a four-dimensional ball with four handles $D^3 \times [0, 1]$ (attached along $D^3 \times \{0\}$ and $D^3 \times \{1\}$). The intersection of the obtained set with $W$ is a neighborhood of the loops in $W$. Therefore, its complement in $W$ is the union of four disks $D_i$ (Corollary 6.2.1). It is clear from the construction that the boundary $\partial Q$ of the set $Q$ and the four two-dimensional orbits containing the disks $D_i$ intersect transversally, i.e., we have four closed curves $l_i$, $i = 1, \dots, 4$, on $\partial Q$. It is possible to choose new small $h_* \geq 0$, $k_* \geq 0$ such that the neighborhoods of the disks $D_i$ in $V_0$ defined by the inequalities $|H| \leq h_*$, $|K| \leq k_*$ are foliated by

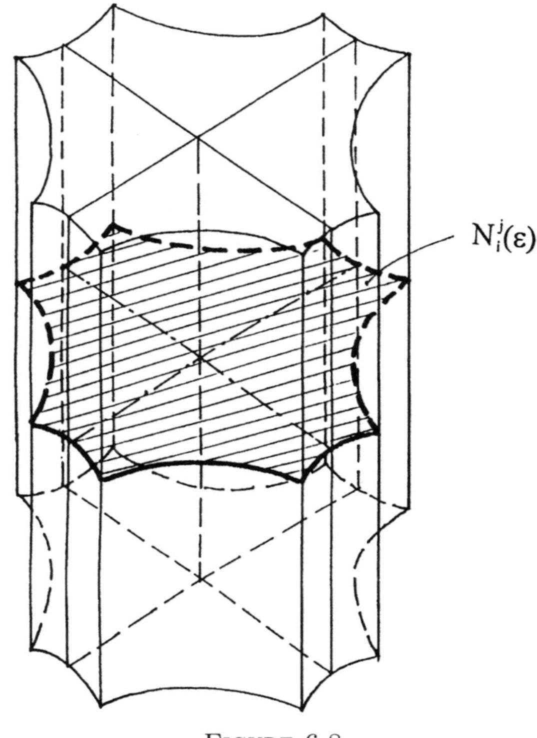


FIGURE 6.8

the pieces of the two-dimensional orbits of the action $\Phi$, which are also transversal to $\partial Q$. Thus, we glue four four-dimensional sets $D_i \times [-h_*, h_*] \times [-k_*, k_*]$ to the set $Q$ along the neighborhood of the closed curves $l_i$ on $\partial Q$ so that each disk $D_i \times \{\epsilon\} \times \{\kappa\}$, $\epsilon \in [-h_*, h_*]$, $\kappa \in [-k_*, k_*]$ is glued to its own closed curve. Now, making $h_*, k_*$ small enough, we obtain an invariant neighborhood $V$ inside the set $V_0 \cap \{|H| \leq h_*, |K| \leq k_*\}$. It is clear, that the neighborhood $V$ contracts to the set $W$ for $h_*, k_* \to 0$.

From now on, the set $V \cap \{H = \epsilon\}$ will be denoted by $V_\epsilon$.

## 6.4. The invariant and formulation of the equivalence theorem

Homeomorphism of the sets $W$ and $W'$ for two IHVFs is not sufficient for isoenergetic equivalence of the fields in their extended neighborhoods $V$ and $V'$. It can be explained by the fact that for $\epsilon \neq 0$, there may be a different number of saddle SPTs on $V_\epsilon$ for given loop types. It follows from the construction of the restriction of the field $X_H$ to $\Sigma_i$ that $V_\epsilon \cap \Sigma_i$ may contain either one long SPT or two short SPTs. Therefore, the following two cases are possible: A) there are four SPTs on $V_\epsilon$ (all of them short) and two SPTs on $V_{-\epsilon}$ (both long); B) there are three SPTs on $V_\epsilon$ (one long and two short), for all $\epsilon \neq 0$. Thus, each of the cases 1–4 describing the loop types (Corollary 6.2.2) is divided into two cases: A and B. Correspondingly, we have eight different cases, which will be enumerated by $n$A and $n$B, $n = 1, \ldots, 4$.

To reduce the number of possibilities we will consider a typical representative for each of the cases $n$A and $n$B, $n = 1, \ldots, 4$, such that every IHVF can be reduced to one of them up to isoenergetic equivalence.

Of the two manifolds $\Sigma_1$ and $\Sigma_2$, the one on which, for a fixed $\epsilon \geq 0$ and therefore for any $0 < \epsilon \leq h_*$ (see the bifurcation diagram in Section 6.3), the function $K$ takes the larger value, is called the upper symplectic manifold. Let us mention that IHVFs $(X_H, K)$, $(X_H, -K)$, $(X_{-H}, K)$ are obviously isoenergetically equivalent. Therefore, changing, if necessary, $(X_H, K)$ to $(X_H, -K)$ in Cases B and $(X_H, K)$ to $(X_{-H}, K)$ in Cases A, we can assume that for $\epsilon > 0$, there are two short SPTs on the upper symplectic manifold on $V_\epsilon$. After that, changing, if necessary, $(X_H, K)$ to $(X_H, -K)$ in Cases A, and $(X_H, K)$ to $(X_{-H}, K)$ in Cases B, we obtain that in Cases 3, 4 the nonorientable loops belong to the upper symplectic manifold. We denote by $(X_H, K)$ the system obtained in this way, and call it a canonical representative of the case. From now on we will only consider the canonical representatives.

As was mentioned earlier, for some local symplectic coordinates, $H$ and $K$ have the form as in Theorem 2.2.1 in $U$ and one can take $\lambda_1 > 0$, $\lambda_2 > 0$ and $\delta > 0$ (see Section 6.1). Therefore, it is easily verified that the upper symplectic manifold is $\Sigma_2$. We also assume that, in Cases 3A–3B the loop $\gamma_2$ is nonorientable. We can always achieve this, preserving all the other conditions, by the local symplectic coordinate change $(x_1, x_2, y_1, y_2) \to (x_1, -x_2, y_1, -y_2)$.

REMARK 6.3. It follows from the construction of the canonical representatives, that for $\epsilon > 0$ there are always two short SPTs on $\Sigma_2$, and therefore, the loop $\gamma_2$ enters $P$ in the direction $x_1 = y_1 = y_2 = 0$, $x_2 > 0$, and $\gamma_4$ in the direction $x_1 = y_1 = y_2 = 0$, $x_2 < 0$. Correspondingly, the loop $\gamma_1$ enters $P$ in the direction $x_2 = y_1 = y_2 = 0$, $x_1 > 0$, in Cases A, and in the direction $x_2 = y_1 = y_2 = 0$, $x_1 < 0$, in Cases B.

Let $(X_H, K)$, $(X_{H'}, K')$ be two IHVFs with Morse Hamilton functions and with the corresponding singular points $p$ and $p'$ of the saddle type. Let $V$ and $V'$ be extended neighborhoods of these points.

THEOREM 6.4.1. *The field $(X_H, K)$ on $V$ is isoenergetically equivalent to the field $(X_{H'}, K')$ on $V'$ if and only if the same Case $nA$ or $nB$ occurs for both fields.*

The "only if" part of the theorem follows from Proposition 6.2.2 and the definition of Cases A and B. The "if" part will be proved in Section 6.7.

Now we consider the equivalence of the Poisson actions. Let $\Phi$ and $\Phi'$ be two Poisson actions of Morse type with singular points $p$ and $p'$ of the saddle-saddle type. Let $V$ and $V'$ be extended neighborhoods of these points.

THEOREM 6.4.2. *Two Poisson actions $\Phi$ and $\Phi'$ are topologically equivalent if and only if they have the same combinations of the orientability types of loops (see Corollary 6.2.2).*

PROOF. The "only if" part of the theorem is obvious. In order to prove the "if" part we show that one can choose one-parameter subgroups in $\mathbb{R}^2$ and an extended neighborhood of the singular point $p$ in such a way that for the corresponding IHVF in $V$ Case $n$A holds, where $n$ is defined as above by the set of loops. Indeed, it follows from the definition of a singular point of the saddle-saddle type that it is possible to choose one-parameter subgroups in $\mathbb{R}^2$ for the action $\Phi$ so that the corresponding Hamilton functions $H, K$ have a singular point of the saddle-saddle type without multiple eigenvalues, i.e., in some neighborhood of this singular point, $H$ and $K$ in symplectic coordinates are as in Theorem 2.2.1. Next we choose an extended neighborhood $V$ for IHVF $(X_H, K)$ as in Section 6.1. If Case $n$A occurs for $(X_H, K)$, then the process ends. If Case $n$B occurs for $(X_H, K)$, then Case $n$A occurs for the IHVF $(X_K, H)$. This can be verified directly by computing the number of SPTs on $\Sigma_1$ and $\Sigma_2$ for a given value $K \neq 0$. It is obvious that for the action $\Phi'$ with the same loop set as $\Phi$ one can choose one-parameter subgroups in $\mathbb{R}^2$ and an extended neighborhood $V'$ of the singular point $p'$ in such a way that Case $n$A hold for the corresponding IHVF $(X', K')$. Then Theorem 6.4.2 follows from Theorem 6.4.1. □

## 6.5. The structure of the auxiliary system and its invariant foliations

Let us consider the auxiliary gradient system $Y$ in $U$ (see Appendix B). To do this, we introduce the Euclidean Riemannian metrics by means of coordinates $(x_1, x_2, y_1, y_2)$. In $U$, the vector field $Y$ has the form

$$\begin{aligned} \dot{x_1} &= y_1(x_2^2+y_2^2)h_2R, \quad \dot{y_1} = x_1(x_2^2+y_2^2)h_2R, \\ \dot{x_2} &= -y_2(x_1^2+y_1^2)h_1R, \quad \dot{y_2} = -x_2(x_1^2+y_1^2)h_1R, \end{aligned} \tag{6.1}$$

where

$$h_i = \frac{\partial h}{\partial \xi_i}, \quad k_i = \frac{\partial k}{\partial \xi_i}, \qquad i = 1, 2,$$

$$R = \frac{(h_1k_2 - h_2k_1)}{h_1^2(x_1^2+y_1^2) + h_2^2(x_2^2+y_2^2)}.$$

We can assume that $R > 0$ in $U$. Therefore, the factor $R$ can be regarded as appeared in the course of a time change. It can be verified directly that the obtained system has three invariant foliations: the $H$-foliations determined by the equation $H$=const, and the $b_i$-foliations, $i = 1, 2$, determined by the equation $b_i = (x_i^2 - y_i^2)/2$ = const, and that the disks $\Sigma^1_{\text{loc}}$, $\Sigma^2_{\text{loc}}$ consist of the singular points of the field $Y$. In order to describe the geometry of the $b_i$-foliations, $i = 1, 2$, we consider a map $\pi$ from $U$ to the coordinate space $(\xi_1, \xi_2, b_1, b_2)$ given by the formula

$$\xi_i = x_i y_i, \quad b_i = (x_i^2 - y_i^2)/2, \qquad i = 1, 2. \tag{6.2}$$

In these coordinates the field (6.1) has the form

$$\begin{aligned} \dot{\xi_1} &= 4h_2(\xi_1^2+b_1^2)^{1/2}(\xi_2^2+b_2^2)^{1/2}R, \qquad \dot{b_1} = 0, \\ \dot{\xi_2} &= -4h_1(\xi_1^2+b_1^2)^{1/2}(\xi_2^2+b_2^2)^{1/2}R, \qquad \dot{b_2} = 0, \end{aligned} \tag{6.3}$$

where $2(\xi_i^2 + b_i^2)^{1/2}$ is substituted for $(x_i^2 + y_i^2)$ in $R$.

We consider the field (6.3) in the domain $|\xi_i| \leq \rho$, $|b_i| \leq \delta$, $i = 1, 2$, with $\rho$ so small that

$$h_i \geq \frac{\lambda_i}{2} > 0, \quad h_1k_2 - h_2k_1 \geq \frac{\Delta}{2} > 0.$$

It is clear that for fixed $\rho$, $\delta$, one can choose $h_*$, $k_*$ so that these inequalities for $\xi_i$, $b_i$ hold in $U$.

The equations (6.3) for $\xi_1$, $\xi_2$ define a Hamiltonian system on the plane $(\xi_1, \xi_2)$ "spoiled" by a time change, and this system depends on the parameters $b_1, b_2$. Since $h$ is a local integral, we get a foliation by the curves $h(\xi_1, \xi_2) = \epsilon$. The parameters $b_1$, $b_2$ affect the velocity of the motion along these curves. For $b_1 \neq 0$, $b_2 \neq 0$ the system does not have singular points in a neighborhood of the point $(0,0)$. For $b_1 = 0, b_2 \neq 0$, there is an interval of singular points $\xi_1 = 0$, and for $b_1 \neq 0, b_2 = 0$, there is an interval of singular points $\xi_2 = 0$. For $b_1 = 0, b_2 = 0$, there is a "cross" of singular points defined by the equation $\xi_1\xi_2 = 0$, and in this case two singular points lie on any curve $h(\xi_1, \xi_2) = \epsilon$, $\epsilon \neq 0$.

Projecting the set $\pi(U)$ to the plane $(\xi_1, \xi_2)$, we get a curvilinear rectangle defined by the inequalities $|h(\xi_1, \xi_2)| \leq h_*$, $|k(\xi_1, \xi_2)| \leq k_*$. From now on, the equations (6.3) will be considered in this rectangle (see Figure 6.9).

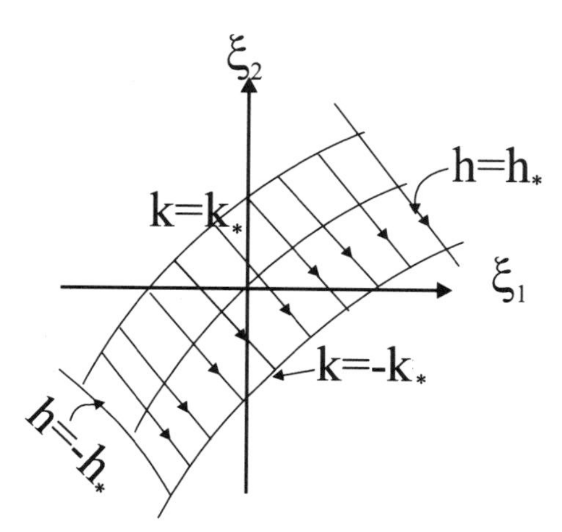


FIGURE 6.9

Now consider the set $G_\epsilon$ in $\pi(U)$ defined by the relations $h(\xi_1, \xi_2) = \epsilon$, $|k(\xi_1, \xi_2)| \leq k_*$, $|b_i| \leq \delta$, $i = 1, 2$, where $|\epsilon| \leq h_*$. This set is invariant with respect to the field (6.3) and is diffeomorphic to a cube. Expressing $\xi_2 = \phi(\xi_1, \epsilon)$ from the equation $h(\xi_1, \xi_2) = \epsilon$ one can see that the coordinates in $G_\epsilon$ are $(\xi_1, b_1, b_2)$, and the integral curves of the field (6.3) in $G_\epsilon$ are the curves $b_1 = \text{const}$, $b_2 = \text{const}$.

The set of singular points of the field (6.3) in $G_\epsilon$ consists of two intervals, $C_1(\epsilon) = \{b_2 = 0, \xi_1 = \xi_1(\epsilon)\}$ and $C_2(\epsilon) = \{b_2 = 0, \xi_1 = 0\}$, where $\xi_1(\epsilon) = \epsilon/\lambda_1 + \cdots$ is the solution of the equation $h(\xi_1, 0) = \epsilon$. It is easy to see that $C_i(\epsilon) = \pi(\Sigma_i \cap U)$.

One can readily see from (6.3) that $\dot{\xi}_1 \geq 0$, and, moreover, the equality holds only on $C_1(\epsilon)$, $C_2(\epsilon)$. The phase portrait of the field (6.3) in $G_\epsilon$ is shown in Figure 6.10. Therefore, for $\epsilon \neq 0$, there is a trajectory defined by the relations $b_1 = 0$, $b_2 = 0$, $0 < \xi_1 < \xi_1(\epsilon)$, for $\epsilon > 0$, and $b_1 = 0$, $b_2 = 0$, $\xi_1(\epsilon) < \xi_1 < 0$, for $\epsilon < 0$. This trajectory goes from one segment of singular points to another. The stable manifold for $C_2(\epsilon)$ is the set $b_1 = 0$, $\xi_1 < 0$ for $\epsilon > 0$, or $b_1 = 0$, $\xi_1 < 0$ for $\epsilon < 0$ without the segment $b_1 = b_2 = 0$, $\xi_1 < \xi_1(\epsilon)$, for $\epsilon < 0$. The unstable manifold for $C_2(\epsilon)$ is the set $b_1 = 0$, $\xi_1 > 0$ for $\epsilon < 0$, and the set $b_1 = 0$, $\xi_1 > 0$ for $\epsilon < 0$, without the interval $b_1 = b_2 = 0$, $\xi_1 > \xi_1(\epsilon)$ for $\epsilon > 0$. Analogously to the case $C_1(\epsilon)$, the stable and unstable manifolds belong to the plane $b_2 = 0$ (see Figure 6.10).

Denote by $\tilde{\Pi}^+(\epsilon)$ the face of the set $G_\epsilon$ defined by the equation $k(\xi_1, \phi(\xi_1, \epsilon)) = k_*$, and by $\tilde{\Pi}^-(\epsilon)$ the boundary defined by the equation $k(\xi_1, \phi(\xi_1, \epsilon)) = -k_*$. The trajectories of the field (6.3) enter $G_\epsilon$ through $\tilde{\Pi}^+(\epsilon)$ and leave through $\tilde{\Pi}^-(\epsilon)$. It is easy to see that due to the conditions $\lambda_1 > 0, \lambda_2 > 0$, $\Delta > 0$, we have $\xi_1 < 0$, $\xi_2 > 0$ on $\tilde{\Pi}^+(\epsilon)$, and $\xi_1 > 0$, $\xi_2 < 0$ on $\tilde{\Pi}^-(\epsilon)$.

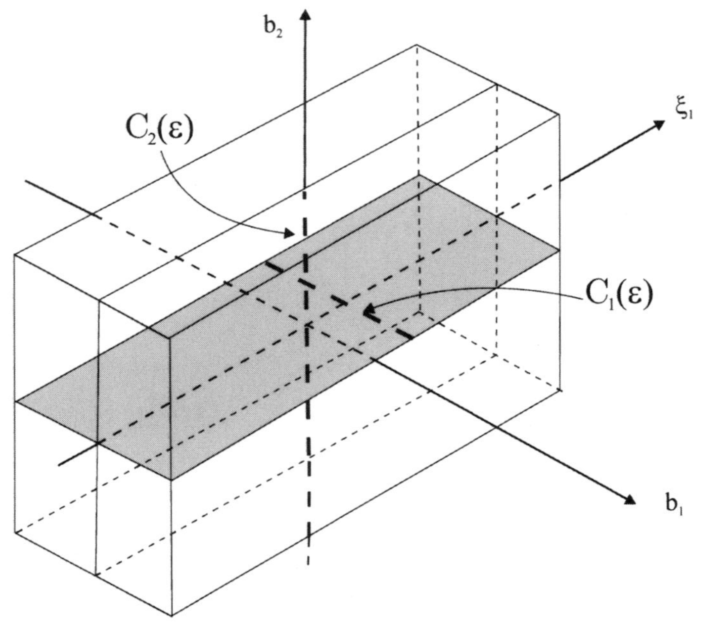


FIGURE 6.10

Now let us return to $U$. Formulas for $\pi$ show that the map $\pi$ is a branched covering, and its branch set is a union of the planes $b_1 = \xi_1 = 0$ and $b_2 = \xi_2 = 0$. There are four preimages over all points from $\pi(U)$ except for the points on these planes, two preimages over the points on these planes, and one preimage (point $p$) over the point $(0, 0, 0, 0)$.

The sets $\tilde{\Pi}^{\pm}(\epsilon)$ do not intersect the curves $C_i(\epsilon)$ (since $\xi_1\xi_2 \neq 0$ on $\tilde{\Pi}^{\pm}(\epsilon)$). Therefore, $\pi^{-1}(\tilde{\Pi}^{+}(\epsilon))$ consists of four curvilinear rectangles, which belong to $U_\epsilon$ on the level set $K = k_*$, and $\pi^{-1}(\Pi^{-1}(\epsilon))$ consists of four curvilinear rectangles, which belong to $U_\epsilon$ on the level $K = -k_*$. We introduce new notation for these rectangles in the following way: $\Pi_1^+(\epsilon)$ is the rectangle on which $x_1 > 0, y_1 < 0$, $x_2 > 0, y_2 > 0$; $\Pi_2^+(\epsilon)$ the rectangle on which $x_1 > 0, y_1 < 0$, $x_2 < 0, y_2 < 0$; $\Pi_3^+(\epsilon)$ the rectangle on which $x_1 < 0, y_1 > 0$, $x_2 > 0, y_2 > 0$; and $\Pi_4^+(\epsilon)$ the rectangle on which $x_1 < 0, y_1 > 0$, $x_2 < 0, y_2 < 0$. In a similar way, we denote by $\Pi_1^-(\epsilon)$ the rectangle on which $x_1 > 0, y_1 > 0$, $x_2 > 0, y_2 < 0$; by $\Pi_2^-(\epsilon)$ the rectangle on which $x_1 > 0, y_1 > 0$, $x_2 < 0, y_2 > 0$; by $\Pi_3^-(\epsilon)$ the rectangle on which $x_1 < 0, y_1 < 0$, $x_2 > 0, y_2 < 0$; and by $\Pi_4^-(\epsilon)$ the rectangle on which $x_1 < 0, y_1 < 0$, $x_2 < 0, y_2 > 0$.

In order to describe the map $\pi$ in a neighborhood of the set $\Sigma^1_{\text{loc}} \cup \Sigma^2_{\text{loc}}$ we restrict $\pi$ to $U_\epsilon$. Then $\pi(U_\epsilon) = G_\epsilon$. We choose a small disk $D_a$ on the plane $b_2 = \text{const}$ around the point $a \in C_2(\epsilon)$ in $G_\epsilon$, $\epsilon \neq 0$, such that $D_a$ does not contain points of the segment $C_1(\epsilon)$. The set $\pi^{-1}(a)$ consists of two points $A_1, A_2$ lying on two curves from $\Sigma^2_{\text{loc}}$ determined by the equation $h(0, \xi_2) = \epsilon$ (these curves form $\pi^{-1}(C_2(\epsilon))$). Thus, it follows from (6.2) that $\pi^{-1}(D_a)$ consists of two disjoint two-dimensional smooth disks: one of them, $D(A_1)$, belongs to a neighborhood of the point $A_1$, and the other, $D(A_2)$, to a neighborhood of the point $A_2$. The restriction of $\pi$ to such a disk is a branched covering over $D_a$ of the type $z \to z^2$ over the disk $|z| \leq 1$, $z \in \mathbb{C}$, with the branching at the point $z = 0$. In this case $D(A_i)$ is foliated by the trajectories of the field $Y$ and the restriction of $Y$ to $D(A_i)$ has a saddle singular point at the point $A_i$. Its two stable separatrices are projected by the map $\pi$ into one semitrajectory of the system (6.3) entering the point $a$, and the two unstable separatrices into a semitrajectory leaving the point $a$ (see Figure 6.11). For $\epsilon \neq 0$, $\pi$ is constructed analogously in a neighborhood of the preimage $\pi^{-1}(C_1(\epsilon))$ of the set $C_1(\epsilon)$ on the fiber $b_1 = \text{const}$. For $\epsilon = 0$ the intervals $C_1(0)$ and $C_2(0)$ in $G_0$ intersect at the point $\pi(p) : b_1 = b_2 = \xi_1 = 0$. For neighborhoods of the points of these intervals different from $\pi(p)$ the map $\pi$ is the same as described above.

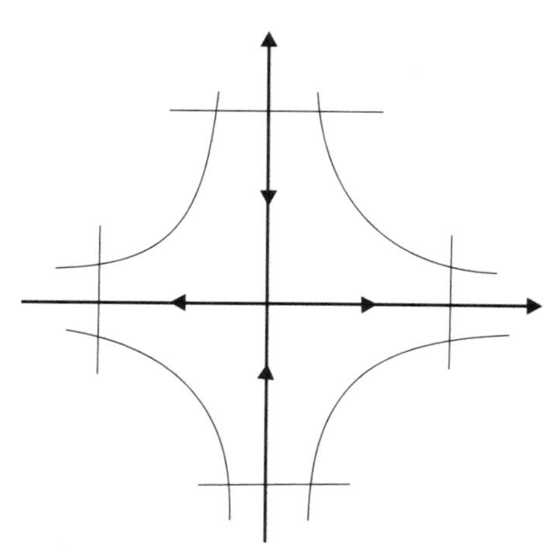

FIGURE 6.11

Now we turn to the description of complete fibers of the $b_1$-foliation in $U_\epsilon$. First we choose $\epsilon = 0$ and consider the fibers $b_1 = 0$ and $b_2 = 0$ in $G_0$. The fiber $b_2 = 0$ contains the interval $C_1(0)$. The first half of the interval $C_1(0)$ on which $b_1 < 0$, is the image of the interval $x_1 = 0$ in $\Sigma^1_{\text{loc}}$. The second half of the interval $C_1(0)$, where $b_1 > 0$, is the image of the interval $y_1 = 0$ in $\Sigma^1_{\text{loc}}$. Thus, $\pi^{-1}(C_1(0))$ is a "cross" on $\Sigma^1_{\text{loc}}$ defined by the equation $x_1 y_1 = 0$. The points $(\pm x_1, 0)$ are mapped into one point on $C_1(0)$. The same holds for the points $(0, \pm y_1)$ under the map $\pi$. There is a semitrajectory $b_1 = 0, b_2 = 0, \xi_1 < 0$ on $G_0$ which enters the point $\pi(p)$. Its preimages in $U_0$ are the four semitrajectories entering the point $p$.

Each of the semitrajectories of this kind goes through the corresponding point $Q_i^+(0)$ of the square $\Pi_i^+(0)$. Through the point $Q_i^+(0)$ passes a smooth curve $\tau_i^+$, and the image of this curve on $\tilde{\Pi}_i^+(0)$ is the trace of the stable manifold of the interval $C_1(0)$ formed by singular points of the field (6.3). Therefore, the set of all semitrajectories of the field $Y$ passing through the points of the curves $\tau_i^+, i = 1, \ldots, 4$, forms the stable manifold of the "cross" on $\Sigma^1_{\text{loc}}$. Let us verify that the $\omega$-limit set of the semitrajectories passing through the curves $\tau_1^+$ and $\tau_2^+$ consists of two intervals of the "cross" $y_1 = 0, x_1 > 0$ and $x_1 = 0, y_1 < 0$, and the $\omega$-limit set of the semitrajectories passing through the curves $\tau_3^+$ and $\tau_4^+$ consists of two intervals of the "cross" $y_1 = 0, x_1 < 0$ and $x_1 = 0, y_1 > 0$ (see Figure 6.12 a)).

Let $L$ be a trajectory of the field $Y$ passing through a point $m \in \tau_1^+$. Since $b_2 = 0$ on $\tau_1^+$, this equality holds along $L$. Therefore, $|x_2| = |y_2|$ on $L$. Along the trajectory $L$, the function $\xi_1$ is negative and it tends to zero as $t \to \infty$. Thus, the function $\xi_2 = -\frac{\lambda_1}{\lambda_2}\xi_1 + O(\xi_1^2)$ also tends to zero as $t \to \infty$. Therefore, $x_2 \to 0$ and $y_2 \to 0$ as $t \to \infty$, i.e., the $\omega$-limit set of $L$ belongs to $\Sigma^1_{\text{loc}}$. From the formulas for

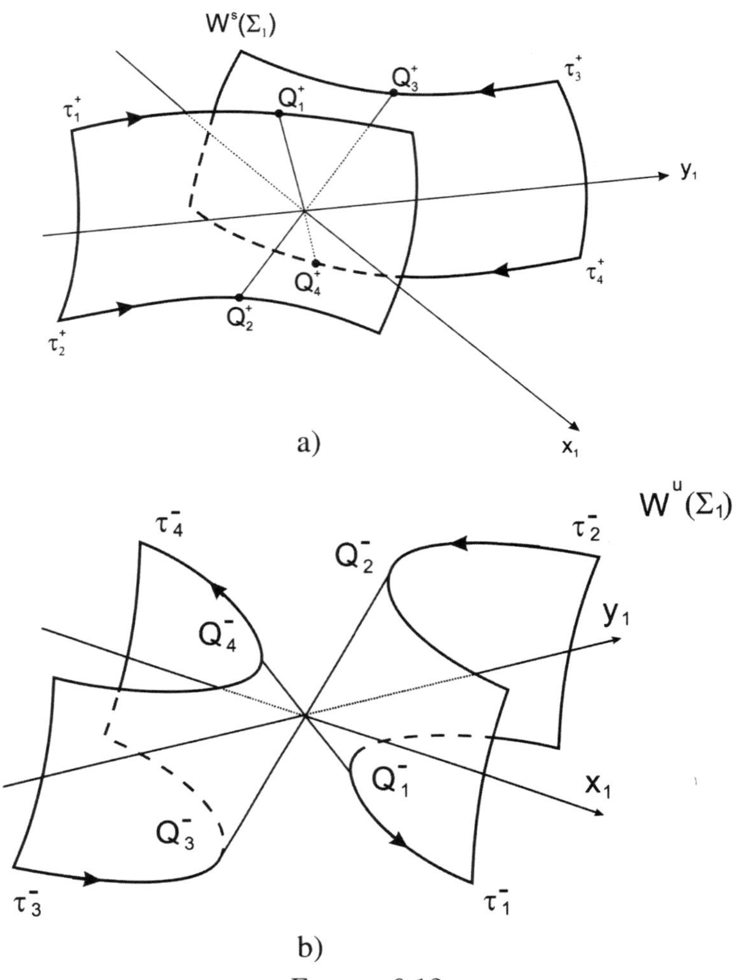


FIGURE 6.12

the map $\pi$, we see that

$$x_1 = \pm\sqrt{b_1 + \sqrt{b_1^2 + \xi_1^2}}, \qquad y = \xi_1 / x_1.$$

Since the function $\xi_1$ does not change sign on $L$, $x_1$ and $y_1$ do not vanish, i.e., the signs of the coordinates $x_1$ and $y_1$ along $L$ are the same as on $\Pi_i^+(0) : x_1 < 0,\ y_1 > 0$. If $b_1 > 0$ at the point $m$, then

$$\lim_{\xi_1 \to 0} x_1 = -\sqrt{2b_1} < 0, \quad \lim_{\xi_1 \to 0} y_1 = 0.$$

If $b_1 < 0$ at the point $m$, then

$$\lim_{\xi_1 \to 0} x_1 = 0, \quad \lim_{\xi_1 \to 0} y_1 = \sqrt{-2b_1} > 0.$$

The same calculations applied to the intervals $\tau_i^+$, $i = 2, 3, 4$, show that analogous statements hold.

Similarly, there are points $Q_i^-(0)$ and curves $\tau_i^-$ on the squares $\Pi_i^-(0)$. The semitrajectories passing through $\tau_i^-$ form the unstable manifold of the "cross" on

$\Sigma^1_{\mathrm{loc}}$ such that the $\alpha$-limit set trajectories passing through $\tau_1^-$ and $\tau_2^-$ are the union of the intervals of the "cross" $y_1 = 0, x_1 > 0$ and $x_1 = 0, y_1 > 0$. Correspondingly, for $\tau_3^-$ and $\tau_4^-$, it is the union of the intervals $y_1 = 0, x_1 < 0$ and $x_1 = 0, y_1 < 0$ (see Figure 6.12 b)).

The stable and unstable manifolds of the "cross" on $\Sigma^2_{\mathrm{loc}}$, which are preimages of the fiber $b_1 = 0$ in $G_0$ are constructed analogously. Related intervals $\zeta_i^\pm$ (analogs of $\tau_i^\pm$) belong to the same squares $\Pi_i^\pm(0)$ and pass through the same points $Q_i^\pm(0)$ such that for a given square, for example $\Pi_i^+(0)$, intervals $\tau_i^+$ and $\zeta_i^+$ intersect transversally at the point $Q_i^+(0)$.

Similar calculations show that the $\omega$-limit set of the semitrajectories passing through the intervals $\zeta_1^+$ and $\zeta_3^+$ is the angle $x_2 > 0, y_2 = 0$ and $x_2 = 0, y_2 > 0$. Correspondingly, for the intervals $\zeta_2^+$ and $\zeta_4^+$ it is the angle $x_2 < 0, y_2 = 0$ and $x_2 = 0, y_2 < 0$. Finally, the $\alpha$-limit set of the semitrajectories passing through the intervals $\zeta_1^-$ and $\zeta_3^-$ is the angle $x_2 > 0, y_2 = 0$ and $x_2 = 0, y_2 < 0$. Correspondingly, for the intervals $\zeta_2^-$ and $\zeta_4^-$ it is the angle $x_2 < 0, y_2 = 0$ and $x_2 = 0, y_2 > 0$ (see Figure 6.13).

REMARK 6.4. The orientation of the curves $\tau_i^\pm$, $\zeta_i^\pm$ is indicated in Figures 6.12–6.13. It shows the direction of increase of the functions $b_i$ on the corresponding curves. Orientation is induced from the related angle on $\Sigma_i$ by the trajectories of the field $Y$. Introduction of this orientation is based on the fact that the sides of the angle are transversal to the $b_i$-foliation, and on the $y_i = 0$ axis the function $b_i$ is positive and tends to $+0$ as it approaches the point $p$, whereas on the $x_i = 0$ axis the function $b_i$ is negative and tends to $-0$ as it approaches the point $p$.

Let us describe the geometry of other fibers of the foliations $b_2 = \text{const}$ and $b_1 = \text{const}$. It follows from the description of $b_1$- and $b_2$-foliations in $G_\epsilon$ and the type of branching that the preimages of the fibers $b_1 = b_1^0 \neq 0$ or $b_2 = b_2^0 \neq 0$ are the same for all $\epsilon$. Consider, for example, the fiber $b_2 = b_2^0 \neq 0$ in $G_\epsilon$. It intersects $C_2(\epsilon)$ at a single point $z$, and does not intersect $C_1(\epsilon)$. It follows from the local description of the branching that the preimage of the fiber $b_2 = b_2^0 \neq 0$ consists of two smooth two-dimensional submanifolds with boundary. The foliation of such a manifold by the trajectories of the vector field is shown in Figure 6.11.

Now we have to study the geometry of the fibers $b_2 = 0$ and $b_1 = 0$. Consider, for example, the fiber $b_2 = 0$ in $G_\epsilon$. This fiber contains the whole segment $C_1(\epsilon)$ and one point $\tilde{z} \in C_2(\epsilon)$. The set $\pi^{-1}(C_1(\epsilon))$ consists of two intervals $d_1(\epsilon)$, $d_2(\epsilon)$, which belong to $\Sigma^1_{\mathrm{loc}}$. The map $\pi$ lifts the trajectory $\tilde{\Gamma}$ connecting the point $\tilde{z}$ and the point $\tilde{z}' \in C_1(\epsilon)$ to four trajectories. It follows from the description of the branching that $\pi^{-1}(\tilde{z})$ consists of two points $z_1$ and $z_2$, which belong to different pieces of the curve $h(0, \xi_2) = \epsilon$ on $\Sigma^2_{\mathrm{loc}}$, and $\pi^{-1}(\tilde{z}')$ consists of two point $z_1'$ and $z_2'$, which belong, respectively, to $d_1(\epsilon)$ and $d_2(\epsilon)$. Straightforward calculations using (6.2) yield the following result.

LEMMA 6.5.1 (the structure of a heteroclinic set). *The set $\pi^{-1}(\tilde{\Gamma})$ forms a contour as follows: for $\epsilon > 0$, there are exactly two trajectories of the filed $Y$ that start at each of the points $z_1$, $z_2$. Trajectories begin at the point $z_i$ and go to different points $z_1'$ and $z_2'$. For $\epsilon < 0$, the trajectories go from the point $z_i'$, $i = 1, 2$, to different points $z_1$, $z_2$.*

Now let us consider an auxiliary interval $\tilde{\Lambda} : \xi_1 = \xi_1^0$ on the plane $b_2 = 0$, which intersects the trajectory $\tilde{\Gamma}$ somewhere between the points $\tilde{z}$ and $\tilde{z}'$. The coordinate

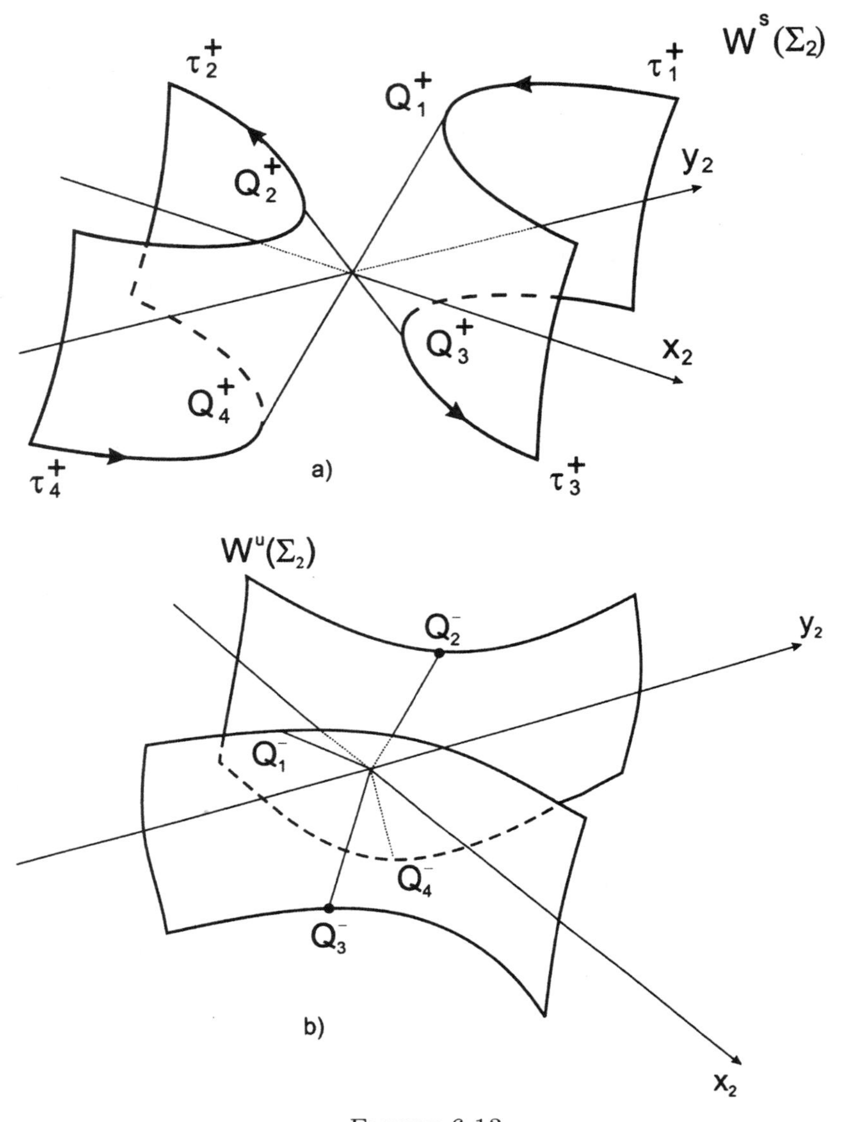


FIGURE 6.13

$b_1$ induces an orientation of this interval. The interval $\tilde{\Lambda}$ divides the fiber $b_2 = 0$ into two rectangles $R_1$, $R_2$ such that $R_1$ contains the point $\tilde{z}$, and $R_2$ contains the interval $C_1(\epsilon)$. In fact, the sets $\pi^{-1}(R_1)$, $\pi^{-1}(R_2)$ are already described: the geometry of the fiber $\pi^{-1}(R_1)$ is the same as that of the fiber $b_2 = b_2^0 \neq 0$, and $\pi^{-1}(R_2)$ consists of two components, each being the union of the stable and unstable submanifolds of the interval $d_i(\epsilon)$, i.e., of two curvilinear rectangles which intersect transversally along $d_i(\epsilon)$. For $\epsilon > 0$, one pair of intervals from $\pi^{-1}(\tilde{\Lambda})$ is a pair of opposite sides of the stable manifold on one of the $d_i(\epsilon)$, and the other pair of the intervals from $\pi^{-1}(\tilde{\Lambda})$ is a pair of opposite sides of the stable manifold on some other $d_j(\epsilon)$, $i \neq j$.

Orientation of $\tilde{\Lambda}$ induces an orientation of each of the intervals $\pi^{-1}(\tilde{\Lambda})$. The boundary of the local stable manifold of the interval $d_i(\epsilon)$, i.e., the boundary of the curvilinear rectrangle can be oriented. Moreover, (6.2) implies that the two orientations induced by the orientations of each of the intervals are opposite to each other.

The same four intervals $\pi^{-1}(\tilde{\Lambda})$, with the same decomposition into pairs, belong to the boundary of each of the components of the set $\pi^{-1}(R_1)$. They are transversal intervals to the unstable separatrix of the saddle singular point on the corresponding component of $\pi^{-1}(R_1)$. The boundary of the component is a simple closed curve. Moreover, its orientations induced by the orientations of each of the two oriented intervals from $\pi^{-1}(\tilde{\Lambda})$ coincide. For $\epsilon > 0$, the description of preimages of rectangles $R_1$, $R_2$ is the same, provided one changes the stable manifolds to unstable, and vice versa.

Lemma 6.5.1 defines rules for gluing the intervals together. We obtain the set shown in Figure 6.14. It consists of a smooth two-dimensional annulus with piecewise smooth boundary and two smooth rectangles which intersect the annulus transversally. These rectangles are the unstable manifolds of the curves $\pi^{-1}(C_1(\epsilon))$.

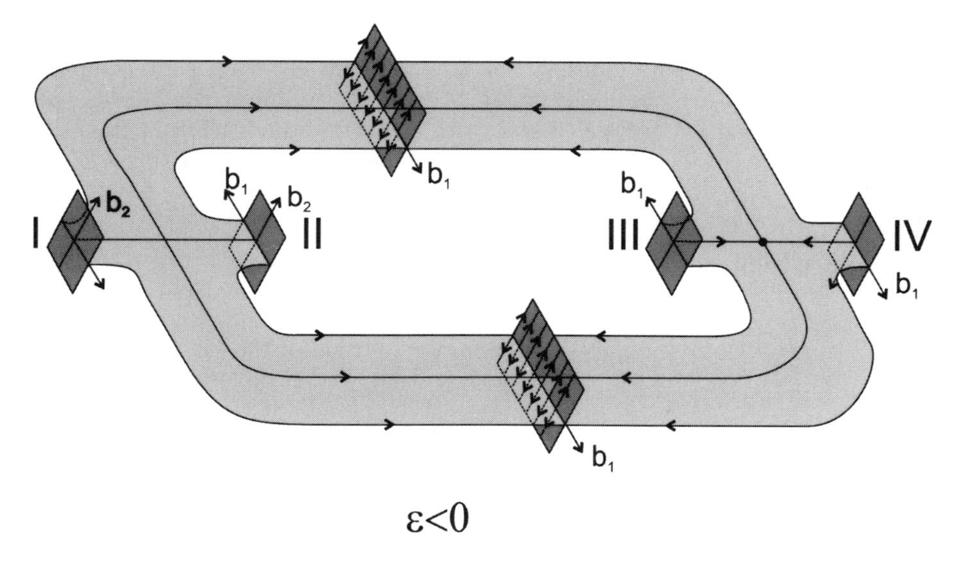


FIGURE 6.14

As $\epsilon \to +0$, the preimage of the fiber $b_2 = 0$ deforms in such a way that the heteroclinic contour contracts to the point $p$, and the two segments $\pi^{-1}(C_1(\epsilon))$ glue together at one point forming the "cross" $\Sigma^1_{\text{loc}}$. Thus, as described above, we obtain the union of the stable and unstable manifolds of the "cross." When passing to $\epsilon < 0$, a heteroclinic contour appears, but now the smooth annulus contains the unstable manifolds of the curves $\pi^{-1}(C_2(\epsilon))$.

REMARK 6.5. There are two mutually transversal foliations of $\tilde{\Pi}^{\pm}(\epsilon)$, by intervals $b_2 = b_2^0$ and $b_1 = b_1^0$. Since the restriction of the map $\pi$ to $\Pi^{\pm}(\epsilon)$ is a diffeomorphism, there are two smooth mutually transversal foliation of these squares as well (they are the traces of the $b_1$-foliation and $b_2$-foliation on the surfaces $K = \pm k_*$, $H = \epsilon$).

In each of the squares $\Pi_i^{\pm}(\epsilon)$ there are two distinguished fibers, $b_1 = 0$ and $b_2 = 0$, which were earlier denoted by $\tau_i^{\pm}$ and $\zeta_i^{\pm}$. The point of their intersection is $Q_i^{\pm}(0)$. For $\epsilon > 0$ ($\epsilon < 0$) the fiber $b_1 = 0$ ($b_2 = 0$) on $\Pi_i^+(\epsilon)$ is the trace of the stable manifold of a CSP which belongs to $\Sigma_2$ ($\Sigma_1$). Moreover, there is a trajectory passing through the point $Q_i^+(0)$, as $t \to \infty$, tends to a point in $\Sigma_2$ ($\Sigma_1$) such that there are two heteroclinic trajectories beginning at that point. Except for the point $Q_i^+(0)$, the fiber $b_2 = 0$ ($b_1 = 0$) is the trace of the stable manifolds of curves of singular points from $\Sigma_1$ ($\Sigma_2$). Distinguished fibers on $\Pi_i^-(\epsilon)$ are described analogously.

Later we will need some results on the orientation at the point $Q_i^+(\epsilon)$ of the 3-frame formed by the tangent vectors to the lines $b_1 = 0$, $b_2 = 0$ on $\Pi_i^+(\epsilon)$ and the vector of the field $Y$ at the point $Q_i^+(\epsilon)$. More precisely, let $e_1^i$ be the unit tangent vector to the line $b_2 = 0$ on $\Pi_i^+(\epsilon)$ directed towards increasing values of $b_1$. Let $e_2^i$ be the unit tangent vector to the line $b_1 = 0$ on $\Pi_i^+(\epsilon)$ directed towards increasing values of $b_2$. Finally, let $e_3^i$ be the normalized vector of the field $Y$ at the point $Q_i^+(\epsilon)$.

LEMMA 6.5.2. *All the 3-frames in $U_\epsilon$ at the points $Q_i^+(\epsilon)$, $i = 1, \dots, 4$, define the same orientation.*

PROOF. In order to determine the orientations of the 3-frames, we compare the orientations of the 4-frames at the points $Q_i^+(\epsilon)$ in $U$, adding to the 3-frames as the first vector the gradient $\nabla H$ calculated with respect to the Euclidean Riemannian metric in $U$. The coordinates $(x_1^i, x_2^i, y_1^i, y_2^i)$ of the points $Q_i^+(\epsilon)$ are determined by the system

$$H = \epsilon, \quad K = k_*, \quad x_1^2 - y_1^2 = 0, \quad x_2^2 - y_2^2 = 0,$$

which yields

$$x_1^i = \sigma_1\sqrt{|u|}, \quad y_1^i = -\sigma_1\sqrt{|u|}, \quad x_2^i = y_2^i = \sigma_2\sqrt{v},$$

where $\xi_1 = u(\epsilon, k_*)$, $\xi_2 = v(\epsilon, k_*)$ is the unique solution of the system

$$h(\xi_1, \xi_2) = \epsilon, \quad k(\xi_1, \xi_2) = k_*,$$

and $\sigma_j = \operatorname{sgn} x_j^i$. We recall that if $\xi_2$ is positive on the $\Pi_i^+(\epsilon)$, then $\xi_1$ is negative. Then tangent vectors to the curves $b_2 = 0$ and $b_1 = 0$ in $\Pi_i^+(\epsilon)$ at the point $Q_i^+$ have the form

$$\left(\frac{\sigma_1}{2\sqrt{|u|}}, 0, \frac{\sigma_1}{2\sqrt{|u|}}, 0\right)^T, \quad \left(0, \frac{\sigma_2}{2\sqrt{|u|}}, 0, \frac{\sigma_2}{2\sqrt{|u|}}\right)^T,$$

respectively. Taking (6.2) into account, we see that the vectors $\nabla H$ and $Y$ have the form

$$\nabla H = (-h_1\sigma_1\sqrt{|u|}, h_2\sigma_2\sqrt{v}, h_1\sigma_1\sqrt{|u|}, h_2\sigma_2\sqrt{v})^T,$$
$$Y = 2R(-\sigma_1 h_2 v\sqrt{|u|}, -\sigma_2 h_1 u\sqrt{|v|}, \sigma_1 h_2 v\sqrt{|u|}, -\sigma_2 h_1 u\sqrt{|v|})^T.$$

Therefore, the signs of the determinants composed from the coordinates of the four vectors are the same at all the points $Q_i^+(\epsilon)$, i.e., these frames induce the same orientation in $U$. For every $\epsilon$, the set $U_\epsilon$ is orientable, and its orientation can be defined by the vectors $\nabla H$ everywhere except for the point $p$. Thus, the 3-frames tangent to $U_\epsilon$ also induce the same orientation. The lemma is proved. □

Therefore, we have described the structure of the field $Y$ in $U$. Now let us consider some properties of the filed $Y$ in $V \setminus U$, assuming that there is a Riemannian metric in $V$ such that it is an extension of the Euclidean metric in $U$. In $V_\epsilon$ the points of $(\Sigma_1 \cup \Sigma_2) \setminus U$ form four disjoint intervals of the curves of singular points

(CSPs) of the field $Y$. It follows from Appendix B that there are stable and unstable manifolds of the CSPs. One has the following statement about these manifolds.

LEMMA 6.5.3. *The stable and unstable manifolds of the curves of singular points of the field $Y$ do not intersect in $V_\epsilon \setminus U$, i.e., all the heteroclinic trajectories of the field $Y$ belong to $U$.*

PROOF. First, consider $V_0$ (i.e., $\epsilon = 0$). The function $K$ vanishes on CSP in $V_0$. Since the function $K$ decreases monotonically as $t$ increases along trajectories of the filed $Y$, those trajectories that leave CSP cannot tend to a point of CSP as $t \to \infty$. There are no other limit points of the filed $Y$ in $V_0 \setminus U$. Thus, as $t$ decreases, all the trajectories that belong to the stable manifolds of CSP come to the upper boundary tori $K = k_*$, $H = 0$. As follows from the description of the stable sets of the "cross", the traces of the trajectories form eight disjoint intervals. Similarly, the trajectories on the unstable manifolds of CSP come to the lower boundary tori $K = -k_*$, $H = 0$.

Let $\epsilon \neq 0$. For all $\epsilon$, the stable and unstable manifolds of CSP in $V \setminus U$ depend on $\epsilon$ smoothly. Therefore, for sufficiently small $h_*$ and $|\epsilon| \leq h_*$, all the trajectories also come to the upper boundary tori and form eight disjoint intervals. Similar statements hold for the unstable manifolds that come to the tori $K = -k_*$, $H = \epsilon$. This implies the statement of the lemma. □

The topology of stable and unstable manifolds of CSP of the field $Y$ in $U$ has been described before (see Figures 6.12–6.13). Let us study their topology in $V$. First, consider the CSP formed by the loops $\gamma_1, \dots, \gamma_4$. The stable and unstable manifolds of the "cross" on $\Sigma_i \cap U$ have been constructed above.

Let us denote by $[p, Q_i^+], i = 1, \dots, 4$, the semitrajectories of the filed $Y$ that enter the singular point $p$, and by $[Q_i^-, p]$ the semitrajectories that leave the point $p$, respectively, where $Q_i^\pm \equiv Q_i^\pm(0)$.

Let $E_k$ be the "figure eight" on $\Sigma_k$ formed by two loops, and let $W^s(E_k)$, $W^u(E_k)$ be its stable and unstable sets, $k = 1, 2$. The intervals $[p, Q_i^+]$ cut each of the sets $W^s(E_k), k = 1, 2$, into two rectangles. Each such rectangle contains the intervals $[p, Q_i^+]$, which are used for gluing. Thus, the topology of $W^s(E_k)$ can be described by pointing out the rules for the identification of the sides of rectangles.

PROPOSITION 6.5.1. *The topological structure of the stable and unstable manifolds of "figures eight" for the standard representatives is as shown in Figures* 6.15, *provided the intervals of the lateral sides are identified with regard to their orientations, and all the points denoted by $p$ are also identified.*

PROOF. For each of the possible cases $n$A, $n$B, we must take their standard representatives and describe the rules of gluing the local pieces of the stable (unstable) manifolds when they are extended along the global parts of the loops $\gamma_i$. To be specific, let us consider the loop $\gamma_1 \in \Sigma_1$ in all the cases $n$A, $n = 1, \dots, 4$. Recall that for canonical representatives the loop $\gamma_1$ near the point $p$ coincides with the semiaxis $x_2 = y_2 = x_1 = 0$, $y_1 > 0$ (initial piece of $\gamma_1$), and semiaxis $x_2 = y_2 = y_1 = 0$, $x_1 > 0$ (final piece of $\gamma_1$). It is only necessary to consider two cases, when $\gamma_1$ is an orientable loop (Cases 1A, 3A, 4A), and when $\gamma_1$ is a nonorientable loop (Case 2A). For $b_1^* > 0$ small enough, for the sections in the level set $H = 0$ we take the disks $N^u = \{x_1^2 - y_1^2 = -2b_1^*, y_1 > 0\}$ and $N^s = \{x_1^2 - y_1^2 = 2b_1^*, x_1 > 0\}$. The coordinates on $N^s$ and $N^u$ are $(x_2, y_2)$. From the equation $h(\xi_1, \xi_2) = 0$ one has

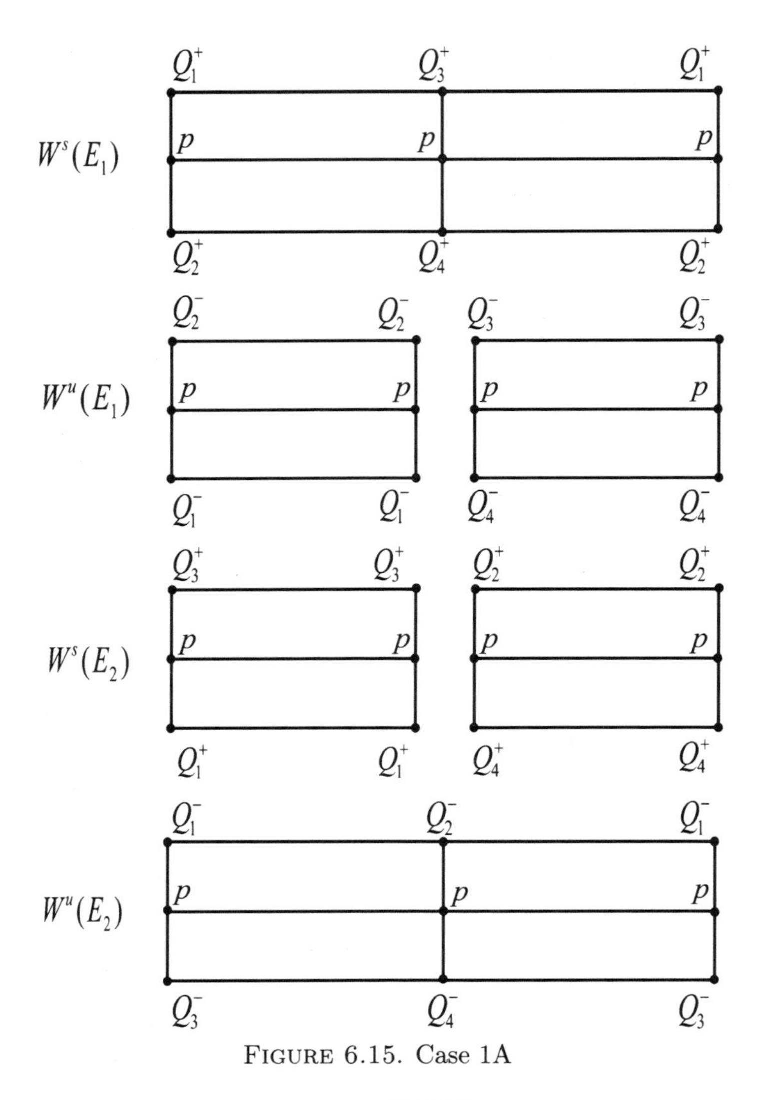


FIGURE 6.15. Case 1A

$\xi_1 = x_1 y_1 = -\frac{\lambda_2}{\lambda_1}\xi_2 + O(\xi_2)$, and $y_1 = \sqrt{x_1^2 + 2b_1^*}$ on $N^u$. Therefore, $x_1 = \phi(\xi_2)$, $y_1 = \sqrt{\phi^2 + 2b_1^*}$. Similarly, on $N^s$ one has $x_1 = \sqrt{y_1^2 + 2b_1^*}$. Therefore, $y_1 = \psi(\xi_2)$, $x_1 = \sqrt{\psi^2 + 2b_1^*}$. Moreover, $N^u$ and $N^s$ are invariant manifolds of the field $Y$ (see Section 6.1). Thus, the traces of $W^s(E_1)$ and $W^u(E_1)$ are two smooth curves passing through the point $m_u = \gamma_1 \cap N^u$ (respectively, $m_s = \gamma_1 \cap N^s$). The curves are formed by two trajectories of the field $Y$ that enter $m_u$ as $t \to \infty$, and by two trajectories that enter $m_s$ as $t \to -\infty$. Notice that on $V$ the manifolds $W^s(E_1)$ and $W^u(E_1)$ intersect $W^u_{\text{loc}}$ and $W^c_1$, as well as their extensions along $\gamma_1$ (in particular, $W^s_{\text{loc}}$ and $W^c_2$) only along the loop $\gamma_1$. Indeed, the only limit sets of the trajectories of the field $Y$ in $V_0$ that belong to $W^s(E_i)$, $W^u(E_i)$ are the "figures eight" $E_1$ and $E_2$, for which $H = K = 0$. Since the function $K$ changes monotonically on the trajectories of the field $Y$ in $W^s(E_i)$, $W^u(E_i)$ on $V_0$, it cannot vanish. Therefore, such a trajectory cannot intersect $W^u_{\text{loc}}$, $W^c_1$, $W^c_2$, $W^s_{\text{loc}}$ where $K = 0$. The disk $N^u$ intersects $W^u_{\text{loc}} = \{x_1 = x_2 = 0\}$ and $W^c_1 = \{x_1 = y_2 = 0\}$ along the intervals that are given by the equations $x_2 = 0$ and $y_2 = 0$, respectively, in coordinates $(x_2, y_2)$.

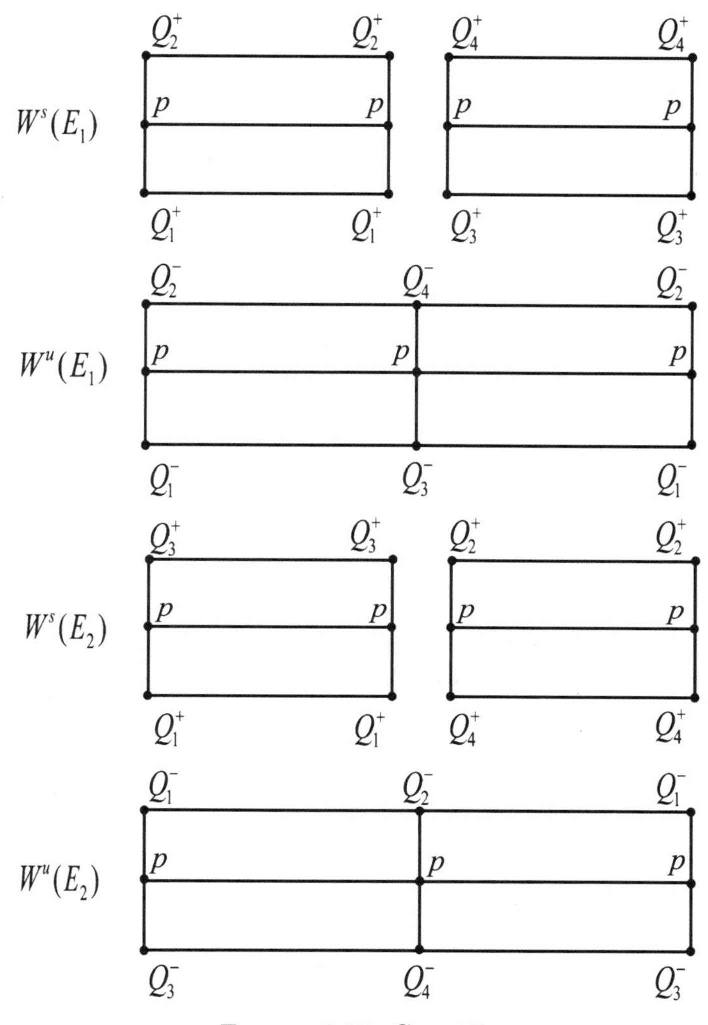


FIGURE 6.15. Case 1B

These intervals divide $N^u$ into four quadrants. We show that two of them intersect the trace of $W^s(E_1)$, and the other two intersect the trace of $W^u(E_1)$.

In order to do this, let us determine the positions of the points of intersection of the curves $\tau_i^+$ and $\tau_j^-$ with $N^u$. Since $\tau_i^+ \in \Pi_i^+(0)$, $\tau_j^- \in \Pi_j^-(0)$, the location of the traces of the curves $\tau_i^+$ and $\tau_j^-$ is determined by the signs of the coordinates $x_2$, $y_2$ corresponding to these curves. It follows from the structure of local stable and unstable manifolds of the "cross" on $\Sigma_1$ (see Figure 6.12) that the curves which intersect $N^u$ are $\tau_3^+$, $\tau_4^+$, $\tau_1^-$, $\tau_2^-$. According to the signs of $x_2$, $y_2$, on $\Pi_i^+(0)$, $\Pi_j^-(0)$, we have the positioning shown in Figure 6.16 a). The location of points of intersection of curves $\tau_1^+$, $\tau_2^+$, $\tau_1^-$, $\tau_2^-$ with $N^s$ can be determined analogously (Figure 6.16 b)). In the figures, the traces of the curves $\tau_i^\pm$ are marked by the same symbols as the curves themselves. As was pointed out before, the points $\tau_i^+$ and $\tau_j^+$ are connected by the traces of the manifolds $W^s(E_1)$, and the points $\tau_i^-$ and $\tau_j^-$ are connected by the traces of the manifolds $W^u(E_1)$. We recall that when

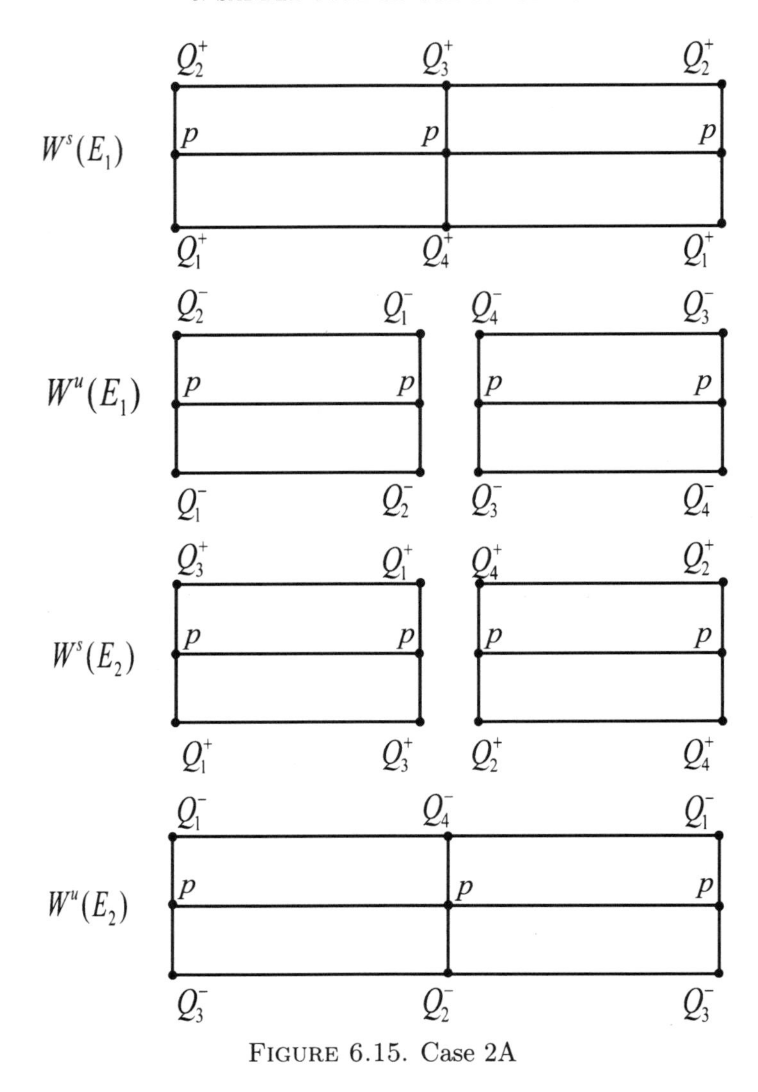


FIGURE 6.15. Case 2A

moving along the loop $\gamma_1$ the trace of $W^u_{\rm loc}$ on $N^u$ merges with the trace of $W_2^c$ on $N^s$, and the trace of $W_c^1$ on $N^u$ merges with the trace of $W^s_{\rm loc}$ on $N^s$ (this results from the transversality along $\gamma_1$ of the intersection of the global stable and unstable manifolds of a saddle singular point $p$ of the field $X_H$ on the level $H = 0$). The definition of the orientability of loops and the fact that the gluing map is symplectic imply that, for an orientable loop, the gluing map preserves the orientation of the axes of the same name, and for a nonorientable loop, the orientation is reversed. Since the traces of $W^s(E_1)$ and $W^u(E_1)$ on $N^u$ map to the traces of $W^s(E_1)$ and $W^u(E_1)$ on $N^s$, the stable and unstable manifolds of the loop $\gamma_1$ look as shown in Figure 6.17.

Similar calculation can be done for the other loops $\gamma_2, \ldots, \gamma_4$ in all the cases. The results of these calculations are presented in Figure 6.17 showing the stable and unstable manifolds of the "figures eight" $E_1$, $E_2$. The proposition is proved. □

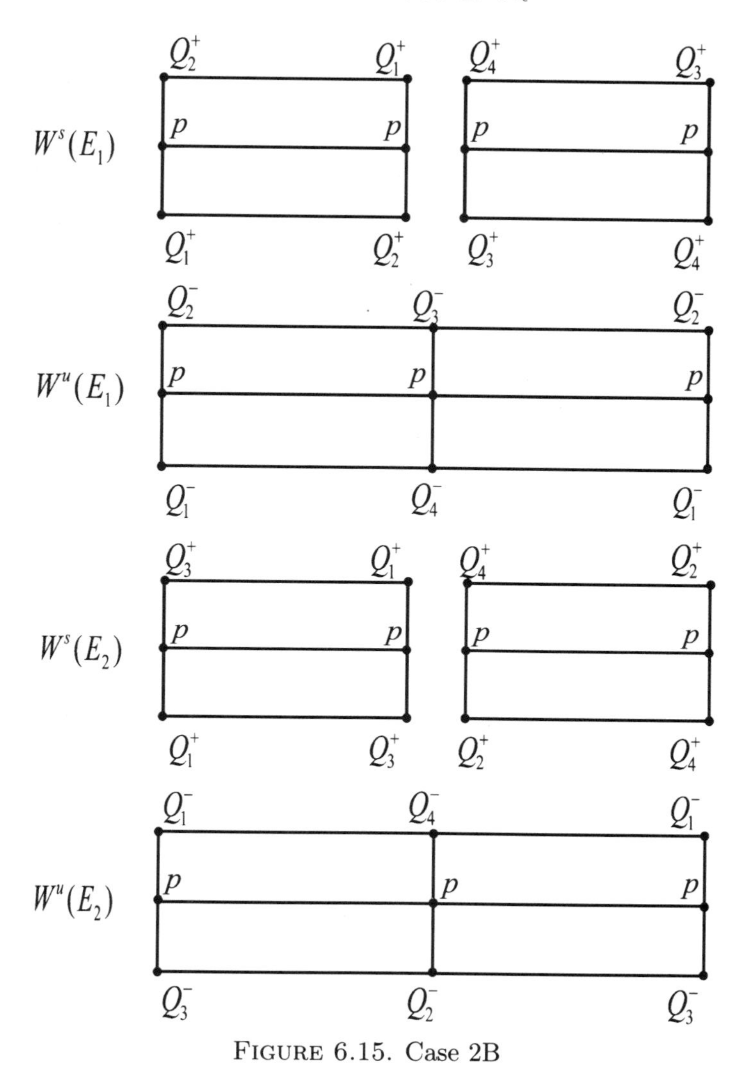


FIGURE 6.15. Case 2B

REMARK 6.6. It is clear from the previous proof that orientations on the curves $\tau_i^{\pm}$, $\zeta_i^{\pm}$ (see Remark 6.4) induce the corresponding orientations on each arc of the curves from the traces $W^s(E_1)$ and $W^u(E_2)$.

Similar arguments prove the following result.

PROPOSITION 6.5.2. *If a closed CSP of the field Y corresponds to an orientable (nonorientable) SPT of the field $X_H$, then the local stable and unstable manifolds of CSP are cylinders (Möbius bands).*

There is a similar statement in [**9**].

## 6.6. The topology of $\partial V_\epsilon$

In this section we will describe the boundary $\partial V_\epsilon$ of the manifold $V_\epsilon$ and, in particular, will find the number of the boundary tori. This boundary is the common level set $H = \epsilon$ and $K = k_*$ of the functions $H$ and $K$. Since the set $K = \pm k_*$ does not intersect $\Sigma_1 \cup \Sigma_2$, the boundary consists of some number of Liouville tori.

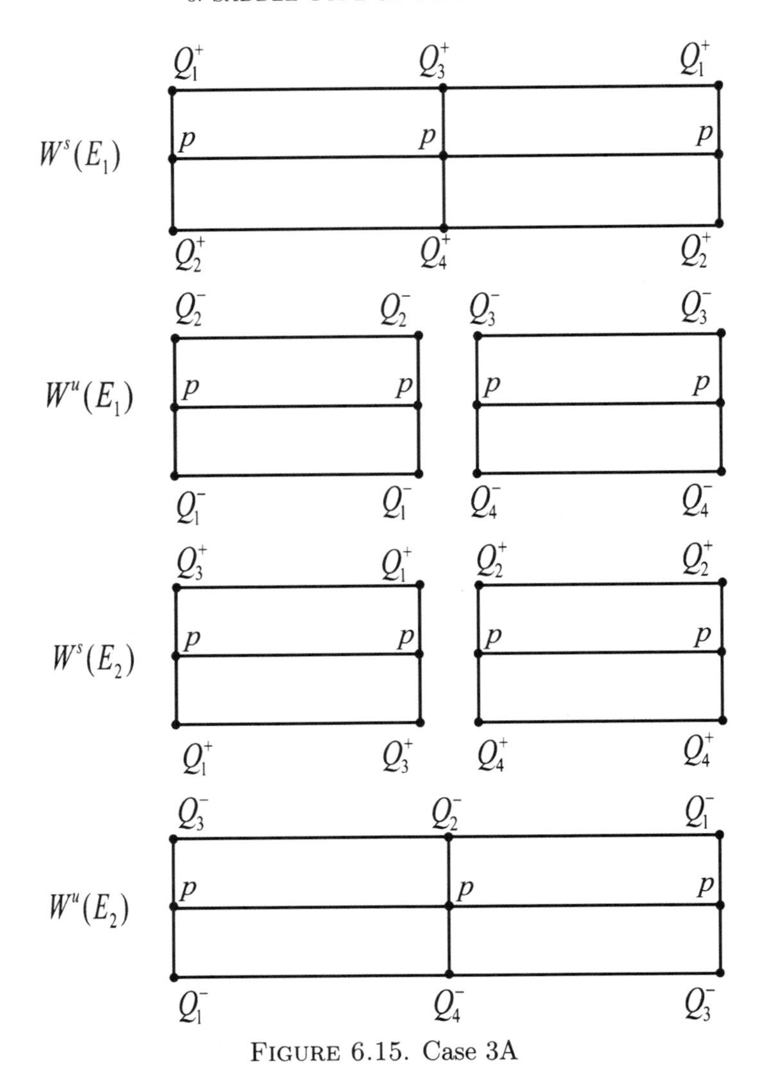


FIGURE 6.15. Case 3A

Moreover, the number of boundary tori is the same for all $\epsilon$, $|\epsilon| < h_*$. Therefore, it is sufficient to find the number of the boundary tori on $\partial V_0$. It is clear that

$$\partial V_0 = \partial V_0^+ \cup \partial V_0^-,$$

where $K = k_*$ on $\partial V_0^+$, and $K = -k_*$ on $\partial V_0^-$.

Let us state the basic properties of trajectories of the field $Y$ on $V_0$ which follow from the fact that the only limit sets of trajectories of the field $Y$ in $V$ are the singular points in $\Sigma_1 \cup \Sigma_2$.

PROPOSITION 6.6.1. 1. *The set of singular points of the field $Y$ on $V_0$ coincides with $E_1 \cup E_2$ (two "figures eight" formed by the loops $\gamma_1, \dots, \gamma_4$).*

2. *All trajectories from $W^s(E_1) \cup W^s(E_2)$ that do not belong to $E_1 \cup E_2$ intersect $\partial V_0^+$, and all trajectories from $W^u(E_1) \cup W^u(E_2)$ intersect $\partial V_0^-$.*

3. *The map along trajectories of the field $Y$ from $\partial V_0^+ \setminus (W^s(E_1) \cup W^s(E_2))$ to $W \setminus (E_1 \cup E_2)$ is a diffeomorphism. Analogously, the map along the trajectories from $W \setminus (E_1 \cup E_2)$ to $\partial V_0^- \setminus (W^u(E_1) \cup W^u(E_2))$ is also a diffeomorphism.*

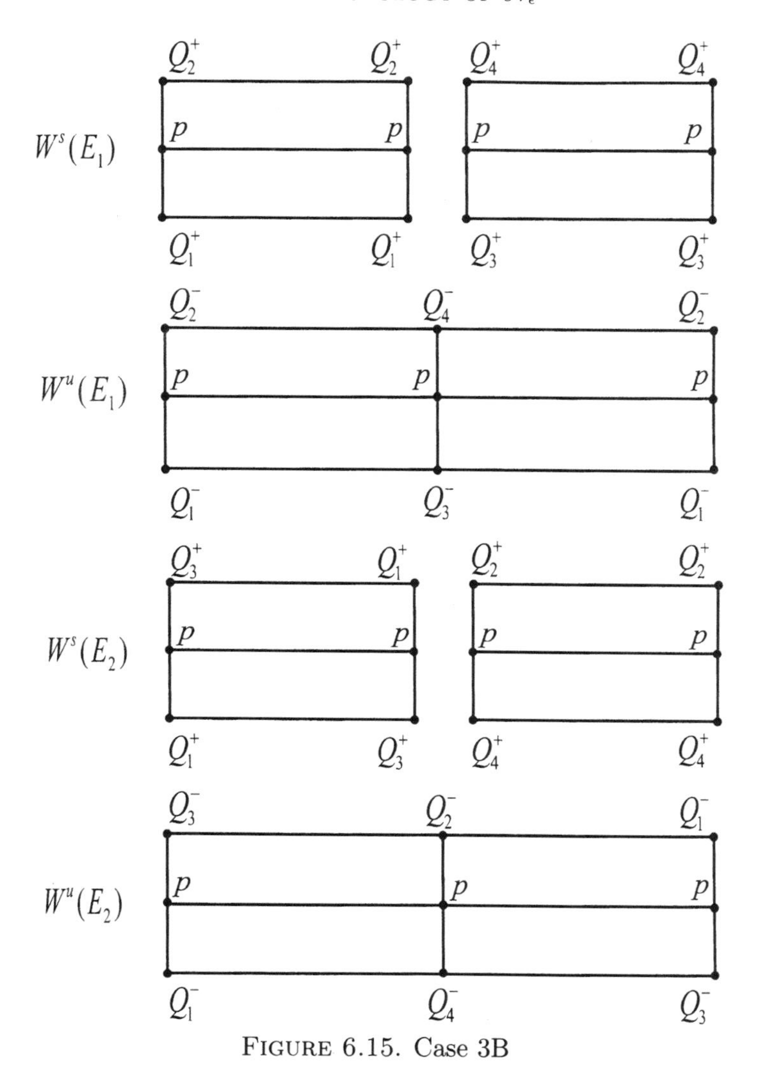


FIGURE 6.15. Case 3B

4. *The sets* $\partial V_0^+ \setminus (W^s(E_1) \cup W^s(E_2))$ *and* $\partial V_0^- \setminus (W^u(E_1) \cup W^u(E_2))$ *consist of four disjoint open disks.*

PROOF. Parts 1–3 follow directly from the structure of the field $Y$. Part 4 can be proved using part 3 and topology of the set $W$. □

Now we describe the structure of the traces of stable manifolds of the "figures eight" on $\partial V_0^+$ and the traces of unstable manifolds of the "figures eight" on $\partial V_0^-$. These traces are smooth curves consisting of some number of simple closed curves. For a given "figure eight" the number of components is determined completely by the topological type of the manifold $W^s(E_i)$ or $W^u(E_i)$ (Proposition 6.5.1, Figures 6.15). The trace of $W^s(E_1)$ on $\partial V_0^+$ intersects the trace of $W^s(E_2)$ at four points $Q_1^+, \dots, Q_4^+$. Analogously, on $\partial V_0^-$ the traces of $W^s(E_1)$ and $W^u(E_2)$ intersect each other at four points $Q_1^-, \dots, Q_4^-$. Therefore, Proposition 6.5.1 implies the following result.

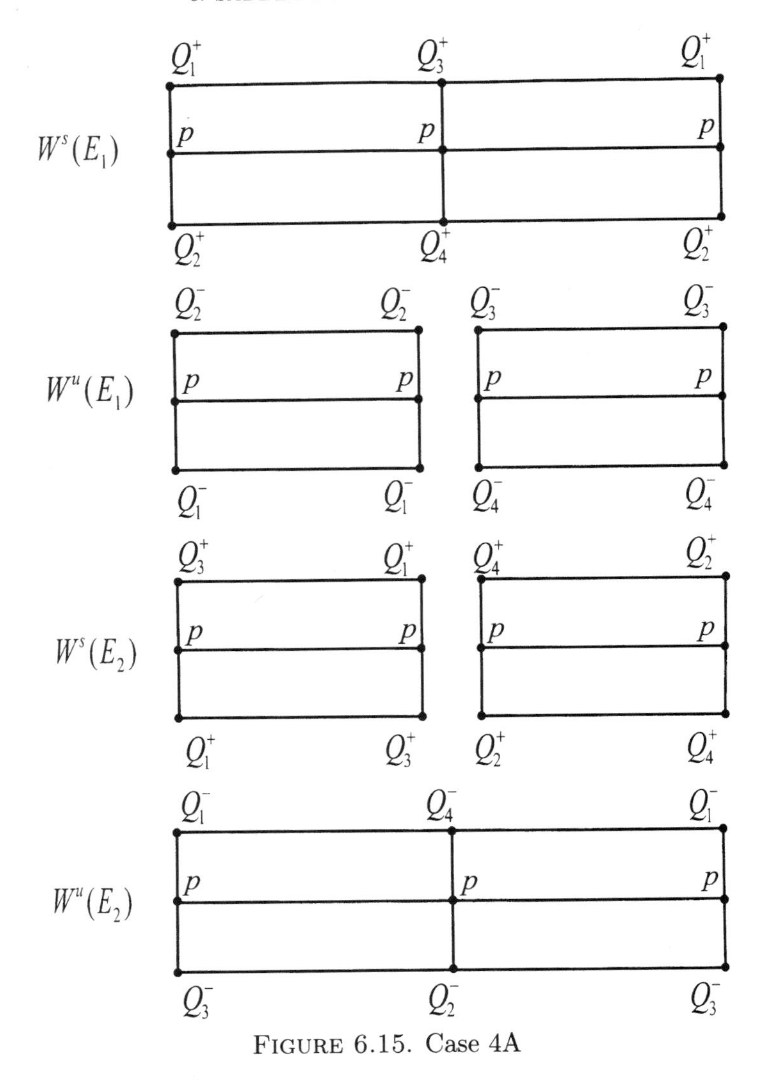


FIGURE 6.15. Case 4A

PROPOSITION 6.6.2. *Mutual position of the traces of stable manifolds of the "figures eight" on $\partial V_0^+$ and the traces of stable manifolds of the "figures eight" on $\partial V_0^-$ is as shown in Figure 6.18.*

The number of tori on $\partial V_0^+$ and on $\partial V_0^-$ and embedding of the described curves into them will be obtained later.

REMARK 6.7. Due to Remark 6.6, all the curves in Figure 6.18 are orientable. We will not indicate this orientation since it does not give us any additional information in any situation, except for the traces of the curves $W^u(E_2)$ in Cases 3A and 3B. The orientation of these curves is determined by the cyclic order $Q_1^-, Q_4^-, Q_3^-, Q_2^-$.

PROPOSITION 6.6.3. *Every torus on $\partial V_0^+$ ($\partial V_0^-$) contains exactly one component of the set*

$$(W^s(E_1) \cup W^s(E_2)) \cap \partial V_0^+ \qquad ((W^u(E_1) \cup W^u(E_2)) \cap \partial V_0^-).$$

PROOF. The existence of a torus on $\partial V_0^+$ that does not intersect the set $W^s(E_1) \cup W^s(E_2)$ contradicts part 4 of Proposition 6.6.1. Now, assume that some

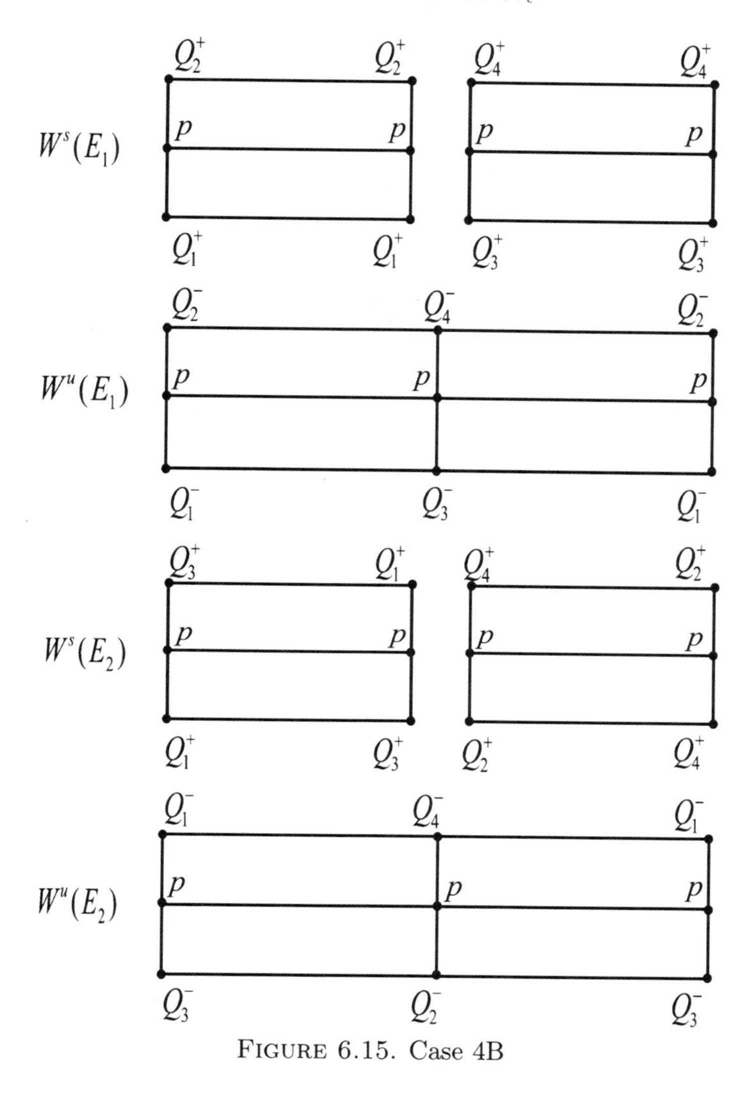


FIGURE 6.15. Case 4B

torus $T$ contains two components $S_1$ and $S_2$. Each of them consists of the union of simple smooth closed curves, some of which intersect each other. Assume that all curves from $S_1$ and $S_2$ are homotopic to zero on $T$. Then $T \setminus (S_1 \cup S_2)$ contains a simple closed curve which is nonhomotopic to zero on $T$. This contradicts part 4 of Proposition 6.6.1. If, for example, $S_1$ contains a curve which is nonhomotopic to zero, then cutting the torus $T$ along all the curves of this component we obtain either annuli or disks, and $S_2$ belongs to the interior of one of these annuli or disks. In any of these cases the complement of $S_2$ contains a non-simply-connected component, i.e., a component which is not a disk. Again, this contradicts part 4 of Proposition 6.6.1. This completes the proof. □

PROPOSITION 6.6.4. *For a fixed $i$, the curves forming the trace of $W^s(E_i)$ on $\partial V_0^+$ (or $W^u(E_i)$ on $\partial V_0^u$) and lying on the same torus are nonhomotopic to zero and isotopic to each other. Two curves lying on the same torus and such that one of them belongs to the trace $W^s(E_1)$ and the other to the trace of $W^s(E_2)$ are nonisotopic. The same is true for the traces of $W^u(E_1)$ and $W^u(E_2)$.*

a)

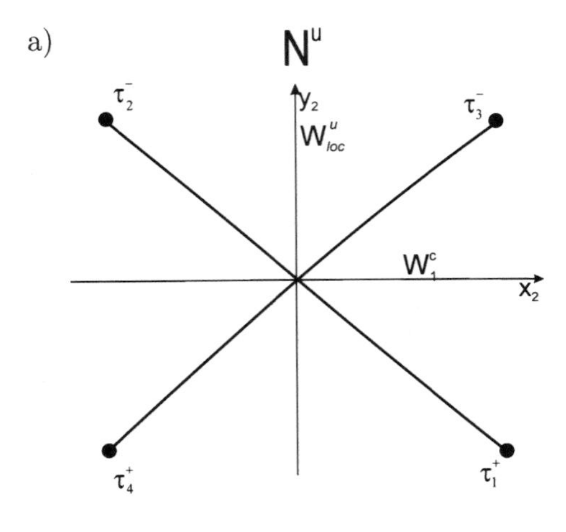


b)

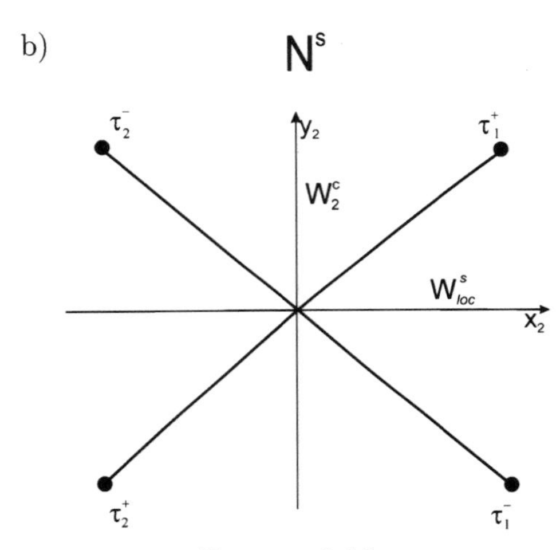


FIGURE 6.16

PROOF. Consider, for example, stable manifolds of "figures eight". Each component of the trace of $W^s(E_1) \cup W^s(E_2)$ belongs to its own torus and consists of a finite set of smooth closed curves which are the traces of $W^s(E_1)$ and $W^s(E_2)$. In all the cases, except 4A and one component of Case 3A (see Figure 6.18), any two curves with a common point (therefore belonging to different $W^s(E_2)$) intersect transversally at a unique point. Therefore, their intersection coefficient is equal to one, and they are nonhomotopic to zero and nonisotopic to each other. Thus, in these cases all the curves are nonhomotopic to zero. If two curves do not intersect (they are the traces of stable manifolds of one "figure eight"), then, since each of them is nonhomotopic to zero, they are isotopic to each other.

Let us consider the remaining situations in Cases 4A and 3A. Now the curves (there are exactly two curves) intersect each other transversally at two points. If their intersection coefficient is equal to zero, then it is easy to show that in the complement of these curves on the torus there is a non-simply-connected component. This contradicts part 4 of Proposition 6.6.4. Therefore, their intersection

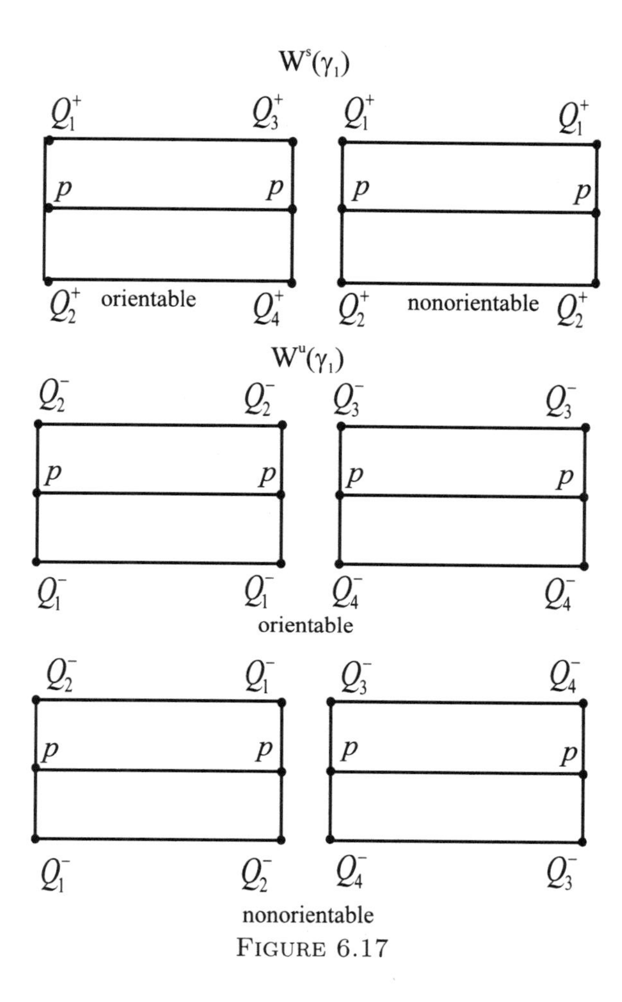


FIGURE 6.17

coefficient is not equal to zero (to be precise, its absolute value is equal to 2), and these curves are nonhomotopic to zero and nonisotopic to each other.

For the unstable manifolds all arguments are similar in all cases except for Case 3B. Let us consider Case 3B. First we choose a curve $l$ which contains all 4 points $Q_1^-, \dots, Q_4^-$ located on it in cyclic order. Each of the other two curves $l_1$ and $l_2$ passes through a pair of nonneighboring points on $l$. Thus, if $l$ is homotopic to zero, then $l_1$ and $l_2$ must intersect each other. Therefore, $l$ is nonhomotopic to zero. If $l_1$ is homotopic to zero, then in the disk bounded by $l_1$ there is an arc of the curve $l$ on which (inside the disk) lies one of the intersection points $l \cap l_2$. The other point lies outside of the disk. Therefore, $l_1 \cap l_2 \neq \varnothing$. Hence, $l_1, l_2$ are nonhomotopic to zero, and therefore, $l_1$ is isotopic to $l_2$. The curves $l_1, l_2$ cut the torus into two annuli, and it follows from the arrangement of the points $Q_1^-, \dots, Q_4^-$ on $l_1, l_2$ that $l$ is not homotopic to $l_1$ and $l_2$. The proposition is proved. □

Thus, we have constructed the traces on $\partial V_0^+$ and $\partial V_0^-$ of stable and unstable manifolds of the "figures eight". Now we would like to construct the traces on $\partial V_0^+$ and $\partial V_0^-$ of the stable and unstable manifolds $W^s(\Sigma_i)$ and $W^u(\Sigma_i)$ of a closed CSP lying on $\Sigma_1$ and $\Sigma_2$. The manifold $\partial V^+$ with boundary is the union of manifolds

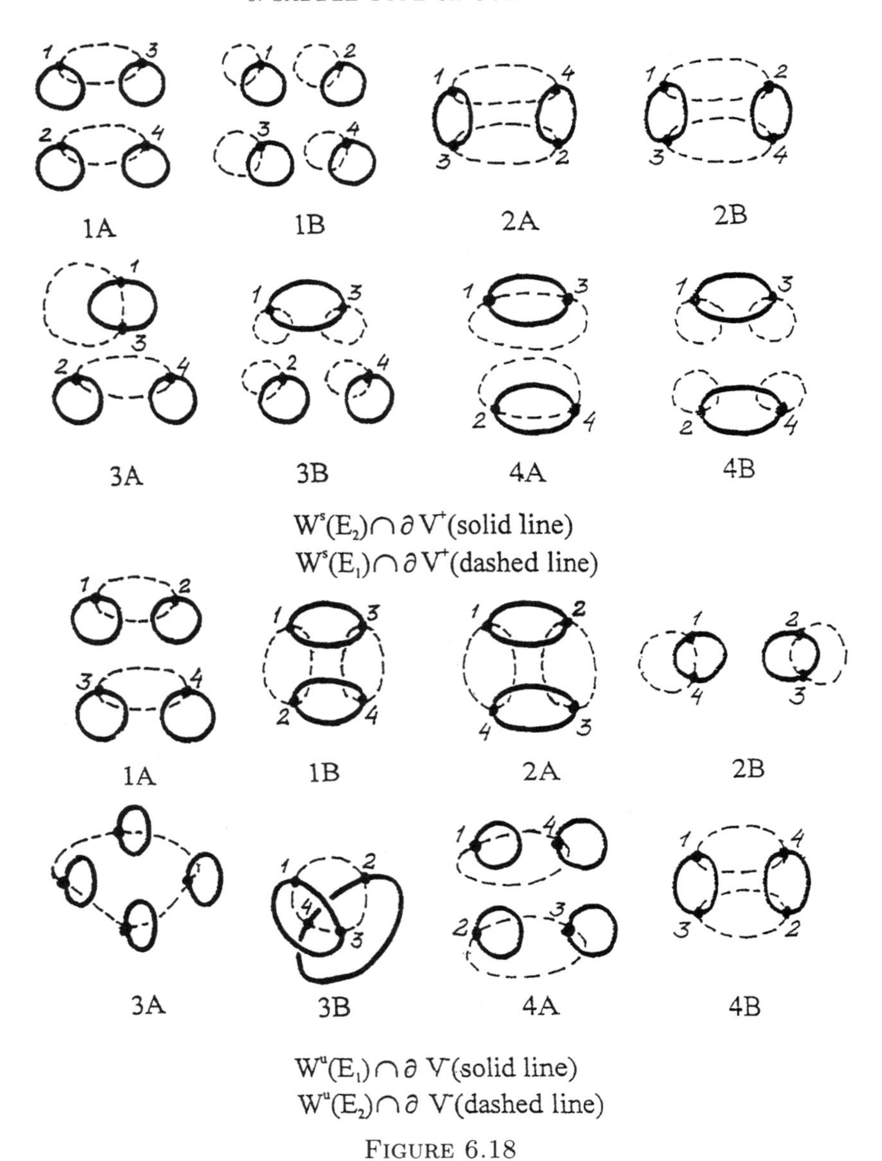


FIGURE 6.18

$\partial V_\epsilon^+$, $|\epsilon| \le h_*$, and it is diffeomorphic to the direct product $\partial V_0^+ \times [-h_*, h_*]$. Let $A_i^s$ denote the closure of the set $W^s(\Sigma_i) \cap \partial V^+$.

PROPOSITION 6.6.5. *The set $A_i^s$ is diffeomorphic to the direct product*

$$(W^s(E_i) \cap \partial V_0^+) \times [-h_*, h_*].$$

*A similar statement is true for $A_i^u = W^u(\Sigma_i) \cap \partial V^-$, $i = 1, 2$.*

PROOF. Consider, for example, $W^s(E_1)$. We have already shown that $W^s(\Sigma_1) \cap \partial V_0^+$ is a union of some number of disjoint closed curves. For $\epsilon < 0$, there are no heteroclinic trajectories of the field $Y$ on $W^s(\Sigma_1) \cap V_\epsilon$. Therefore, $W^s(\Sigma_1) \cap \partial V_\epsilon^+$ is also a union of some number of smooth closed curves. For $\epsilon > 0$, there are four heteroclinic trajectories on $W^s(\Sigma_1) \cup V_\epsilon$; all others intersect $\partial V_\epsilon^+$. Thus, $W^s(\Sigma_2) \cap \partial V_\epsilon$ is a union of four nonclosed smooth curves with boundary points

$Q_i^+(\epsilon), i = 1, \ldots, 4$. Trajectories from $W^s(\Sigma_1) \cap \partial V_\epsilon^+$ pass through these points. The trace of the fiber $b_2 = 0$ passing through the point $Q_i^+(\epsilon)$ is a segment such that all its points except $Q_i^+(\epsilon)$ belong to $W^s(\Sigma_1)$. Thus, the closure of $W^s(\Sigma_1) \cup \partial V_\epsilon^+$ is a union of a certain number of closed curves. Therefore, $A_1^s$ is a union over $\epsilon$ of closed smooth curves. These curves are constructed from intervals of the local fibers $b_2 = 0$ near the points $Q_i^+(\epsilon)$ and the global parts, which are traces of $W^s(E_1/U)$ on $\partial V_\epsilon^+$. It follows from Proposition B.5 that the traces on $\partial V_\epsilon^+$ of the fibers from $W^s(E_1/U)$ depend smoothly on $\epsilon$. Due to smooth dependence on $\epsilon$ of the $b_2$-foliation of the set $\partial V^+$, the local parts depend smoothly on $\epsilon$. Consequently, $A_1^s$ is a smooth two-dimensional submanifold with boundary smoothly foliated by smooth closed curves. This implies the proposition. □

## 6.7. Proof of the equivalence theorems

In this section we prove that the conditions of Theorem 6.4.1 are sufficient. Let $(X_H, K)$, $(X_{H'}, K')$ be two IHVFs with Morse Hamilton functions with saddle singular points $p$ and $p'$, respectively. Suppose that $V$ and $V'$ are their extended neighborhoods, $\sigma = \mu(V)$, $\sigma' = \mu'(V')$, where $\mu$, $\mu'$ are the moment maps. We assume that the canonical representatives of the fields $(X_H, K)$, $(X_{H'}, K')$ are chosen. We will construct a homeomorphism $f : V \to V'$ defining the isoenergetic equivalence of the fields. Let $I = [-h_*, h_*]$, $h_* > 0$, be the range of $H$ in $V$, and $I' = [-h'_*, h'_*]$, $h'_* > 0$, the corresponding range of $H'$. Similarly, $J = [-k_*, k_*]$ and $J' = [-k'_*, k'_*]$ are the ranges of $K$ and $K'$.

Now we construct the auxiliary homeomorphism $q : \sigma \to \sigma'$. First we define an arbitrary homeomorphism $q_1 : I \to I'$, $q_1(0) = 0$, $q_1(h_*) = h'_*$. Then we define a function $q_2(h, k)$ with the following properties:

1. $q_2(h, \pm k_*) = \pm k'_*$;
2. $q_2(h, k_i(h)) = k'_i(q_1(h))$,

where $k_i(h)$ and $k'_i(h)$ are the functions that define $\mu(\Sigma_i)$ and $\mu'(\Sigma'_i)$. The homeomorphism $q$ is given by the formula $q(h, k) = (q_1(h), q_2(h, k))$.

At the next step we construct homeomorphisms $\rho_i : \Sigma \to \Sigma'$. The "figure eight" $E_i$ ($E'_i$) divides $\Sigma$ ($\Sigma'$) into two regions. In one of them $H > 0$ ($H' > 0$), in the other $H < 0$ ($H' < 0$). One of the regions contains long SPTs and it is connected, the other contains short SPTs and it is disconnected. We work with the canonical representatives of the cases, and, therefore, in the regions of the same type on $\Sigma_i$, $\Sigma'_i$ (i.e., connected or disconnected) the functions $H$ and $H'$ are of the same sign. Namely, the region $H > 0$ on $\Sigma_2$ (respectively, the region $H' > 0$) is always disconnected. On $\Sigma_1$ (and $\Sigma'_1$) the region $H > 0$ ($H' > 0$) is disconnected in Case A and connected in Case B. The set $\Sigma_i \cup U$ has a $b_i$-foliation $x_i^2 - y_i^2 = 2b_i$, $|b_i| \leq b_i^*$ such that the fiber $b_i = 0$ consists of two smooth intervals intersecting transversally at the point $p$ (simlarly for $\Sigma'_i \cap U'$). One of these intervals, $d_i$ ($d'_i$), belongs to the connected region and divides $\Sigma_i$ ($\Sigma'_i$) into two components. The other, $n_i$ ($n'_i$), belongs to the nonconnected region and does not divide $\Sigma_i$ ($\Sigma'_i$). In the local coordinates the interval $d_2$ is defined by the equation $x_2 + y_2 = 0$. The interval $d_1$ is defined by the equation $x_1 + y_1 = 0$ in Case A and by the equation $x_1 - y_1 = 0$ in Case B. This follows from the agreement about the numbering of the loops for the canonical representatives of the cases (see Section 6.4). Therefore, $d_2$ ($d'_2$) always belongs to the region $H < 0$ ($H' < 0$), whereas the segment $d_1$ ($d'_1$)

belongs to the region $H < 0$ ($H' < 0$) in Case A, and to the region $H > 0$ ($H' > 0$), in Case B.

We now extend the local $b_i$-foliation to the whole $\Sigma_i$. For this we consider one component of the set $\Sigma_i \setminus d_i$ and construct a smooth foliation which is transversal to the curves $H = \text{const}$ and coincides with the $b_i$-foliation in $\Sigma_i \cap U$. In particular, its leaves are transversal to the loop. As the parameter $b_i$ on the loop we take a cyclic coordinate on the circle, $b_i \in [-\pi, \pi] (\text{mod} 2\pi)$, extended from $\Sigma_i \cap U$. In this way the foliation itself is also parametrized. The analogous construction works for the other component of $\Sigma_i \setminus d_i$.

In order to define $\rho_i$ we first establish a correspondence between the components of the sets $\Sigma_i \setminus d_i$ and $\Sigma'_i \setminus d'_i$ using local coordinates on $\Sigma_i \cap U$. Namely, we put the component of the set $\Sigma_2 \setminus d_2$ on which $x_2 + y_2 \geq 0$, in correspondence with the component of the set $\Sigma'_2 \setminus d'_2$ on which $x'_2 + y'_2 \geq 0$. In Case A, we put the component of the set $\Sigma_1 \setminus d_1$ on which $x_1 + y_1 \geq 0$, in correspondence with the component of the set $\Sigma'_1 \setminus d'_1$ on which $x'_1 + y'_1 \geq 0$. In Case B we put the component of the set $\Sigma_1 \setminus d_1$ on which $x_1 - y_1 \geq 0$, in correspondence with the component of the set $\Sigma'_1 \setminus d'_1$ on which $x'_1 - y'_1 \geq 0$. It should be noted that the orientation type of the loops lying in the components corresponding to each other (for the same fixed cases) is preserved. Now we note that we have defined the coordinates $(\epsilon, b_i)$ on the components. Let us define a homeomorphism $\alpha_1 : [-\pi, \pi] \to [-\pi, \pi]$ such that $\alpha_1(0) = 0$, $\alpha_1(\pi) = \pi$, $\alpha_1(\pm b_1^*) = \pm' b_1^*$. Then $\rho_1(\epsilon, b_1) = (q_1(\epsilon), \alpha_1(b_1))$. Let us extend $\rho_1$ to $d_1$. Consider a point $x \in d_1$. Then a long SPT passes through $x$. It also intersects $d_1$ at another point $x_1$. The same value $\epsilon = H(x) = H(x_1)$ corresponds to the points $x_1$ and $x_2$. However, in the component considered the function $b_1$ has different signs near $x_1$ and near $x_2$. Therefore, the map $\rho_1$ is extended to the points of $d_1$ by continuity. Using the same function $\alpha_1$, a similar construction can be carried out for the other component of the set $\Sigma_1 \setminus d_1$. Later we will use the notation $\rho_1$ for the union of these homeomorphisms (coinciding on $d_1$).

The same construction gives $\rho_2$. On the component of $\Sigma_2 \setminus d_2$ the homeomorphism is defined by the same formula $\rho_2(\epsilon, b_2) = (q_1(\epsilon), \alpha_2(b_2))$, where $\alpha_2 : [-\pi, \pi] \to [-\pi, \pi]$, $\alpha_2(0) = 0$, $\alpha_2(\pi) = \pi$, $\alpha_2(\pm b_2^*) = \pm' b_2^*$.

REMARK 6.8. It should be noted that $\operatorname{sgn} b_i = \operatorname{sgn} \alpha_i(b_i)$. Therefore, it is easy to verify that the restriction of the homeomorphism $\rho_i$ to $\Sigma_i \cap U$ has the following property: $\rho_i$ maps oriented intervals of the axis of the local coordinate system $(x_i, y_i)$ into oriented intervals of the corresponding axis of the local coordinate system $(x'_i, y'_i)$ in $U'$ preserving the orientations.

Our next goal is to define the conjugating homeomorphism $f_u : U \to U'$. To do this we first establish the correspondence between the sets $\Pi_i^+ = \bigcup_{\epsilon \in I} \Pi_i^+(\epsilon)$, $\Pi_i^- = \bigcup_{\epsilon \in I} \Pi_i^-(\epsilon)$, $i = 1, \dots, 4$, and those of the second system. It is clear that it will be sufficient to define the correspondence for $\epsilon = 0$. Let us choose, for example, the set $\Pi_1^+(0)$. In it there is a point $Q_1^+ = \tau_1^+ \cap \zeta_1^+$ corresponding to the fibers $b_2 = 0$ and $b_1 = 0$ with respect to the local coordinates $(b_1, b_2)$ near the point $Q_1^+$ in $\Pi_1^+$ (see Section 6.5). It follows from Proposition B.5 (see also Figure 6.12) that the $\omega$-limit points of trajectories of the field $Y$ passing through the points of the curve $\tau_1^+$ form a set of points in $\Sigma_1$ which is the angle $A_1 : x_1 = 0, y_1 < 0$ and $x_1 > 0, y_1 = 0$. Trajectories of the field $Y$ passing through points of the curve $\zeta_1^+$ tend to points which form the angle $A_2 : x_2 = 0, y_2 < 0$ and $x_2 > 0, y_2 = 0$ on $\Sigma_2$.

Homeomorphisms $\rho_i$ take these angles into the angles $A_1'$, $A_2'$ in $\Sigma_1'$, $\Sigma_2'$ with the same sign combinations (see Remark 6.9).

According to Proposition B.5 applied to the IHVF $(X_{H'}, K')$, the angle $A_1'$ is $\omega$-limit points of the trajectories of the field $Y'$ passing through the intervals $'\tau_1^+$ and $'\tau_2^+$, and the angle $A_2'$ is the $\omega$-limit of the trajectories passing through the intervals $'\zeta_1^+$ and $'\zeta_3^+$. Therefore there is a unique rectangle $'\Pi_1^+(0)$ containing the curves $'\tau_1^+$ and $'\zeta_1^+$ such that the angles $A_1'$ and $A_2'$ are the $\omega$-limit sets. Let us put the rectangle $'\Pi_1^+(0)$ into correspondence with the rectangle $\Pi_1^+(0)$. The correspondence of the remaining rectangles $\Pi_i^{\pm}(0)$ (and, therefore, the sets $\Pi_i^{\pm}$) is defined similarly. It follows from the choice of $\rho_i$ and Remark 6.8 that this correspondence simply preserves the numeration.

We will construct the homeomorphism $f_u$ in the closure of the regions into which the neighborhood is divided by unions of stable and unstable manifolds of singular points of the field $Y$ lying on $\Sigma_1 \cap U$ and $\Sigma_2 \cap U$. In the coordinates on $U$ these manifolds precisely coincide with the sets defined by the equations $b_1 = 0$ or $b_2 = 0$. Thus, the regions into which $U$ is divided can be described in the following way.

Each $\Pi_k^+$, $\Pi_j^-$ is divided by the sets $b_1 = 0$ and $b_2 = 0$ into four parts $\Pi_{ks}^+$, $\Pi_{jl}^-$, $s, l = 1, \dots, 4$. Each of these parts is defined by the signs of the functions $b_1$ and $b_2$. We will call the regions $\Pi_{ks}^+$ and $\Pi_{jl}^-$ incident whenever all the trajectories from $\Pi_{ks}^+$ arrive to $\Pi_{jl}^-$. It is obvious that for each region $\Pi_{ks}^+$ there is a unique region $\Pi_{jl}^-$ incident to $\Pi_{ks}^+$. Moreover, the following lemma is true.

LEMMA 6.7.1. *A pair* $(\Pi_k^+, \Pi_j^-)$ *uniquely defines a pair of incident regions* $(\Pi_{ks}^+, \Pi_{jl}^-)$, *and vice versa.*

PROOF. It is sufficient to show that for a given $\Pi_k^+$ each region $\Pi_{ks}^+$ is incident to a unique region $\Pi_{jl}^-$, and the correspondence $s \to j$ is one-to-one, for $j, s = 1, \dots, 4$. Let us fix $\epsilon$ and consider, for example, the rectangle $\Pi_1^+(\epsilon)$ defined by the inequalities $x_1 > 0$, $y_1 < 0$, $x_2 > 0$, $y_2 > 0$. We will show that the quadrants $N_1 : b_1 > 0$, $b_2 > 0$ and $N_2 : b_1 > 0$, $b_2 < 0$ are mapped along the trajectories into two distinct rectangles $\Pi_j^-(\epsilon)$, $\Pi_k^-(\epsilon)$, $j \neq k$. Indeed, $\xi_1 < 0$, $\xi_2 > 0$ on $\Pi_1^+(\epsilon)$. Along the trajectory of the field $Y$, $\xi_1$, $\xi_2$ change the sign once (see Section 6.5), i.e, on $\partial V_\epsilon^-$ we have $\xi_1 > 0$, $\xi_2 < 0$. On $N_1$ the inequalities $|x_1| > |y_1|$, $|x_2| > |y_2|$ hold since $b_1 > 0$, $b_2 > 0$. Since $b_1, b_2$ are integrals of the field $Y$, these inequalities are preserved along the trajectory. Hence, along the trajectory, $y_1$ and $y_2$ change their signs, and $x_1$ and $x_2$ preserve the signs. Therefore, the trajectories from $N_1$ get into the square where $x_1 > 0$, $y_1 > 0$ and $x_2 > 0$, $y_2 > 0$, i.e., into $\Pi_1^-(\epsilon)$. Analogously, on $N_2$, $|x_1| > |y_1|$, $|x_2| < |y_2|$, hence $y_1$ and $x_2$ change their signs along the trajectory, and $x_1$ and $y_2$ preserve the signs. Thus, $N_2$ is mapped into the square where $x_1 > 0$, $y_1 > 0$ and $x_2 < 0$, $y_2 > 0$, i.e., into $\Pi_2^-(\epsilon)$. Similar considerations can be applied to the remaining squares. The lemma is proved. □

Let us denote by $R_{kj}$ the closure of the set consisting of the pieces of trajectories of the field $Y$ passing through a pair of incident regions on $\Pi_k^+$, $\Pi_j^-$. We obtain 16 distinct regions $R_{kj}$.

REMARK 6.9. The pervious lemma and the constructed correspondence between $\Pi_i^{\pm}$ and $'\Pi_i^{\pm}$ give a one-to-one correspondence between the regions $R_{kj}$ and the respective regions $'R_{kj}$.

PROPOSITION 6.7.1. *The homeomorphisms $\rho_1, \rho_2$ can be extended to a homeomorphism $f_u : U \to U'$ that defines an isoenergetic equivalence of the IHVFs $(X_H, K)_{|U}$ and $(X_{H'}, K')_{|U'}$.*

PROOF. Each $R_{kj}$ is foliated by the level sets $V_\epsilon$ into the sets $R_{kj}(\epsilon) = R_{kj} \cap V_\epsilon$. It can be easily seen from the description of geometry of $b_1$- and $b_2$-foliations in Section 6.3 that the sets $R_{kj}(\epsilon)$ are some of the preimages of a part of the set $G_\epsilon$ under the map $\pi$, with a fixed set of signs of coordinates $b_1, b_2$ coinciding with the signs of $b_1, b_2$ on $\Pi^+_{ks}$, $\Pi^-_{jl}$ defining the region $R_{kj}$. Therefore, $R_{kj}(\epsilon)$ looks as shown in Figure 6.19. In each $R_{kj}$ unique continuous coordinates $(\epsilon, k, b_1, b_2)$ are defined. The correspondence between $R_{kj}$ and $R'_{kj}$ is established above (Remark 6.9).

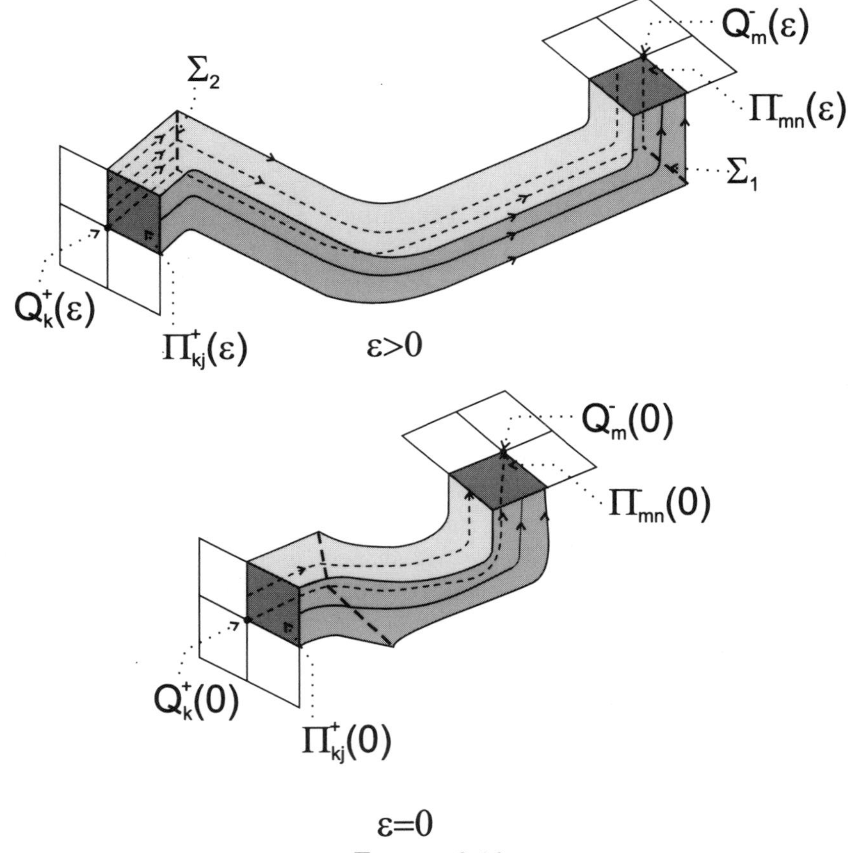


FIGURE 6.19

It follows from the structure of the field (6.3) in $G_\epsilon$ that there is a unique simple arc $L(b_1^0, b_2^0)$ in the intersection of two fibers $b_1 = b_1^0$ and $b_2 = b_2^0$ such that the function $\xi_1$ decreases monotonically along this arc, and consequently (for a fixed $h = \epsilon$) $K$ decreases monotonically from $k_*$ to $-k_*$. Every such curve intersects $\bar{\Pi}^+_{ks}$ at a unique point. If $b_1^0 b_2^0 \neq 0$, then the arc $L(b_1^0, b_2^0)$ coincides with the piece of a trajectory of the field $Y$. If $b_1^0 = 0$, $b_2^0 \neq 0$, or $b_2^0 = 0$, $b_1^0 \neq 0$, then the arc $L(b_1^0, b_2^0)$ consists of three pieces: a semitrajectory from $W^s(\Sigma_i)$, its limit point in $\Sigma_i$, and a semitrajectory from $W^u(\Sigma_i)$ going from the same point on $\Sigma_i$. If $b_1^0 = b_2^0 = 0$, $\epsilon \neq 0$, then the arc $L(b_1^0, b_2^0)$ consists of five pieces: a semitrajectory from $W^s(\Sigma_i)$, its limit point $c_1$ on $\Sigma_i$, heteroclinic trajectory going from $c_1$ into $c_2 \in \Sigma_j$, $i \neq j$, the point $c_2$, and a semitrajectory from $W^u(\Sigma_i)$ such that its $\alpha$-limit point is $c_2$. For $\epsilon = 0$, the points $c_1$ and $c_2$ coincide with the point $p$, and the heteroclinic trajectory contracts to the point $p$.

Now, let us construct maps $f_U^{kj} : R_{kj} \to R'_{kj}$. Let $m \in R_{kj}$ be a point with coordinates $(x_1, x_2, y_1, y_2)$. Then the coordinates $\epsilon = H(m)$, $\kappa = K(m)$, $b_1 = b_1(m)$, $b_2 = b_2(m)$, where $b_i(m) = (x_i^2 - y_i^2)/2$ are defined. Let $f_U^{kj}$ be the point in $R'_{kj}$ with coordinates $(q_1(\epsilon), q_2(\epsilon, \kappa), \alpha_1(b_1), \alpha_2(b_2))$. It is clear that $f_U^{kj}$ is a homeomorphism in $R_{kj}$.

Now we will verify that the restriction of the homeomorphism $f_U^{kj}$ to the sets $\Sigma_1 \cap U$ and $\Sigma_2 \cap U$ coincides with the homeomorphisms $\rho_1$ and $\rho_2$, respectively. To this end note that the intersection $R_{kj} \cap \Sigma_i \cap U$ is one of the closed sectors into which $\Sigma_i \cap U$ is divided by the fiber $b_i = 0$. It follows from the proof of the previous proposition that the signs of the functions $b_1$ and $b_2$ are preserved on each $R_{kj}$, and in the pair $(x_1, y_1)$ ($(x_2, y_2)$, respectively) the sign of one of the coordinates is preserved. For example, since $b_1 > 0$ and $b_2 > 0$, we have $x_1 > 0$ and $x_2 > 0$ on $R_{11}$. Then on $\Sigma_1 \cap U$ the sector we want to be $R_{11}$ is the intersection of the region where $b_1 > 0$ and the region $x_1 > 0$. On $\Sigma_2 \cap U$ it is the region where $b_2 > 0$ and $x_2 > 0$. The sectors for the remaining $R_{kj}$ can be defined analogously. It follows from the construction of $\rho_1$ and $\rho_2$ that $\rho_i(R_{kj} \cap \Sigma_i) = (R'_{kj} \cap \Sigma'_i)$, $i = 1, 2$.

Now let $m \in R_{kj} \cap \Sigma_1$. Then $f_U^{kj}(m) = (q_1(\epsilon), q_2(\epsilon, \phi_1(\epsilon)), \alpha_1(b_1), \alpha_2(b_2))$, where the curve $(\epsilon, \phi_1(\epsilon))$ defines the set $\mu(\Sigma_1)$ on the moment plane, and $\alpha_2(0) = 0$. By the construction of the map $q$, we have $q_2(\epsilon, \phi_1(\epsilon)) = \phi'_1(q_1(\epsilon))$, which means that the point $(q_1(\epsilon), q_2(\epsilon, \phi_1(\epsilon))$ belongs to the set $\mu(\Sigma'_1)$ on the moment plane of IHVF $(X_{H'}, K')$. Since $b'_2 = 0$ and the system of equations $h'(\xi'_1, \xi'_2) = \epsilon'$, $k'(\xi'_1, \xi'_2) = \kappa(\epsilon')$ has a unique solution $\xi'_2 = 0$, $\xi'_1 = \phi'_1(\epsilon')$, we have $b'_2 = 0$ and $\xi'_2 = 0$. Therefore, $x'_2 = y'_2 = 0$, and we have $\rho_1(m) \in \Sigma'_1$. Note that a point with coordinates $(\epsilon, b_1)$ in $\Sigma_1 \cap U$ has coordinates $(\epsilon, \phi_1(\epsilon), b_1, 0)$ in $R_{kj}$, and vice versa. The same is true for the second system. Therefore, $f_U^{kj}(m) = \rho_1(m) = ((\epsilon'), q_2(\epsilon', \phi'_1(\epsilon')), \alpha_1(b_1), 0)$. Analogously, $f_U^{kj}|_{\Sigma_2} = \rho_2$.

In order to complete the proof we have to verify that the homeomorphisms constructed in $R_{kj}$ are compatible on their common boundaries. Let $m \in R_{kj} \cap R_{ln}$. Then

$$m \in W^s(\Sigma_1) \cup W^s(\Sigma_2) \cup W^u(\Sigma_1) \cup W^u(\Sigma_2).$$

If $m \in \Sigma_1 \cup \Sigma_2$, then the compatibility of the homeomorphisms is already proved, since in this case they coincide with $\rho_1$ or $\rho_2$. Suppose $m$ does not belong to $\Sigma_1 \cup \Sigma_2$. Then $m \in W^s(\Sigma_1) \cup W^s(\Sigma_2)$, or $m \in W^u(\Sigma_1) \cup W^u(\Sigma_2)$. Denote $\epsilon = H(m)$, $x = K(m)$, $b_1 = b_1(m)$, $b_2 = b_2(m)$. For example, let $m \in W^s(\Sigma_1)$ and assume that $m$ does not belong to a heteroclinic trajectory. Then the trajectory $L(m)$ of the field $Y$ passing through $m$ intersects $\Pi_k^-$ and $\Pi_l^+$, i.e., $k = l$. We will denote by $m_1$ the $\omega$-limit point of the trajectory $L(m)$, $m_1 \in \Sigma_1$. It follows from what was proved above that the images of the point $m_1$ under the maps $f_U^{kj}$ and $f_U^{kn}$ coincide with $m'_1 = \rho_1(m_1) \in \Sigma'_1$ where $m'_1 \in R'_{kj} \cap R'_{kn}$. Since $L(m)$ is not a heteroclinic trajectory, either $b_1 \neq 0$, $b_2 = 0$ or $b_2 \neq 0$, $b_1 = 0$, and $\epsilon \leq 0$. The same is true for the point $m_1$. By the construction of the map $\rho_1$, at the point $m'_1$ either $b'_1 \neq 0$, $b'_2 = 0$ and $b'_1$ is of the same sign as $b_1$ or $b'_1 = b'_2 = 0$, $\epsilon' \leq 0$. Thus, in every $R'_{kj}$ and $R'_{kn}$ there is a unique (not heteroclinic) trajectory of the field $Y'$ for which the point $m'_1$ is the $\omega$-limit point. Since $R'_{kj}$ and $R'_{kn}$ intersect the same set $'\Pi^+_k$, these trajectories coincide in $R'_{kj}$ and $R'_{kn}$. Therefore, on this trajectory the points $f^{kj}_U(m)$ and $f^{kn}_U(m)$ belong to the same level set $K' = \kappa' = q_2(\epsilon, \kappa)$, i.e., due to the monotonicity of $K'$, they coincide. Analogously, one can verify that $f^{kj}_U(m)$ and $f^{ln}_U(m)$ are compatible at all points which do not belong to the heteroclinic set. Note that for the points $m \in W^u(\Sigma_j)$ we work with $\Pi^-_j$.

Now let $m$ belong to the heteroclinic trajectory $L(m)$. Then $m$ determines two points $m_1 \in \Sigma_1$ and $m_2 \in \Sigma_2$, which are $\omega$- and $\alpha$-limit points of the trajectory $L(m)$. On the trajectory $L(m)$ the functions $b_1$ and $b_2$ are equal to zero, and the value of the function $K$ is between $K(m_1)$ and $K(m_2)$. Denote by $m'_i \in \Sigma'_i$ the points $\rho_s(m_i) = f^{kj}_V(m_i) = f^{ln}_U(m_i)$, $s = 1, 2$. By the construction of $\rho_s$, we have $b'_1 = b'_2 = 0$ at the points $m'_i$. Therefore, these points are limit points of heteroclinic trajectories. It follows from Lemma 6.5.1 that there is a unique heteroclinic trajectory of the field $Y'$ with $\alpha$- and $\omega$-limit points $m'_1$ and $m'_2$. Therefore, heteroclinic trajectories coincide in $R'_{kj}$ and $R'_{ln}$. Finally, the points $f^{kj}_U(m)$ and $f^{ln}_U(m)$ coincide, since they belong to this heteroclinic trajectory, and the values of the function $K'$ at these points coincide and lie between $K'(m'_1)$ and $K'(m'_2)$, due to the construction of $f^{kj}_U(m)$. The proposition is proved. □

Our next task is to extend $f_v$ to the whole $V$. Recall some facts proved earlier. On $\partial V^+_\epsilon$, which is a union of a finite number of tori, there are two sets $\Omega^+_1(\epsilon)$ and $\Omega^+_2(\epsilon)$ of smooth closed curves, which are the traces of $W^s(\Sigma_1)$ and $W^s(\Sigma_2)$, respectively. The curves from the same family do not intersect each other, and the curves from different families intersect transversally at four points $Q^+_i(\epsilon)$. The point $Q^+_i(\epsilon)$ belongs to the rectangle $\Pi^+_1(\epsilon)$, on which there are coordinates $(b_1, b_2)$. The set $\Pi^+_1(\epsilon)$ is divided by the intervals $b_1 = 0$ and $b_2 = 0$ into four quadrants $\Pi^+_{is}$, $s = 1, \dots, 4$. For a fixed $j$, the curves from $\Omega^+_j(\epsilon)$, $j = 1, 2$, are divided by the points $Q^+_i(\epsilon)$ into four open arcs $\lambda^j_1, \dots, \lambda^j_4$. Consider an arc $\lambda^j_n$. Note that its closure $\bar\lambda^j_n$ may be a simple closed curve or a simple segment. In the first case the boundary of the arc $\lambda^j_n$ consists of one point $Q^+_i(\epsilon) \in \Pi^+_i(\epsilon)$, and the smooth closed curve $\bar\lambda^j_n$ is embedded into a torus in such a way that it divides its tubular neighborhood into two half-neighborhoods. Let us fix one of them. There are two quadrants, $\Pi^+_{ip}(\epsilon)$ and $\Pi^+_{im}(\epsilon)$, with the common side transversal to the arc $\lambda^j_n$ in these half-neighborhoods. We will say that these quadrants lie on the same side of $\lambda^j_n$. If the curve $\bar\lambda^j_n$ is not closed, then its boundary consists of two distinct points $Q^+_i(\epsilon) \in \Pi^+_i(\epsilon)$, $Q^+_m(\epsilon) \in \Pi^+_m(\epsilon)$. Two quadrants from $\Pi^+_i(\epsilon)$ and two quadrants from $\Pi^+_m(\epsilon)$ have common points with $\lambda^j_n$ (see Figure 6.20).

Let us specify the side of the curve $\lambda^j_n$ by an oriented normal. Then the above four quadrants will couple into pairs lying on the same side of $\lambda^j_n$ (one from $\Pi^+_i(\epsilon)$ and one from $\Pi^+_m(\epsilon)$). Therefore, every quadrant is uniquely determined by the signs of $b_1$ and $b_2$. Note that $\lambda^j_n \cap \Pi^+_m(\epsilon)$ is given by the equation $b_k = 0, k \neq j$ on $\Pi^+_m(\epsilon)$.

LEMMA 6.7.2. *Let two quadrants $\Pi^+_{ip}(\epsilon)$ and $\Pi^+_{mq}(\epsilon)$ lie on the same side of the arc $\lambda^j_n$. Then the signs of the coordinate $b_j$ are different on these quadrants, and the signs of the other coordinate $b_k$ are the same.*

PROOF. If $\bar\lambda^j_n$ is a closed curve, then $i = m$, and our lemma follows from the fact that both quadrants belong to the same $\Pi^+_i(\epsilon)$ and have the common side $b_j = 0$.

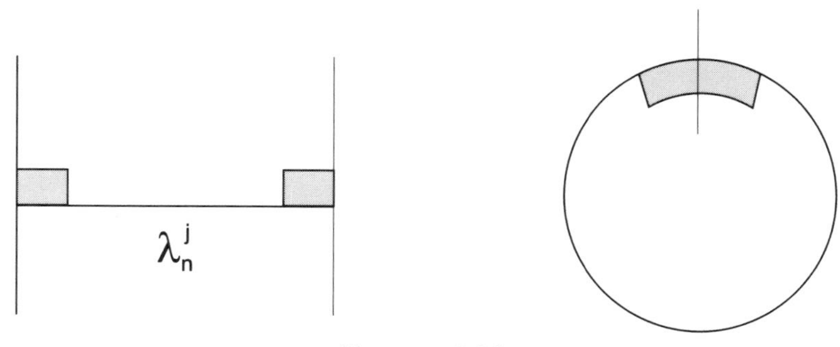


FIGURE 6.20

Now, let $\bar{\lambda}^j_n$ be a nonclosed curve, and $Q^+_i(\epsilon)$ and $Q^+_m(\epsilon)$ its endpoints. Consider the following 2-frame at the point $Q^+_i(\epsilon)$: it is defined by unit vectors tangent to the torus such that one of them, $e^i_1$, is tangent to the curve $b_2 = 0$ and is directed towards the increase of the coordinate $b_1$, and the other, $e^i_2$, is tangent to the curve $b_1 = 0$ and is directed towards the increase of the coordinate $b_2$. The analogous frame at the point $Q^+_m(\epsilon)$ is denoted by $e^m_1, e^m_2$. It follows from Lemma 6.5.2 that the frames obtained at the points $Q^+_i(\epsilon)$ and $Q^+_m(\epsilon)$ determine the same orientation of the torus. For definiteness, let $j = 1$. Then $b_1$ is a coordinate on $\lambda^1_n$. We define a smooth field of frames $(e_1, e_2)$ on $\bar{\lambda}^1_n$, where $e_1$ is a field of unit vectors tangent to $\lambda^1_n$ and containing $e^i_1$, and $e_2$ is a field of unit vectors transversal to $\lambda^1_n$ and containing the vector $e^i_1$ and a vector tangent to the line $b_1 = 0$ at the point $Q^+_m(\epsilon)$. Then a frame $(e^*_1, e^*_2)$ from the same equivalence class as $(e^i_1, e^i_2)$ and $(e^m_1, e^m_2)$ is defined at the point $Q^+_m(\epsilon)$. Translating the curve $\lambda^1_n$ down to $\Sigma_1$ along the trajectories of the field $Y$, we can see that the signs of $b_1$ on $\lambda^1_n$ are different near the points $Q^+_i(\epsilon)$ and $Q^+_m(\epsilon)$. Therefore, $e^*_1 = e^m_1$, and, due to the equivalence of the frames, $e^*_2 = e^m_2$. Since the vector $e^*_2$ belongs to the collections defining the side of the arc $\lambda^1_n$ and coincides with $e^m_2$, the quadrants lying on the same side of $\lambda^1_n$ have the same signs of $b_2$. The lemma is proved. □

Recall that the curves $\Omega^+_1(\epsilon)$, $\Omega^+_2(\epsilon)$ divide the tori $\partial V^+_\epsilon$ into four open disks $N^+_i(\epsilon)$ (Proposition 6.6.1). There are similar disks $N^-_i(\epsilon)$ on $\partial V^-_\epsilon$. Trajectories through the points of the same disk $N^+_i(\epsilon)$ get into the same disk $N^-_i(\epsilon)$.

Let us consider one of the tori $T^+$ on $\partial V^+_\epsilon$ ($T^-$ on $\partial V^-_\epsilon$, respectively), and an open disk $N \subset T^+ \setminus (\Omega^+_1(\epsilon) \cup \Omega^+_2(\epsilon)$ ($N \subset T^- \setminus (\Omega^-_1(\epsilon) \cup \Omega^-_2(\epsilon))$. Let $[N]$ denote the closure of the disk $N$ in the sense of Carathéodory. We want to describe the set $[N]$. To do this we need the following result.

LEMMA 6.7.3. *Up to a diffeomorphism of the torus $T^+$ ($T^-$), the embedding of the orientable components of the set $\Omega^+_1(\epsilon) \cup \Omega^+_2(\epsilon)$ ($\Omega^-_1(\epsilon) \cup \Omega^-_2(\epsilon)$) into $T^+$ ($T^-$) looks like shown in Figure 6.21.*

The proof of this lemma follows immediately from Propositions 6.6.2 and 6.6.4, Lemma 6.7.2, and the topology of a torus. Figure 6.21 shows the embedding of one component of each type into the torus. Note that the four upper parts in Figure 6.21 describe the $W^s$-components and the botton two describe the $W^u$-compponents for the corresponding cases. For other components the letters have to be changed according to Figure 6.18 and with regard to the orientation. This lemma clarifies the structure of the set $[N]$. It is a curvilinear rectangle. Its vertices correspond to

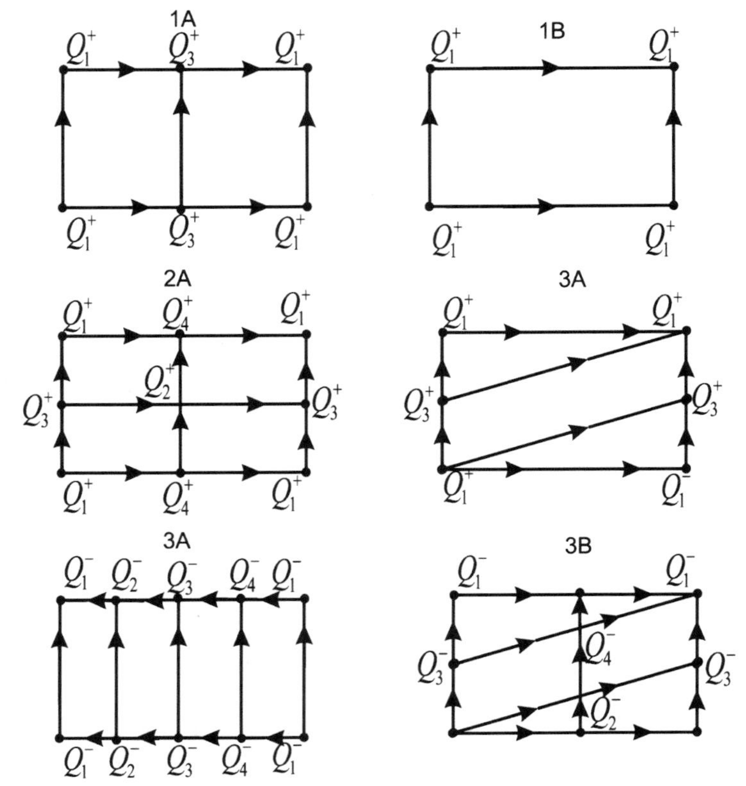


FIGURE 6.21

the points $Q_i^+(\epsilon)$ (for the torus $T^+$), one pair of its opposite sides corresponds to the arcs from $\Omega_1^+(\epsilon)$, and the other pair corresponds to the arcs from $\Omega_2^+(\epsilon)$. For each vertex of the set $[N]$ there is a unique quadrant from $N$, i.e., from some $\Pi_i^+(\epsilon)$ adjacent to it. Notice that some vertices and sides of $[N]$ may correspond to the same vertex or side on $T^+$. The opposite sides of the rectangle $[N]$ are oriented in the same way. It follows from the description of $[N]$ that the set $N$ itself is obtained from $[N]$ by gluing together some of its vertices and sides.

Let us construct two foliations $B_1$ and $B_2$ on each torus $T^+$ such that their leaves are simple closed smooth curves that contain the system of curves $\Omega_1^+(\epsilon)$ (for $B_1$) and $\Omega_2^+(\epsilon)$ (for $B_2$). Any two leaves of different foliations will intersect each other transversally. Moreover, these foliations are extensions of the local $b_1$- and $b_2$-foliations that exist near the points $Q_i^+(\epsilon)$. More precisely, if a leaf of the $B_i$-foliation contains two local leaves of the $b_i$-foliation, then the values of the function $b_i$ on these leaves are equal. The points $Q_i^+(\epsilon)$ divide the system of curves $\Omega_k^+(\epsilon)$, $k = 1, 2$, into four open arcs $\lambda_1^k, \dots, \lambda_4^k$. Two points $a \in \lambda_i^k$ and $b \in \lambda_j^k$ are called congruent if they belong to the trajectories of the vector field $Y$ that tend to the same point of the CSP as $t \to \infty$. Then it is clear that for each point of $\lambda_i^k$ there is a congruent point on $\lambda_j^k$. The arcs $\lambda_i^k$ and $\lambda_j^k$ are called congruent.

Let $N_i^+(\epsilon)$ be a disk and $[N_i^+(\epsilon)]$ its Carathéodory closure.

LEMMA 6.7.4. *If opposite sides of the rectangle $[N_i^+(\epsilon)]$ correspond to different arcs on a torus, then there exists another rectangle $[N_j^+(\epsilon)]$ such that some parts of its opposite sides are the same arcs.*

PROOF. Since we consider the canonical representatives of the cases, in each of the cases $n$A, $n$B, our statement follows from the analysis of mutual location on tori of curves from the collections $\Omega_1^+(\epsilon)$ and $\Omega_2^+(\epsilon)$ (see Lemma 6.7.3 and Figure 6.18). □

REMARK 6.10. We can see from Figure 6.18 that for $[N_j^-(\epsilon)]$ this statement is not true in general.

Thus, there are two possible cases for $N_1^+$. Namely, if a pair of opposite sides of the rectangle $[N_i^+(\epsilon)]$ is one arc on a torus, then, identifying them, we get an annulus. In this annulus one can easily construct a foliation by the closed curves transversal to the arc under consideration, which extends the local fibers and contains the other pair of opposite sides of the rectangle as leaves.

In the case described in Lemma 6.7.4, the annulus is formed by gluing the rectangles $[N_i^+(\epsilon)]$ and $[N_j^+(\epsilon)]$ along the sides corresponding to the same arcs on the torus.

The opposite sides in $[N_i^+(\epsilon)]$ and $[N_j^+(\epsilon)]$ are oriented identically by the corresponding parameter $b_k$, which is induced from $\Sigma_k$ (by this we mean the function $b_k$ extended to $\Sigma_k$). Therefore, we can join by the same leaf of $B_k$-foliation the points on the sides of the rectangles which correspond to equal values of the function $b_k$. This guarantees that a pair of congruent points will belong to the same leaf of foliation. Moreover, let us consider two annuli such that each of them is glued from two rectangles, and two opposite sides of the rectangles from one annulus are congruent to two opposite sides of the second annulus. Let us choose a leaf of the $B_k$-foliation in one annulus and a pair of points $a_1, a_2$ belonging to this leaf and to two opposite sides of one rectangle. Then there is a pair of congruent points $\hat{a}_1, \hat{a}_2$ on the second annulus, which also belong to the opposite sides of the rectangle. By the construction of $B_k$-foliation, these points belong to one leaf.

We constructed the foliations on a fixed $\partial V_\epsilon^+$. It is easy to see that these foliations can be constructed simultaneously for all $\epsilon$. Therefore, we may assume that there are two continuous foliations consisting of smooth leaves on $\partial V^+$, and any two leaves from different foliations are transversal.

We now extend the homeomorphism $f_U$ to $V$. We will assume that $B_1'$- and $B_2'$-foliations on $\partial' V_\epsilon$ are also constructed for the system $(X_{H'}, K')$. Let us choose a disk $N$ on $\partial V_0^+$.

PROPOSITION 6.7.2. *There is a unique disk $N'$ on $\partial' V_0^+$ and a homeomorphism $F_N : N \to N'$ such that the restriction of the map $F_N$ to $U$ coincides with $f_U$ and $F_N$ maps the $B_i'$-foliation of $N$ into the $B_i'$-foliation of $N'$, $i = 1, 2$. Moreover, $F_N$ extends uniquely to the closure $\bar{N}$.*

PROOF. Consider the disk $N$ and one of the oriented arcs of the set $\Omega_1(\epsilon)$ belonging to its boundary. Suppose that this arc $\lambda_1$ connects the points $Q_i^+$ and $Q_j^+$ (possibly coinciding). For example, assume that it is oriented from $Q_i^+$ to $Q_j^+$. It follows from the description of the structure of $N$ that there exists a quadrant $\Gamma_1$ from $\Pi_i^+(\epsilon)$ such that one of its boundary segments belongs to $\lambda_1$. Then there is a unique quadrant from $\Pi_j^+(\epsilon)$ such that it lies on the same side of $\lambda_1$ with $\Gamma_1$

and belongs to $N$. One of the boundary segments of this quadrant belongs to $\lambda_1$ and the other determines the second oriented arc $\lambda_2$ belonging to the boundary of $N$. This arc connects the point $Q_j^+$ and some point $Q_k^+$ (possibly coinciding with $Q_i^+$ or $Q_j^+$). Continuing this process, we obtain four arcs $\lambda_1$, $\lambda_2$, $\lambda_3$, $\lambda_4$ (some of them may coincide) and four quadrants $\Gamma_1$, $\Gamma_2$, $\Gamma_3$, $\Gamma_4$.

We consider the identical cases and, therefore, in accordance with the structure of the traces of $W^s(\Sigma_1')$ and $W^s(\Sigma_2')$ on $\partial V_0^+$ (see Figure 6.18), there exists a torus on $\partial' V_0^+$ containing the same set of oriented arcs which connect the points $'Q_s^+$ with the same indices. The local map $f_U$ transforms the points $Q_i^+$ into the points $'Q_i^+$ with the same indices and preserves the signs of the functions $b_j$, $j = 1, 2$, and, therefore, $f_U$ transforms the quadrant $\Gamma_1$ adjoining the point $Q_i^+$ in $N$, to the quadrant $\Gamma_1'$ adjoining the point $'Q_i^+$. This quadrant belongs to the unique disk $N'$. Since the signs of $b_i$ are preserved under the local map, Lemma 6.7.2 shows that the quadrant $\Gamma_2$ is mapped into the quadrant $'\Gamma_2$ which adjoins the point $'Q_j^+$ and belongs to the same disk $N'$. Continuing this argument, we come to the conclusion that the images of all four quadrants $\Gamma_1, \dots, \Gamma_4$ from $N$ belong to $N'$.

There are also $B_1'$- and $B_2'$-foliations on the disk $N'$. Now, the homeomorphism $f_V$ can be extended in the following way.

We have already established correspondence between vertices and the arcs of the boundary of the disk $N$ and the corresponding elements in $N'$.

Let $m \in N$ be a vertex. Then the local map $f_U$ maps it into the corresponding vertex in $N'$. If $m$ belongs to the boundary of $N$ and does not coincide with a vertex, then on the boundary arc $\lambda$, $m \in \lambda$, a certain value of the parameter $b_1 = b_1^0$ (if $\lambda \in W^s(\Sigma_1)$) or $b_2 = b_2^0$ (if $\lambda \in W^s(\Sigma_2)$) corresponds to $m$. Let $\lambda' \subset N'$ be the arc corresponding to the arc $\lambda \subset N$. Let us put the point $m' \in \lambda'$ such that $b_1' = \alpha_1(b_1)$ (or $b_2' = \alpha_2(b_2)$) in correspondence with the point $m$. Suppose that $m$ does not belong to the boundary of the disk $N$. Then a leaf of $B_1$-foliation that intersects the boundary of the disk $N$ at either one or two points corresponding to the value of the parameter $b_1 = b_1^0 \neq 0$, and a leaf of $B_2$-foliation that intersects the boundary of the disk $N$ at the points corresponding to the unique value $b_2 = b_2^0 \neq 0$, both pass through $m$. Let $m' \in N'$ be the intersection point of the leaf of $B_1'$-foliation and the leaf of $B_2'$-foliation intersecting the boundary of the disk $N'$ at the points corresponding to the value $b_1' = \alpha_1(b_1^0)$ and $b_2' = \alpha_2(b_2^0)$, respectively. We put $m' \in N'$ in correspondence with the point $m \in N$. It is evident that the constructed mapping $F_N$ is a homeomorphism $\bar{N} \to \bar{N}'$. The proposition is proved. □

REMARK 6.11. It follows from the construction of the homeomorphism $F_N$ that on two disks $N_1$, $N_2$ with common boundary points, the mappings $F_{N_1}$ and $F_{N_2}$ are compatible at these common points, and, therefore, one can define a homeomorphism $F_0 : \partial V_0^+ \to \partial' V_0^+$ which is an extension of $f_U$.

REMARK 6.12. Recall that the set $(W^s(\Sigma_1) \cup W^s(\Sigma_2)) \cap \partial V^+$ has the structure of a direct product of the trace of this set on $\partial V^+$ and $I$ (Proposition 6.6.5). Therefore, there is an induced correspondence between the disks $N$ and $N'$ on all $\partial V_\epsilon^+$ and $\partial' V_\epsilon^+$. Since $B_1$-, $B_2$-foliations exist on the entire $\partial V^+$, the constructed homeomorphism $F_0$ extends to the homeomorphism $F : \partial V^+ \to \partial' V_+$.

Now we extend the homeomorphisms $f_U$, $F$ to the entire $V$. We represent $V$ as the union of four closed sets invariant with respect to the field $Y$. Each of these sets is a closure in $V$ of the components $\mathcal{N}_1$, $\mathcal{N}_2$, $\mathcal{N}_3$, $\mathcal{N}_4$ of the set $V \setminus (W^s(\Sigma_1) \cup$

$W^s(\Sigma_2) \cup W^u(\Sigma_1) \cup W^u(\Sigma_2))$. The intersection of the components with $V_\epsilon$ is a union of pieces of trajectories passing through the fixed open disk $N_i^+(\epsilon) = \mathcal{N}_i \cap \partial V_\epsilon^+$, $i = 1, \ldots, 4$. Therefore, the disk $N_i^-(\epsilon) = \mathcal{N}_i \cap \partial V_\epsilon^-$ is also defined. Let us extend the homeomorphism to $\mathcal{N}_i \setminus U$. For each point $m$ of the set $\mathcal{N}_i$ there is a unique trajectory of the field $Y$ through $m$ that intersects $N_i^+(\epsilon)$, $\epsilon = H(x)$, and a unique disk $N_i^-(\epsilon)$. If the point $m$ belongs to the set $(\bar{\mathcal{N}}_i \setminus \mathcal{N}_i) \setminus U$ and does not belong to $\Sigma_1 \cup \Sigma_2$, then two situations are possible:

1. $m \in W^s(\Sigma_k \setminus U)$;
2. $m \in W^u(\Sigma_k \setminus U)$.

In the first case there is a unique trajectory of the field $Y$ that passes through $m$ and intersects $\partial V_\epsilon^+$ at the point $m_1$ which belongs to the arc $\lambda$ from $\Omega_k^+(\epsilon)$. According to the definition of the side of a curve from $\Omega_k^+(\epsilon)$ (see above) and Lemma 6.7.2, a side of a curve will be called right (left) if the function $b_j$, $j \neq k$, is positive (negative) in two quadrants from the corresponding rectangles $\Pi_s^+(\epsilon)$ and $\Pi_k^+(\epsilon)$ adjacent to this curve. A leaf of the $B_j$-foliation transversal to the arc $\lambda$ passes through the point $m_1$. In the neighborhood of the point $m_1$ on $\partial V_\epsilon^+$ this leaf splits into two parts: left and right. By Lemma B.5 from Appendix B, each part of the leaf determines a unique curve that is made of a semitrajectory of the field $Y$ passing through the point $m_1$, its $\omega$-limit point $m_* \in \Sigma_k$, and one semitrajectory of the field $Y$ for which $m_*$ is the $\alpha$-limit point. Hence, there are two curves, left $L_l(m_1)$ and right $L_r(m_1)$, associated with the point $m$. Note that there are two possibilities: either both curves $L_l(m_1)$ and $L_r(m_1)$ belong to the set $\bar{\mathcal{N}}_i$ (if the whole neighborhood of the point $m_1$ on $\partial V_\epsilon^+$ belongs to $\bar{\mathcal{N}}_i$), or only one of them does. Moreover, the curves $L_l(m_1)$ and $L_r(m_1)$ have a common part between the points $m_1$ and $m_*$.

In the second case (when $m \in W^u(\Sigma_k \setminus U)$) there is a unique trajectory of the field $Y$ from $W^u(\Sigma_k \setminus U)$ passing through the point $m$. Its $\alpha$-limit point is a point $m_* \in \Sigma_k$. Two different trajectories of the field $Y$ exist which start from $m_*$, belong to $W^s(\Sigma_k)$, and intersect $\partial V_\epsilon^+$ at two distinct points $m_1$ and $m_2$. As above, the point $m_i$, $i = 1, 2$, determines a pair of curves $L_l(m_i)$ and $L_r(m_i)$. It follows from Lemma B.5 of Appendix B that two curves out of the above-mentioned four contain the point $m$: the curve associated with the point $m_1$, and the curve associated with the point $m_2$. It follows from the preservation of signs of the functions $b_i$ in $U$ that one of these curves is left, the other is right, and they have a common part which coincides with the trajectory of the field $Y$ that contains the point $m$. Also, note that two cases are possible: either both curves belong to the set $\bar{\mathcal{N}}_i$ (if $m_1$ and $m_2$ belong to the boundary of $\bar{N}_i$), or only one of them does.

Let us start constructing the homeomorphism $f_i$ on $\bar{\mathcal{N}}_i \setminus U$. We assume here that corresponding objects exist in the second system as well. Let $m \in \bar{\mathcal{N}}_i \setminus U$. For the points $m \in \bar{\mathcal{N}}_i$ the mapping $f_i$ is defined in the following way. For any point $m$ there is a unique point $m_1$ on $N_i^+(\epsilon)$, $\epsilon = H(m)$, that belongs to the same trajectory of the field $Y$ as the point $m$. Then $F(m_1) = m_1' \in {'N_i^+}(\epsilon'), \epsilon' = q_1(\epsilon)$. Let us take a point $m' = f_i(m)$ such that $K'(m') = q_2(\epsilon, K(m))$ and $m'$ belongs to the trajectory of the field $Y'$ passing through the point $m_1'$. Since $K$ and $K'$ vary strictly monotonically on trajectories of the fields $Y$ and $Y'$ respectively, we obtain a homeomorphism $f_i$ defined on $\mathcal{N}_i$. It can be easily seen from the construction that $f$ is an extension of $f_U$ and $F$ to $\mathcal{N}_i$.

Now let $m \in (\bar{\mathcal{N}}_i \backslash \mathcal{N}_i)\backslash U$ and $m \in W^s(\Sigma_k)$. Then we have found the trajectory of the field $Y$ which contains $m$, the point $m_1$ being the intersection of this trajectory with $\bar{N}_i^+(\epsilon)$, and the pair of curves $L_l(m_1)$ and $L_r(m_1)$ on $\bar{\mathcal{N}}_i$. Note that $K$ also varies strictly monotonically on these curves, and $K(m_*) \leq K(m) \leq K(m_1)$, where $m_*$, as above, is an $\omega$-limit point of the trajectory of the field $Y$ through the point $m$. Hence, we obtain a point $m_1' = F(m_1)$ such that, due to the construction of $F$, a trajectory from $W^s(\Sigma_k')$ passes through this point so that there are two curves $L_l'(m_1')$ and $L_r'(m_1')$ passing through $m_1'$. The property of a curve to be right or left is preserved under the homomorphism $F$, since it is determined by the sign of the function $b_i$, which is preserved by the mapping $F$. Therefore, we put the curve $L_r'(m_1')$ into correspondence with the curve $L_r(m_1)$, and the curve $L_l'(m_1')$ into correspondence with the curve $L_l(m_1)$. The correspondence between the points of the curves is determined by the function $K$: we put the point $m'$ with $K'(m') = q_2(\epsilon, K(m))$ in correspondence with the point $m$. From the construction of the mapping $q$ we have the inequality $K'(m_*') \leq q_2(\epsilon, K(m)) \leq \kappa_*$, where $m_*' = \rho_k(m_*)$. Therefore, the point $m'$ belongs to $W^s(\Sigma_k')$.

Now let $m \in W^u(\Sigma_k)$. Then two congruent points $m_1$ and $m_2$ on $\bar{N}_i^+(\epsilon)$ and two curves, for example, $L_l(m_1)$ and $L_r(m_2)$ are defined. The images $m_1'$ and $m_2'$ of these points are also congruent. We put the curve $L_r(m_2')$ into correspondence with the curve $L_r(m_2)$, and the curve $L_l(m_1')$ into correspondence with the curve $L_l(m_1)$. We recall that the curves $L_l(m_1')$ and $L_r(m_2')$ have a common part, which coincides with a semitrajectory from $W^u(\Sigma_k')$ of the field $Y'$. We put the point $m'$ with $K'(m') = q_2(\epsilon, K(m))$ in correspondence with the point $m$. At the point $m$ we have the inequality $K(m_*) \geq K(m) \geq -\kappa_*$, where $m_*$ is the $\alpha$-limit point of the semitrajectory of the field $Y$ through $m$. From the construction of the mapping $q$ we have $K'(m_*') \geq q_2(\epsilon, K(m)) \geq -x_*$, where $m_*' = \rho_k(m_*)$. Therefore, $m' \in W^u(\Sigma_k)$.

To complete the construction of the homeomorphism $f : V \to V'$ it remains to verify the following two facts.

1. The constructed mapping $f_i : \bar{\mathcal{N}}_i \to \bar{\mathcal{N}}_i'$ is a homeomorphism.
2. The homeomorphisms $f_i$ are compatible on their common boundaries.

Let us start from the verification of the first fact. We use the construction of $f_i$ and Lemma B.5 of Appendix B. It is evident from the construction that $f_i$ is a homeomorphism in $\mathcal{N}_i$. Let $m \in \partial\mathcal{N}_i = \bar{\mathcal{N}}_i \setminus \mathcal{N}_i$, $m \notin \Sigma_1 \cup \Sigma_2$, and the sequence of points $m^{(n)} \in \bar{\mathcal{N}}_i$ converges to $m$. There are two curves defined above which contain a common piece passing through the point $m$. Denote them by $L_l(m)$ and $L_r(m)$. We will assume that both curves belong to $\bar{\mathcal{N}}_i$ (the case when only one of the curves belongs to $\bar{\mathcal{N}}_i$ is simpler and can be treated similarly). For example, assume that the common piece of $L_l(m)$ and $L_r(m)$ belongs to $W^u(\Sigma_k)$. Then a semitrajectory of the field $Y$ through the point $m$ exists and it intersects $\partial V_\epsilon^-$, $\epsilon = H(m)$. Denote by $m_1$ the trace of this trajectory on $\partial V_\epsilon^-$. Being the trace of the manifold $W^u(\Sigma_k) \cap V^\epsilon$, the point $m_1$ belongs to some smooth arc $\lambda$. The arc $\lambda$ divides a neighborhood of the point $m_1$ on $\partial V_\epsilon^-$ into two half-neighborhoods, which will be called, as above, right and left. In our case both half-neighborhoods belong to $N_i^-(\epsilon)$. Denote by $m_1^{(n)}$ the sequence of points that are traces of the trajectories of the field $Y$ on $\partial V_\epsilon^-$ passing through the points $m^{(n)}$. It follows from the theorem on the continuous dependence of solutions of a differential equation on the initial conditions and parameters on a finite interval of time that the points $m_1^{(n)}$ converge to $m_1$. Let us break the sequence $m_1^{(n)}$

into two subsequences. We say that a point belongs to the right subsequence if it lies in the right half-neighborhood, and to the left subsequence if it lies in the left half-neighborhood. We will double the points $m_1^{(n)}$ belonging to the arc $\lambda$, and put one of them into the right half-neighborhood and the other into the left half-neighborhood. Then we say that one of the curves $L_l(m)$ and $L_r(m)$ is defined by the right half-neighborhood, and the other by the left one. Due to Lemma B.5 from Appendix B, a subsequence of the curves $L(m_1^{(n)})$ passing through the points $m_1^{(n)}$ in the right half-neighborhood of the point $m_1$ converges to $L_r(m)$, and a subsequence of curves $L(m_1^{(n)})$ passing through the points $m_1^{(n)}$ from the left half-neighborhood of the point $m_1$ converges to $L_l(m)$. The sequence of the curves $L(m_1^{(n)})$, for the points from the right half-neighborhood of the point $m_1$, defines a sequence of the points $m_2^{(n)} \in \partial V^+$ converging to the point $m_2$ which is the trace of the curve $L_r(m)$ (Lemma B.5 from Appendix B). Therefore, we obtain a converging sequence of points $'m_2 = F(m_2)$. The right half-neighborhood of the point $'m_2$ defines a unique curve $L_r'('m_2)$. The point $'m_2$ determines the curves $L('m_2^{(n)})$ composed from trajectories of the field $Y'$ and converging to $L_r'('m_2)$ in the topology of the uniform convergence of the curves parametrized by the value $K$. On each curve $L_r'('m_2^{(n)})$, the map $f_i$ specifies a point $'m^{(n)} = f_i(m^{(n)})$ for which $K'('m_1^{(n)}) = q_2(\epsilon_n, K(m_n))$, $\epsilon_n = H(m_n)$. Moreover, from the construction of $f_i$ we have $K'('m^{(n)}) \to q_2(\epsilon, K(m)) = K'(f_i(m))$. Therefore, the sequence $f_i(m^{(n)})$ converges to the point $f_i(m)$. The proof for the other half-neighborhood of the point $m_1$ is similar.

Now let $m \in \bar{\mathcal{N}_i} \cap \bar{\mathcal{N}_j}$ (it is evident that $\mathcal{N}_i \cap \mathcal{N}_j = \varnothing$, $i \neq j$). Then the following cases are possible:

a) $m \in W^s(\Sigma_k) \setminus \Sigma_k$;

b) $m \in W^u(\Sigma_k) \setminus \Sigma_k$;

c) $m \in \Sigma_k$, for some $k = 1, 2$.

In the first case two curves $L_r(m)$ and $L_l(m)$ pass through $m$, and then the coincidence of the homeomorphisms $f_i$ and $f_j$ follows from their construction, since they coincide on the common piece of the curves $L_r(m)$ and $L_l(m)$ that belongs to $W^s(\Sigma_k)$. Similarly, in the second case there exist two curves $L_r(m)$ and $L_l(m)$ with the common piece that belongs to $W^u(\Sigma_k)$. On this common piece the construction of $f_i$, $f_j$ gives us the same point. At the points of $\Sigma_1 \cap \Sigma_2$ the maps simply coincide with $\rho_1$ or $\rho_2$. Thus the conjugating homeomorphism $f : V \to V'$ is constructed.

## 6.8. The topology of the extended neighborhood $V$

We conclude this chapter by describing the topological structure of the level sets $V_\epsilon$, $|\epsilon| \leq h_*$. For $\epsilon \neq 0$, the sets $V_\epsilon$ are smooth submanifolds with boundary. They are glued from some standard blocks indicated in [**9**] and can be described as in [**9**], but such a description does not agree with the partition into common level sets $H = \epsilon$, $K = \kappa$. Therefore, it is more convenient to construct another partition of $V_\epsilon$ into standard pieces using manifolds $P_1, \ldots, P_5$ that will be defined below.

Let $R$ be a smooth two-dimensional manifold, and $g$ a smooth function on $R$. Suppose that there exists a compact connected component $L$ of the level line $g = 0$. Denote by $B$ a connected submanifold containing $L$, which is defined, for a positive $c$ small enough, by the inequality $|g| \leq c$. Moreover, we assume that a smooth involution $G : B \to B$, $G^2 = \mathrm{id}_B$ acts on $B$, and $g$ is invariant with respect to $G$.

Let us define a three-dimensional manifold $P$ with boundary as a smooth bundle over the circle $S^1$ with fiber $B$ and structural group $\{\mathrm{id}_B, G\}$ (if $G = \mathrm{id}_B$, then $P$ is a direct product $S^1 \times B$). The manifold $P$ has a foliation generated by a function $F$ which is a natural extension of $g$ to $P$. Indeed, the bundle over $S^1$ can be regarded as a bundle over the interval $I$ in which the boundary fibers are identified by the map $G$. Since the level set of the function $g$ in mapped into itself by the map $G$ and the function $g$ is extended to $B \times I$ by the formulas $F(x, s) = g(x)$, $s \in I$, we get a smooth function $F$ on $P$.

Now we define five model manifolds (see Figure 6.22):

$$P_1 : R = \mathbb{R}^2 = \{(x, y)\}, \quad g = y^2 - x^2(1 - x^2), \quad c = 1/4, \quad G = \mathrm{id},$$

$$P_2 : R = \mathbb{R}^2, \quad g = y^2 - (4 - x^2)(1 - x^2)^2, \quad c = 1, \quad G = \mathrm{id},$$

$$P_3 : R = S^2 = \{(x, y, z) \in \mathbb{R}^3 \mid x^2 + y^2 + z^2 = 1\}, \quad g = yz, \quad c = 1/2, \quad G = \mathrm{id},$$

$$P_4 : R = \mathbb{R}^2, \quad g = y^2 - x^2(1 - x^2), \quad c = 1/4, \quad G : (x, y) \to (-x, -y),$$

$$P_5 : R = S^2 = \{(x, y, z) \in \mathbb{R}^3 \mid x^2 + y^2 + z^2 = 1\}, \quad g = yz, \quad c = 1/2,$$
$$G : (x, y, z) \to (-x, -y, -z).$$

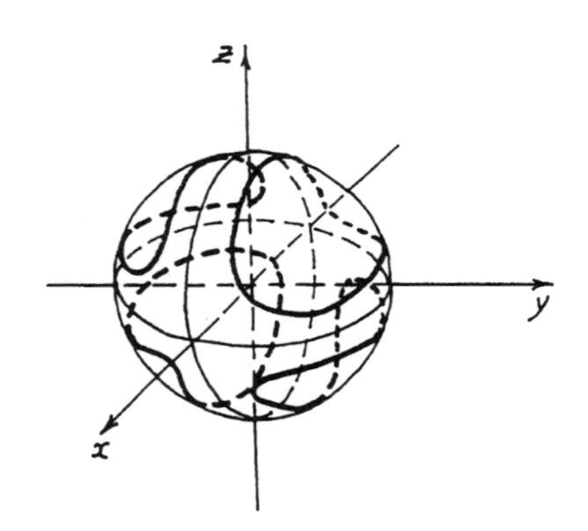


B for $P_3$,$P_5$

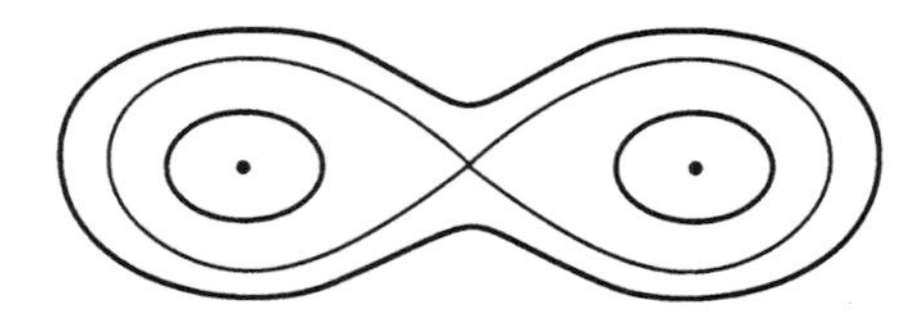

B for $P_1$,$P_4$

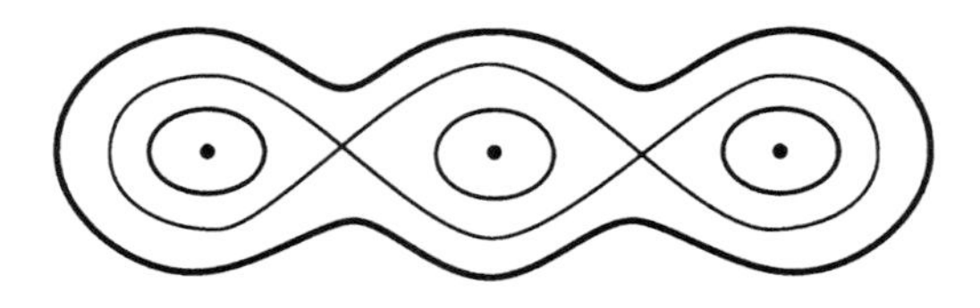

B for $P_2$

FIGURE 6.22

In addition, we define two foliations of the manifolds $P_1, \dots, P_5$. One of these foliations is a natural partition into fibers when the point $s$ is fixed in the bundle base of $S^1$. For the construction of the other foliation we consider, in Cases 1, 2, 4, a gradient vector field of the function $g$ with respect to the Riemannian metric $dx^2 + dy^2$ in $R$. It is easy to verify that this field is invariant with respect to the action of the involution $G$. In other words, if $(x(t), y(t))$ is a trajectory of this field, then $G(x(t), y(t))$ is also its trajectory. For Cases 3 and 5 we consider a vector field of the form $-\nabla g + (\nabla g, \nabla S)\nabla S$ in $\mathbb{R}^3$, where the gradient is defined under the metric $dx^2+dy^2+dz^2$, and $S = (x^2+y^2+z^2)/2$. This field is invariant with respect to the involution $G$, and the sphere $x^2 + y^2 + z^2 = 1$ is its invariant submanifold.

The invariance of the constructed vector fields in $R$ allows us to extend them naturally to a gradient-like vector fields $\operatorname{gr} F$ in the bundles $P_1, \dots, P_5$ in such a way that each fiber of this bundle is an invariant submanifold of the field, the boundary tori are transversal to the field, and the limit sets of the field trajectories are closed curves of singular points obtained from the critical points of the corresponding function $g$. In particular, stable manifolds of each CSP of this kind transversally intersect the upper boundary of the manifold $P_i$ (where the function $F$ is equal to $c$) along smooth closed curves (two in the case $G = \mathrm{id}$ and one in the case $G \neq \mathrm{id}$). These closed curves are not homotopic to zero, and, in the case when both of them belong to the same boundary torus, they are homotopic to each other. Since stable manifolds of different CSPs do not intersect, all the traces of stable manifolds of the CSP lying on the same torus are homotopic to each other. This allows us to construct a smooth foliation of the upper tori of the manifold $P_i$ by closed curves containing traces of stable manifolds of CSP as fibers. Each leaf of this foliation must be transversal to all the curves of the foliation $s = \text{const}$ on the upper tori. Let us extend the constructed foliation along the trajectories of the vector field $\operatorname{gr} F$ in $P_i$. Due to Lemma B.5 from Appendix B, we obtain a foliation in $P_i$.

To describe the structure in the level sets of $V$ we cut the manifold $V_\epsilon$ into two pieces along the level set

$$K = k_m(\epsilon) = \frac{1}{2}(k_1(\epsilon) + k_2(\epsilon)),$$

where $k_i(\epsilon)$ is a value of the function $K$ on $\Sigma_i \cap V_\epsilon$. We denote these pieces by $V_\epsilon^1(K \geq k_m(\epsilon))$ and $V_\epsilon^2(K \leq k_m(\epsilon))$.

First we find the number of Liouville tori on the intermediate level set $K = k_m(\epsilon)$, i.e., on the boundary part of the submanifold $V_\epsilon^i$ that is different from $\partial V_\epsilon^+$ and $\partial V_\epsilon^-$. To this end we find, similarly to Proposition 6.6.3, the number of connected components formed on the level set $K = k_m(\epsilon)$ by the traces of $W^s(\Sigma_1)$ and $W^u(\Sigma_2)$ for $\epsilon > 0$, and by the traces of $W^s(\Sigma_2)$ and $W^u(\Sigma_1)$ for $\epsilon < 0$. Since on the level set $K = k_m(\epsilon)$ these traces intersect each other at the points through which the heteroclinic trajectories pass, we choose local cross-sections $\Pi_i^m(\epsilon)$, $i = 1, \dots, 4$, to the heteroclinic trajectories of the field $Y$ as preimages under the map $\pi : V_\epsilon \to G_\epsilon$ of the rectangle $\tilde{\Pi}^m(\epsilon) : \xi = \xi_1^m$, where $0 < \xi_1^m < \xi_1(\epsilon)$ for $\epsilon > 0$, and $\xi_1(\epsilon) < \xi_1^m < 0$ for $\epsilon < 0$ (see Section 6.5). We have $\xi_1 < 0$, $\xi_2 > 0$ on $\tilde{\Pi}^+(\epsilon)$. For $\epsilon > 0$, the sign of $\xi_2$ is preserved and the sign of $\xi_1$ changes only once between $\tilde{\Pi}^+(\epsilon)$ and $\tilde{\Pi}_m(\epsilon)$. For $\epsilon < 0$, the sign of $\xi_1$ is preserved and the sign of $\xi_2$ changes. Therefore, for a given $\epsilon$, the regions $\Pi_i^m(\epsilon)$ differ by combinations of signs of the coordinates $x_1, x_2, y_1, y_2$. Let us fix the following numeration of the regions:

1. For $\epsilon > 0$,

$$\begin{aligned}
\Pi_1^m(\epsilon) &: x_1 > 0, \quad y_1 > 0, \quad x_2 > 0, \quad y_2 > 0;\\
\Pi_2^m(\epsilon) &: x_1 > 0, \quad y_1 > 0, \quad x_2 < 0, \quad y_2 < 0;\\
\Pi_3^m(\epsilon) &: x_1 < 0, \quad y_1 < 0, \quad x_2 > 0, \quad y_2 > 0;\\
\Pi_4^m(\epsilon) &: x_1 < 0, \quad y_1 < 0, \quad x_2 < 0, \quad y_2 < 0.
\end{aligned}$$

2. For $\epsilon < 0$,

$$\begin{aligned}
\Pi_1^m(\epsilon) &: x_1 < 0, \quad y_1 < 0, \quad x_2 < 0, \quad y_2 > 0;\\
\Pi_2^m(\epsilon) &: x_1 < 0, \quad y_1 > 0, \quad x_2 > 0, \quad y_2 < 0;\\
\Pi_3^m(\epsilon) &: x_1 > 0, \quad y_1 < 0, \quad x_2 < 0, \quad y_2 > 0;\\
\Pi_4^m(\epsilon) &: x_1 > 0, \quad y_1 < 0, \quad x_2 > 0, \quad y_2 < 0.
\end{aligned}$$

In each region $\Pi_i^m(\epsilon)$, the traces of the corresponding stable and unstable manifolds for $\Sigma_1, \Sigma_2$ are intervals intersecting transversally at the point $Q_i^m(\epsilon)$ through which a heteroclinic trajectory passes. On the level set $K = k_m(\epsilon)$ these intervals are connected by smooth arcs forming a set of closed curves which is the complete trace of $W^s(\Sigma_1)$ or $W^s(\Sigma_2)$. Similar statement holds for $W^u(\Sigma_1)$ or $W^u(\Sigma_2)$. Determination of the adjoining rules for the connecting smooth arcs for $W^s(\Sigma_1)$, for $\epsilon > 0$, is based on the fact that the trajectories of the field $Y$ passing through these arcs intersect $\partial V_\epsilon^+$ forming curves which belong to the boundaries of the disks $N^+(\epsilon)$. Therefore, using the known structure of traces of $W^s(\Sigma_1)$ on $\partial V_\epsilon^+$ and preservation of functions $b_1, b_2$ along the trajectories of the field $Y_m$, one can follow the transformation of quadrants from $\Pi_i^+(\epsilon)$ to $\Pi_j^m(\epsilon)$, hence, a transformation of the arcs from $W^s(\Sigma_1)$ composing the boundary. Analogously, using $\partial V_\epsilon^-$, one can construct traces of $W^u(\Sigma_2)$ on the level set $K = k_m(\epsilon)$. In the same way one can construct traces of $W^u(\Sigma_1)$ and $W^s(\Sigma_2)$, for $\epsilon < 0$.

Let us monitor, for example, the movement of one arc from $W^s(\Sigma_1) \cap \partial V_\epsilon^+$ in Case 1A (see Figure 6.18). On the torus $T_1^+(\epsilon)$ we consider an arc $l$ from $W^s(\Sigma_1)$ connecting the points $Q_1^+(\epsilon)$ and $Q_3^+(\epsilon)$ and oriented from 1 to 3 (the second arc is oriented from 3 to 1). Then $b_1$ is positive on $l$ near the point 1. By Lemma 6.7.2, $b_1 < 0$ near the point 3. Consider a quadrant $\Gamma_1^+$ on $\Pi_1^+(\epsilon)$ whose boundary contains $l$, for example, the quadrant $b_1 \geq 0$, $b_2 \geq 0$. By Lemma 6.7.2, the quadrant $\Gamma_2^+$, $b_1 \leq 0$, $b_2 \geq 0$ from $\Pi_3^+(\epsilon)$ also contains $l$ in the boundary. While passing from $\Pi_1^+(\epsilon)$ to $K = k_m(\epsilon)$, the quadrant $\Gamma_1^+$ transfers to the quadrant $b_1 > 0$, $b_2 > 0$ on a certain region $\Pi_j^m(\epsilon)$. Let us determine $j$. During the translation, the sign of $\xi_1$ changes and the sign of $\xi_2$ remains the same. Since the functions $b_1, b_2$ are local integrals of the field $Y$, we have $x_1^2 - y_1^2 = 2b_1 > 0$, $x_2^2 - y_2^2 = 2b_2 > 0$, i.e., $|x_1| > |y_1|$, $|x_2| > |y_2|$ along the trajectories of the field. Thus, the change of the sign of $\xi_1$ occurs due to the change of the sign of $y_1$. On $\Pi_1^+(\epsilon)$ we have $x_1 > 0$, $y_1 < 0$, $x_2 > 0$, $y_2 > 0$. Therefore, on the level $K = k_m(\epsilon)$ we obtain the region with $x_1 > 0$, $y_1 > 0$, $x_2 > 0$, $y_2 > 0$, i.e., $\Pi_1^m(\epsilon)$. Since on $\Pi_3^+(\epsilon)$ we have $x_1 < 0$, $y_1 > 0$, $x_2 > 0$, $y_2 > 0$ and the sign of $x_1$ changes along the trajectory, the image of $\Gamma_2^+$ is the quadrant $b_1 < 0$, $b_2 > 0$ on $\Pi_1^m(\epsilon)$. Thus, we obtain a closed curve on the level set $K = k_m(\epsilon)$ containing the point 1.

The traces of $W^s(\Sigma_1)$ and $W^u(\Sigma_2)$ for $\epsilon > 0$, and of $W^u(\Sigma_1)$ and $W^s(\Sigma_2)$ for $\epsilon < 0$, obtained in this way, are depicted in Figure 6.23, according to Cases $n$A and $n$B. Similarly to Proposition 6.6.3, one can prove the following result.

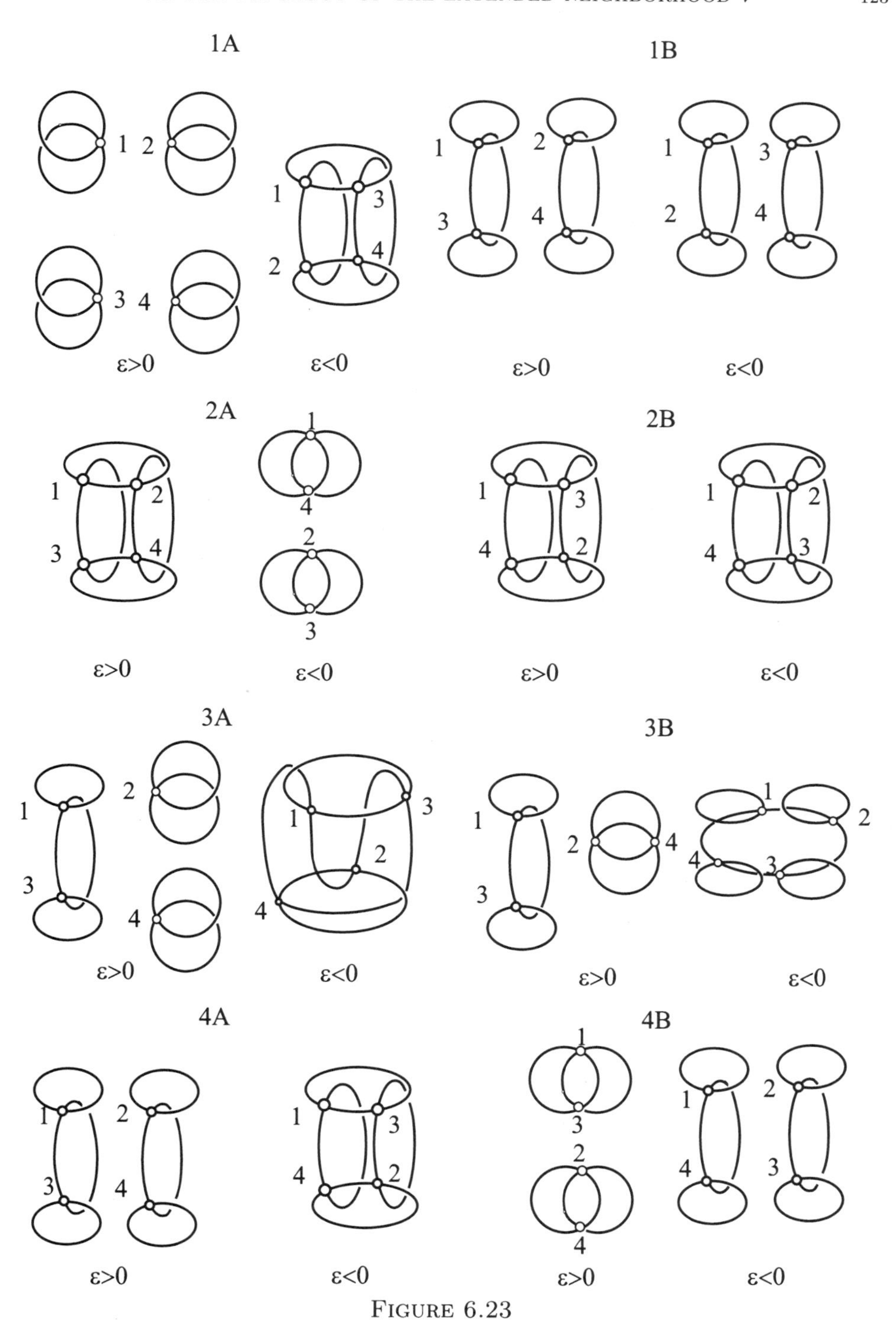


FIGURE 6.23

PROPOSITION 6.8.1. *The number of tori on the level $K = k_m(\epsilon)$ is equal to the number of connected components of the constructed sets.*

Let us determine the number of components of the sets $V_\epsilon^i$ in every Case $n$A, $n$B. The boundary of $\partial V_\epsilon^i$ consists of two pieces. One of them is $\partial V_\epsilon^+$ (for $V_\epsilon^1$) or

$\partial V_\epsilon^-$ (for $V_\epsilon^2$). We denote the other by $\partial V_\epsilon^m$, and it is equal to $V_\epsilon \cap \{K = k_m(\epsilon)\}$. For example, let us consider $V_\epsilon^1$. We call the tori $T_j^+(\epsilon) \subset \partial V_\epsilon^+$ and $T_k^m(\epsilon) \subset \partial V_\epsilon^m$ coupled if there exists a trajectory of the field $Y$ intersecting $T_j^+(\epsilon)$ and $T_k^m(\epsilon)$. It is equivalent to the fact that there is a disk $N^+(\epsilon) \subset T_j^+(\epsilon)$ which is transformed along trajectories of the field $Y$ into some disk from $T_k^m(\epsilon)$. The movement of disks can be determined by the movement of their quadrants. Therefore, one can determine the coupled pairs of the tori. It should be noted that under the mapping of $V_\epsilon^+$ to $\partial V_\epsilon^m$ the images of two disks whose common boundary is the trace of $W^s(\Sigma_1)$ (for $\epsilon > 0$) or of $W^s(\Sigma_2)$ (for $\epsilon < 0$) are two adjacent disks with the same common boundary. Thus, the tori on $\partial V_\epsilon^+$ are cut into annuli by smooth homotopic curves which are the traces of $W^s(\Sigma_2)$ (for $\epsilon > 0$) or of $W^s(\Sigma_1)$ (for $\epsilon < 0$). Under the mapping these annuli are glued into tori on $\partial V_\epsilon^m$.

There is a similar description for $V_\epsilon^2$. In this case the annuli on $\partial V_\epsilon^m$ can be obtained by cutting along the traces of $W^u(\Sigma_1)$ for $\epsilon > 0$, or of $W^u(\Sigma_2)$ for $\epsilon < 0$.

For the description of the regluing of the annuli we define a graph $\Gamma_\epsilon^1$ such that its vertices correspond to the tori on $\partial V_\epsilon^+$ and $\partial V_\epsilon^m$, and the edges connecting these vertices correspond to the annuli which begin at $\partial V_\epsilon^+$ and end at $\partial V_\epsilon^m$. The number of annuli may be different for the same case depending on the sign of $\epsilon$. All possible types of graphs $\Gamma_\epsilon^1$ for the cases $n$A, $n$B coincide with the upper halves of the graphs in Figure 6.24. Analogously, one can define graphs $\Gamma_\epsilon^2$ which coincide with the lower halves of the graphs in Figure 6.24. They describe the regluing of the annuli $\partial V_\epsilon^-$ when we move to $\partial V_\epsilon^m$.

PROPOSITION 6.8.2. *The number of arcwise connected components of the set $V_\epsilon^i$ is equal to the number of components of the graph $\Gamma_\epsilon^i$.*

PROOF. According to Figure 6.24, $\Gamma_\epsilon^i$ consists of at most two components. For example, let us consider $V_\epsilon^1$, $\epsilon > 0$. Assume that $\Gamma_\epsilon^1$ consists of a single component. Then any two vertices of the graph are connected by a path consisting of edges and vertices. We will show that any two points $a_1$ and $a_2$ can be connected by a path.

Without loss of generality we may assume that these points do not belong to stable and unstable manifolds of CSP. Therefore, these points are connected by segments of trajectories of the field $Y$ with a pair of points $m_1$, $m_2$ on $\partial V_\epsilon^+$. The points $m_1$, $m_2$ belong to two tori $T_1$, $T_2$. If $T_1 = T_2$, then the points $m_1$, $m_2$ can be connected by a path on this torus, hence the points $a_1$ and $a_2$ are connected by a path on $V_\epsilon^1$. Now suppose that the tori are different. Let us consider two vertices $c_1$ and $c_2$ corresponding to the tori $T_1$, $T_2$ in the graph $\Gamma_\epsilon^1$, and the path connecting them. It follows from the form of the graph $\Gamma_\epsilon^1$ that $c_1$ and $c_2$ are always connected by a path formed by two edges which have a common vertex at the intermediate level set. This means that in $V_\epsilon^1$ there are trajectories $l_i$ beginning at the torus $T_i$, $m_i \in T_i$, and ending at the torus $T \subset \partial V_\epsilon^m$. On $T_1$, let us join the point $m_1$ and the point $l_1 \cap T_1$, the point $m_2$ and the point $l_2 \cap T_2$, and, on $T$, let us connect the points $l_1 \cap T$ and $l_2 \cap T$. We obtain a path in $V_\epsilon^1$ which connects the points $m_1$ and $m_2$, and, thus, the points $a_1$ and $a_2$.

Now let us assume that $\partial V_\epsilon^1$ is connected, and $\Gamma_\epsilon^1$ is not connected. Therefore, $\Gamma_\epsilon^1$ consists of two subgraphs. Let us denote by $R_1$ the closure of the set of points lying on the trajectories passing through the annuli on the tori which correspond to one component of the graph $\Gamma_\epsilon^1$. Let us denote by $R_2$ the second such set. The graph $\Gamma_\epsilon^1$ is disconnected, so it consists of two component which together form the complete graph of the connected manifold $V_\epsilon$. Therefore, it follows from the

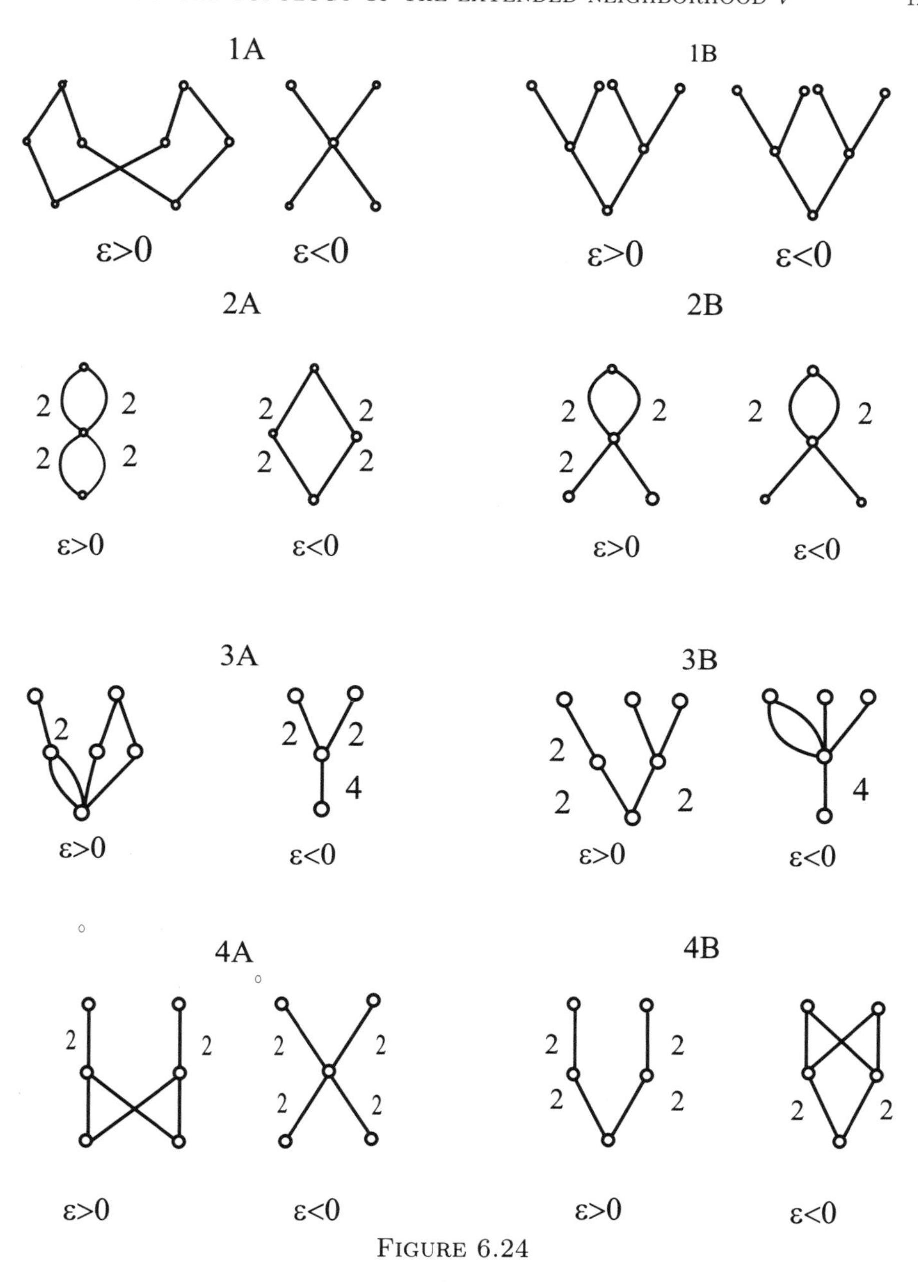


FIGURE 6.24

construction of the graph $\Gamma_\epsilon^1$ that $R_1 \cap R_2$ is nonempty, invariant with respect to the field $Y$, and contains only the points from $(W^s(\Sigma_2) \cup W^u(\Sigma_2)) \cap V_\epsilon$. Let $x \in R_1 \cap R_2$, $x \in W^s(\Sigma_2) \setminus \Sigma_2$. Then in any neighborhood of the point $x$ there exist points from $R_1$ and $R_2$ that do not belong to $W^s(\Sigma_2) \cup W^u(\Sigma_2)$. Then the trajectories of the field $Y$ passing through these points intersect the same torus on $\partial V_\epsilon^+$ (the one which is intersected by the trajectory through $x$). This means that this torus corresponds to the same vertex in both subgraphs, contradicting the discontinuity of $\Gamma_\epsilon^1$. The proof is similar for $x \in W^u(\Sigma_2) \setminus \Sigma_2$, if one moves from the point $x$ to $\partial V_\epsilon^m$. Therefore, the only possibility left is that $x \in \Sigma_2$, i.e., that $R_1$ and $R_2$ intersect each other only along $\Sigma_2 \cap V_\epsilon$, i.e., along a CSP. This contradicts the fact that the set $R_1 \cup R_2 = V_\epsilon^1$ is a smooth manifold with the boundary $\partial V_\epsilon^+ \cup \partial V_\epsilon^m$, because a finite set of one-dimensional piecewise smooth curves cannot divide a three-dimensional manifold. The proposition is proved. $\square$

Now we can assign a component of the graph $\Gamma_\epsilon^i$ to each component of the set $\partial V_\epsilon^i$, $i = 1, 2$. We obtain a collection of such components. It follows from Figure 6.24 that there are 5 different types of components. Similar graphs can be constructed also for the model manifolds $P_1, \dots, P_5$. The correspondence to $P_i$ is shown in Figure 6.25. In order to do this, we consider a vector field $v = \operatorname{gr} F$ on $P_i$ that is transversal to the boundary of $\partial P_i$ (for some Riemannian metric on $P_i$). In this way one can determine stable and unstable manifolds of saddle CSP which cut the tori into annuli and determine the movement of the annuli through the critical levels.

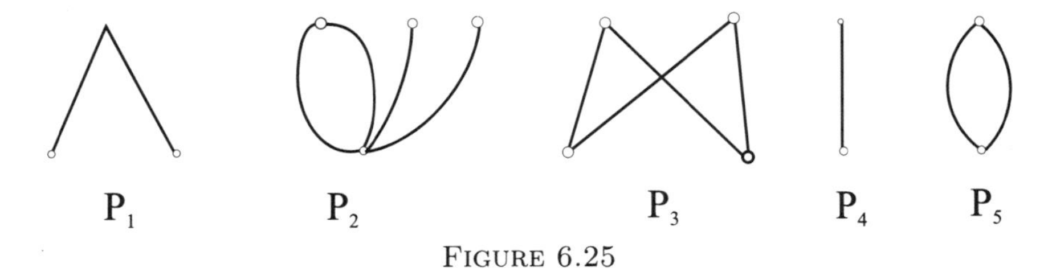


FIGURE 6.25

PROPOSITION 6.8.3. *Let $Z$ be a connected component of the set $V_\epsilon^i$, $\epsilon \neq 0$. Then $Z$ is homeomorphic to the model manifold $P_k$ with the same graph.*

PROOF. According to Proposition 6.8.1, $V_i^\epsilon$ consists of at most two components. If it consists of two components, then each of them contains one SPT, oriented or nonoriented. In the orientable case we will associate to such a component the manifold $P_1$, in the nonorientable one the manifold $P_4$ (for these cases the homeomorphism between $V_\epsilon^i$ and $P_1, P_4$ follows from [**9**]). If $V_\epsilon^i$ consists of one component containing one SPT, then we have the previous case. Now, assume that $V_\epsilon^i$ is connected and contains two SPTs. Note that $V_\epsilon^i$ cannot contain oriented and nonoriented SPTs simultaneously, because we can only have such a pair of SPTs in Cases 3A, 3B, for $\epsilon > 0$, on $V_\epsilon^1$, but then $V_\epsilon^1$ is not connected. Therefore, there are three possible cases, according to the types of graphs and the properties of the canonical representatives.

1. If $V_\epsilon^i$ contains two nonorientable SPTs, then $V_\epsilon^i$ corresponds to $P_5$.
2. If $V_\epsilon^i$ contains two orientable SPTs, and $\partial V_\epsilon^m$ consists of one or three tori, then $V_\epsilon^i$ corresponds to $P_2$.
3. If $V_\epsilon^i$ contains two orientable SPTs, and $\partial V_\epsilon^m$ consists of two tori, then $V_\epsilon^i$ corresponds to $P_3$.

Let us show that indeed, there is a homeomorphism between $V_\epsilon^i$ and the corresponding $P_i$. Consider, for example, a component $Z \in V_\epsilon^1$, and use the constructed $B_1$-, $B_2$-foliations on $Z$ and the analogous foliations on the corresponding $P_i$. Similarly to Theorem 6.4.1, one can construct a homeomorphism $F : Z \cap \partial V_\epsilon^+ \to \partial P_i$ such that the chosen boundary must contain the same number of tori as $Z \cap \partial V_\epsilon^+$. For $\epsilon > 0$, the role of $B_2$-foliation on $P_i$ is played by the foliation $s = \text{const}$, where $s \in S^1$. For $\epsilon < 0$, the foliation $s = \text{const}$ in $P_i$ plays the role of $B_2$-foliation. $\square$

The homeomorphism $F$ must preserve foliations. It also must transform the traces of stable manifolds of CSP of the field $F$ into traces of stable manifolds of CSP of the field $\operatorname{gr} F$. The homeomorphism we construct possesses the following property. Suppose trajectories of the field $Y$ which tend to the same point on CSP pass through two points in $V_\epsilon^i$. Then the images of these points have the same property with respect to the field $\operatorname{gr} F$. Proceeding as in Theorem 6.4.1, we get the following result.

COROLLARY 6.8.1. *If $V_\epsilon$ is connected and contains two SPTs of the field $X_H$, then their stable and unstable manifolds merge.*

This statement follows from the construction of the model manifolds $P_2, P_3, P_5$ and the preceding proposition.

The entire manifold $V_\epsilon$ is obtained by gluing $V_\epsilon^1$ and $V_\epsilon^2$ along $\partial V_\epsilon^m$. This gluing is defined uniquely by the transformation of the disks $N_i$ along the trajectories of the field $Y$. The topological types of the manifolds $V_\epsilon^1$ and $V_\epsilon^2$ for every case $n$A, $n$B, and the way they are glued are defined by the corresponding graph in Figure 6.24.

Now we describe the topology of the set $V_0$. It is a union of two closed subsets $V_0^1$ and $V_0^2$ corresponding to $K \geq 0$ and $K \leq 0$ which are glued together along the common boundary $W = \{H = 0, K = 0\}$. The set $V_0^1 \setminus W$ is diffeomorphic to the direct product $\partial V_0^+ \times (0, K_*]$. When $k \to +0$, the tori $H = 0, K = k$ are pasted to $W$, the open disks $N_i^+(0)$ on the tori converge smoothly to the disks from $W \setminus (E_1 \cup E_2)$, the curves $W^s(E_1 \cup E_2)$ converge smoothly to the corresponding "figures eight" $E_1$ and $E_2$, and the points $Q_1^+ - Q_4^+$ merge with the point $p$. Let us remark that the trace of $W^s(E_i)$ consists of a certain number of smooth nonintersecting closed curves, which either themselves form pairs of incident curves or are divided by points into pairwise incident pieces lying on different curves. In any case a pair of incident pieces are pasted to the same loop. Note that $V_0^2$ is organized similarly. The gluing of $V_0^1$ and $V_0^2$ is defined uniquely by the correspondence of the disks $N_i$.

In conclusion we describe the topological structure of the extended neighborhood $V$. Let us represent $V$ as a union of two closed subsets: $V^+ = \{0 \leq H \leq h_*\}$ and $V^- = \{-h_* < H \leq 0\}$. Consider, for example, $V^+$. The set $V^+ \setminus V_0$ is a smooth four-dimensional submanifold diffeomorphic to $V_{h_*} \times (0, h_*]$. As $\epsilon \to +0$, the manifold $V_\epsilon$ is deformed into $V_0$. Let us divide $V_\epsilon$ into three pieces, $Z_\epsilon^1 = \{k_2(\epsilon) \leq K \leq k_*\}$, $Z_\epsilon^2 = \{k_1(\epsilon) \leq K \leq k_2(\epsilon)\}$, $Z_\epsilon^3 = \{-k_* \leq K \leq k_1(\epsilon)\}$, where $k_1(\epsilon) < k_2(\epsilon)$ are the critical values of the function $K$ on $V_\epsilon$. The boundary $\partial Z_\epsilon^2$ consists of two pieces, $D_1 = \{H = \epsilon,\ K = k_2(\epsilon)\}$ and $D_2 = \{H = \epsilon,\ K = k_1(\epsilon)\}$. Each of them is a merged stable and unstable manifold of the corresponding SPT of the field $X_H$. In the set $Z_\epsilon^2$ there is a piece of the stable manifold of CSP from $\Sigma_1$ of the field $Y$ and a piece of the unstable manifold of CSP from $\Sigma_2$ of the field $Y$. Stable manifold of each CSP in $Z_\epsilon^2$ is a smooth two-dimensional manifold, diffeomorphic to a cylinder in the case of orientable CSP, and to a Möbius band in the case of nonorientable CSP. When this manifold approaches the boundary $D_1$, it is transversal to $D_1$ at all points except for the points of CSP. Therefore, it intersects $D_1$ along smooth open arcs. The end points of these arcs (they belong to the stable manifold) lie on CSP. These end points are limit points for heteroclinic trajectories. In particular, there are two such points on $D_1$. Notice also that if we remove a CSP from $D_1$ and the traces of the stable manifolds of CSP from $\Sigma_1$, then $D_1$ splits into four open disks such that the disks $N_i^+(\epsilon) \subset \partial V_\epsilon^+$ are mapped into them along the trajectories of the field. The unstable manifold of CSP from $\Sigma_2$ has a similar structure.

As $\epsilon \to +0$, the topological limit of $Z_\epsilon^2$ is $W$. It is convenient to describe this passage to the limit along the path $\lambda$ that lies between the curves $K = k_1(h)$ and $K = k_2(h)$ on the moment plane $(h, (k_1(h) + k_2(h))/2)$ and connects a point on the interval $H = h_*$ with the point $(0, 0)$. Let us observe the deformation of the Liouville tori in $M$. On every such torus there is a system of curves described in Proposition 6.6.5, and the deformation is described exactly as on $V_0$ for $k \to +0$. As $\epsilon \to +0$, the sets $Z_\epsilon^1$ and $Z_\epsilon^2$ deform into $V_0^1$ and $V_0^2$, respectively, and also $D_1$ and $D_2$ deform into $W$. Thus, we have described the topological structure of the set $V$.

CHAPTER 7

# Saddle-Focus Type Singular Points

In this chapter we consider the structure of IHVF in the extended neighborhood of a simple singular point of the saddle-focus type. We will assume that both integrals, i.e., the Hamiltonian $H$ and the second integral $K$, are equal to zero at the singular point. Then the stable and unstable manifolds of the saddle-focus point belong to the level set $H = K = 0$. Locally they are two-dimensional disks in the three-dimensional level set of the Hamilton function $H = 0$. Since we assume that there are no other singular points in the connected component of the common level set $H = K = 0$ containing the point $p$ (Morse Hamiltonian condition), in the general case stable and unstable manifolds of the singular point merge when continued along trajectories of the field $X_H$, provided that the level set $H = K = 0$ is compact. However, as was mentioned in Proposition 3.3.5, other situations are also possible. For example, a hyperbolic SPT, oriented or nonoriented, lying in the same three-dimensional level set $H = 0$ may be contained in the closure of the unstable manifold of the singular point. In this case the following condition must hold: this hyperbolic SPT must also belong to the level set $K = 0$. This imposes an additional restriction. Therefore, generally speaking, these situations occur when one considers not individual IHVFs, but families of IHVFs that depend on at least one parameter. We may observe an even more degenerate situation that occures mainly in two-parameter families of IHVFs, in the case when the unstable (or stable) manifold of a saddle-focus contains in its closure a parabolic SPT. In this chapter we restrict ourselves to considering the first three situations, which we regard as the basic ones. Let us remark that an SPT contained in the closure of the unstable (or stable) manifold of a saddle-focus point can contain other hyperbolic or parabolic SPTs in the closure of its unstable (or stable) manifold, i.e., the process of "capturing" new SPTs may multiply. However, the condition that these new SPTs belong to the common level set $H = K = 0$ (in the absence of special symmetries) is a restriction. Moreover, the greater is the number of SPTs belonging to this level, the stronger this restriction is. Since the study of this situation does not give us any conceptually new information, we will consider the above-mentioned cases only.

## 7.1. The structure of a garland and its separatrix set

Let $p$ be a singular saddle-focus type point of the IHVF $(X_H, K)$. Without loss of generality, we will assume that $H(p) = K(p) = 0$. Note that for the induced action $\Phi$ the point $p$ is a singular point of the focus-focus type. Consider the garland of the point $p$. Recall that in some neighborhood of the point $p$, there are only two orbits of the action (cylinders) that contain the point $p$ in their closure. For IHVF $(X_H, K)$ these orbits coincide locally with the stable $W^s(p)$ and unstable $W^u(p)$ manifolds of the point $p$. It follows from the Morse property of $H$ (Proposition 3.3.5) that the limit set of a cylindrical orbit can be either a singular point or a closed one-dimensional orbit of the hyperbolic type. In other words, the unstable two-dimensional manifold of the saddle-focus point of an integrable system of the Morse type can end either on the saddle-focus point itself or on a hyperbolic SPT. It was mentioned above that in the second case the process may multiply: the hyperbolic SPT $\gamma$, in turn, has its own unstable manifolds which can end on other SPTs, etc. Therefore, here we restrict ourselves to the main cases, when there is at most one hyperbolic SPT in the closure of the manifolds of the point $p$. Thus, we have the following possibilities.

1. The garland consists of the point $p$ itself.
2. The garland consists of the point $p$ and one oriented hyperbolic SPT.
3. The garland consists of the point $p$ and one nonoriented hyperbolic SPT.

First we describe the separatrix set of the garland in each of these three cases. To do this we define three standard sets.

1. Let $(D_i, x_i)$, $i = 1, 2$, be two closed smooth disjoint two-dimensional disks with distinguished interior points $x_i \in D_i$. Then we denote by $Q_1$ the union of the two disks $D_1, D_2$ with the points $x_1$ and $x_2$ identified.

2. Consider the set defined by the relations $u^2 + v^2 \leq p^2, uv = 0, p > 0$, on the plane $(u, v)$. We call it the "cross". Let $Q_2$ denote the direct product of the cross and a circle. The boundary $\partial Q_2$ consists of four circles $s_1^+, s_2^+, s_1^-, s_2^-$, where $s_1^+, s_2^+$ are obtained from the boundaries of the interval $v = 0$, and $s_1^-, s_2^-$, from the boundaries of the interval $u = 0$.

3. $Q_3$ is a fiber bundle over a circle, with the "cross" as a fiber, such that its structural group is the antipodal map of the "cross" $(u, v) \to (-u, -v)$. The boundary $\partial Q_3$ consists of two closed curves $s^+, s^-$.

PROPOSITION 7.1.1. *The separatrix set of the garland is homeomorphic to one of the following sets.*

a) *In Case 1, to the set obtained from $Q_1$ by gluing the disks $D_1, D_2$ along some homeomorphism of the boundary circles, i.e., it is homeomorphic to a sphere with two identified points.*

b) *In Case 2, to the set obtained by gluing the set $Q_2$, in which the circles $s_2^+, s_2^-$ are glued in the orientable way, and the set $Q_1$. The circle $s_1^+$ is glued to $\partial D_1$, and $s_1^-$ to $\partial D_2$.*

c) *In Case 3, to the set, obtained from $Q_1$ and $Q_3$ by gluing $s^+$ to $\partial D_1$, and $s^-$ to $\partial D_2$.*

PROOF. a) On the stable manifold $W^s(p)$, let us choose a smooth closed curve $s_1$ transversal to trajectories of the field $X_H$. Let us denote by $(D_1, p)$ the part of $W^s(p)$ containing the point $p$ and bounded by the curve $s_1$. Similar disk on $W^u(p)$ bounded by the curve $s_2$ is denoted by $(D_2, p)$. In Case 1 trajectories of the field $X_H$ leaving $s_1$ return to $s_2$, defining a gluing diffeomorphism for $\partial D_1$ and $\partial D_2$. Since $D_1$ and $D_2$ are only identified at one point, it is clear that the topological type of the set does not depend on the gluing map.

b) In Case 2 the curve $\gamma$ is a hyperbolic SPT of the field $X_H$. Extension of the stable manifold of the point $p$ along trajectories of the field $X_H$ merges with one of the components of the set $W^u \backslash \gamma$, where $W^u$ is the unstable manifold of SPT $\gamma$, and extension of the unstable manifold of the point $p$ merges with one of the components of the set $W^s \setminus \gamma$ of the stable manifold of SPT $\gamma$. Gluing homeomorphisms are defined by mapping the curves $s_1$ and $s_2$ along trajectories into the corresponding closed curves $l_1(\gamma)$ and $l_2(\gamma)$ which lie on the indicated components of invariant manifolds of the SPT $\gamma$, and are transversal to trajectories of the field $X_H$. Two remaining components merge with each other. Thus, we obtain the desired set.

c) In this case we proceed analogously keeping in mind that each set $W^s \setminus \gamma$ and $W^u \setminus \gamma$ consists of one component. The proposition is proved. □

## 7.2. The structure of the auxiliary gradient system

Let $U_0$ be a neighborhood of the singular point $p$ in which Theorem 2.2.1 holds. This means that $H = h(\xi, \eta)$, $K = k(\xi, \eta)$, where $\xi = x_1y_1 + x_2y_2$, $\eta = x_1y_2 - x_2y_1$ in $U_0$. Then the field $Y$ (in the Euclidian Riemannian metric with respect to coordinates in $U_0$) has the form

$$\begin{aligned} \dot{x_1} &= (y_1h_\eta - y_2h_\eta)G, \quad \dot{x_2} = (y_1h_\eta - y_2h_\eta)G, \\ \dot{y_1} &= (x_1h_\eta - x_2h_\xi)G, \quad \dot{y_2} = (x_2h_\eta - X_1h_\xi)G, \\ G &= (h_\xi k_\eta - h_\eta k_\xi)/(h_\xi^2 + h_\eta^2). \end{aligned} \tag{7.1}$$

We assume that the neighborhood $U_0$ is chosen in such a way that the value $G = \Delta(\alpha^2 + \beta^2) + O(r)$ is not equal to zero in $U_0$, $r^2 = x_1^2 + x_2^2 + y_1^2 + y_2^2$, and $\pm(\alpha \pm i\beta)$ are the eigenvalues of the field $X_H$ at the singular point $p$. The multiplier $G$ in (7.1) may be omitted, and, after that, the system becomes Hamiltonian with respect to the symplectic structure $dx_1 \wedge dx_2 + dy_2 \wedge dy_1$ with the Hamilton function $H$. It is integrable and its additional integral is the function $Q = x_1^2 + x_2^2 - y_1^2 - y_2^2$.

Below we will study the system (7.1) in the neighborhood $U \subset U_0$ which is invariant with respect to the field $Y$ and is defined, for sufficiently small $h_*, Q_0, k_*$, by the inequalities $|H| \leq h_*$, $|Q| \leq Q_0$, $|K| \leq k_*$. Here, the invariance means that a segment of a trajectory of the field $Y$ is contained in $U$ for $|K| \leq k_*$.

The point $p$ is a singular point of the saddle type of the field (7.1) with the eigenvalues $(\alpha, -\alpha)$, $\alpha > 0$, of multiplicity two, and it lies on the level $H = 0$. The two-dimensional stable and unstable manifolds of this point form the entire set of solutions of the system of equations $H = 0$, $Q = 0$.

To construct a local homeomorphism which realizes isoenergetic equivalence of two IHVFs in a neighborhood of a singular point of the saddle-focus type we consider the following construction. The cylinder $C_\epsilon^+$ in $U$ defined by the equations $H = \epsilon$, and $K = k_*$ is foliated by circles $Q = \text{const}$ ($Q$-foliation). To define the point on the cylinder uniquely we consider another foliation $\tau = \text{const}$ into segments which are transversal to the leaves of the $Q$-foliation. Then we observe how this foliation is mapped along trajectories of the field $Y$ on the cylinder $C_\epsilon^- : H = \epsilon$, $K = -k_*$. Since $p$ is the only singular point of the field $Y$ in $U$ and all trajectories of the field $Y$ move from $C_\epsilon^+$ to $C_\epsilon^-$ in bounded time, this map is a diffeomorphism, for $\epsilon \neq 0$. Moreover, for $\epsilon = 0$, this map has a discontinuity on the curve $Q = 0$, since this curve is the trace of the stable manifold of the point $p$. The fiber $\tau = \text{const}$ on $C_\epsilon^+$ is broken under mapping at the point $Q = 0$ into two parts. One part transforms into a curve on $C_0^-$ lying in one of the two semiannuli into which the circle $Q = 0$ divides the annulus $C_0^-$. This curve has a limit point $a_1$ on the circle $Q = 0$. The other part transforms into the curve lying in the second semiannulus and has a limit point $a_2$ on the circle $Q = 0$. The point $a_2$ is shifted by the angle $\pi$ relative to the point $a_1$ (Figure 7.1). For $\epsilon \neq 0$, the fiber $\tau = \text{const}$ is mapped on $C_\epsilon^-$ without discontinuities, but as $\epsilon \to 0$, the image of the fiber approaches the curve formed by the half of the circle $Q = 0$ on $C_0^-$ between the points $a_1$ and $a_2$ and by the images of the fiber $\tau = \text{const}$ on $C_0^-$. Let us construct a foliation with the described properties.

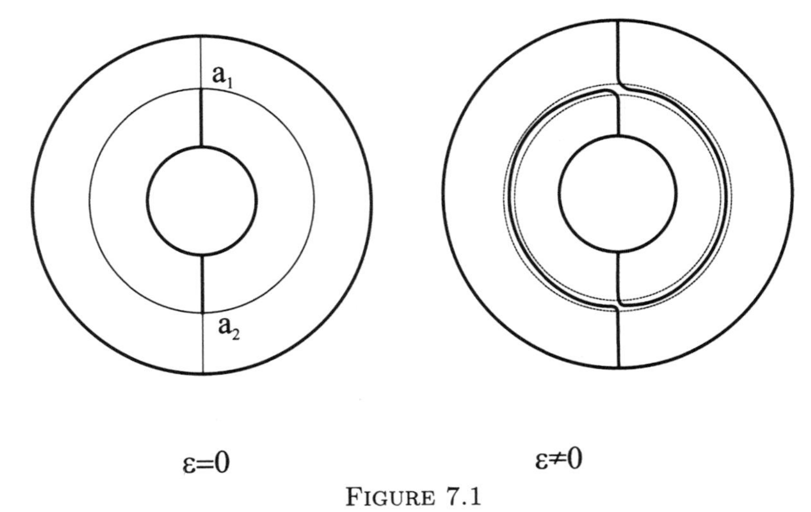


FIGURE 7.1

Let us introduce the coordinates $(\zeta, \eta, Q, \tau)$ in $U$ by the formulas

$$x_1 = ((Q^2 + \xi^2 + \eta^2)^{1/2} + Q)\cos\tau, \qquad x_2 = ((Q^2 + \xi^2 + \eta^2)^{1/2} + Q)\sin\tau,$$
$$y_1 = ((Q^2 + \xi^2 + \eta^2)^{1/2} - Q)\cos\tau, \qquad y_2 = ((Q^2 + \xi^2 + \eta^2)^{1/2} - Q)\sin\tau,$$

where $\xi, \eta, Q$ are defined above. In these coordinates the field (7.1) can be written in the form

$$\xi' = Gr^2 h_\eta, \qquad \eta' = -Gr^2 h_\xi, \qquad Q' = 0,$$
$$\tau' = G(\xi h_\xi + \eta h_\eta)/((Q^2 + \xi^2 + \eta^2)^{1/2} + Q), \tag{7.2}$$

where $r^2 = 2(Q^2 + \zeta^2 + \eta^2)^{1/2}$.

The Lie derivative of the function $K$ with respect to the vector field (7.2) is equal to $K' = -Gr^2[h,k]$, $[h,k] = h_\xi k_\eta - h_\eta k_\xi$. Therefore, it is convenient to introduce a new time parameter $K$. Then

$$d\xi/dK = -h_\eta/[h,k], \quad d\eta/dK = h_\xi/[h,k], \quad dQ/dK = 0,$$
$$d\tau/dK = -(\eta h_\eta + \xi h_\xi)/(2(\xi^2 + \eta^2)[h,k]) + Q(\xi h_{dz} + \eta h_\eta)/((\xi^2 + \eta^2)[h,k]r^2). \tag{7.3}$$

The first two equations in (7.3) have the integral $h(\xi, \eta) = \epsilon$. For a given $\epsilon$, they define the functions $\xi(K, \epsilon)$, $\eta(K, \epsilon)$ that give the dependence of $\xi$ and $\eta$ on $K$. Substitution of these expressions on the right-hand side of the last equation in (7.3) and integration with respect to $K$ from $-k_*$ to $k_*$ determines the function $T(Q, \epsilon)$. This function defines the image of the interval $\tau = 0$ from the cylinder $C_\epsilon^+$ when it is mapped to the cylinder $C_\epsilon^-$ along the trajectories (7.3).

The first term in the expression for $d\tau/dK$ is $(d/dK)(-\arctan(\xi/\eta)/2)$. Therefore, the properties of the function $T(Q, \epsilon)$ are determined by the integral

$$\tau_1(Q, \epsilon) = Q \int_{-k_*}^{k_*} \frac{\xi h_\xi + \eta h_\eta}{(\xi^2 + \eta^2)[h,k]r^2} dK.$$

Introducing the polar coordinates $\xi = \rho\cos\theta$, $\eta = \rho\sin\theta$, and using the first two equations in (7.3), one can rewrite the integral in the form

$$\tau_1(Q,\epsilon) = \frac{Q}{2}\int \frac{d\theta}{(Q^2+\rho^2)^{1/2}}.$$

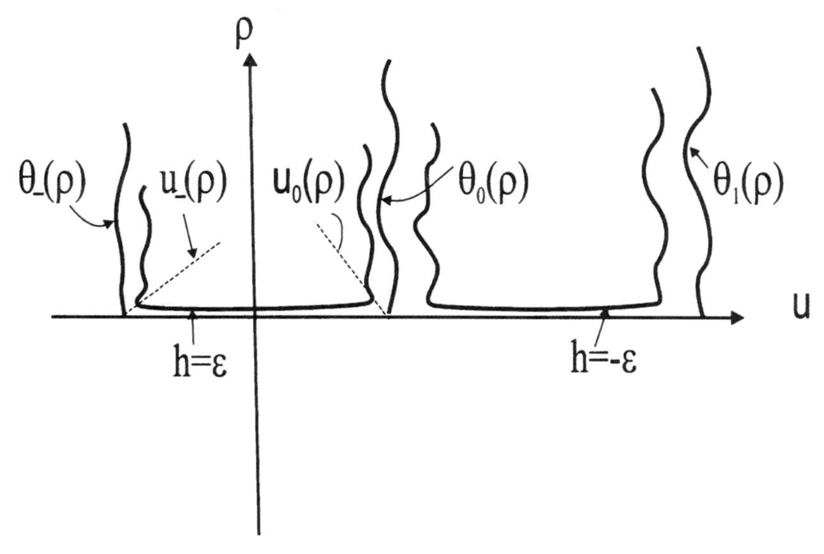


FIGURE 7.2

The integral is taken from $\theta_1(\epsilon)$ to $\theta_2(\epsilon)$, along the curve $\hat{h}(\rho,\theta) = h(\rho\cos\theta, \rho\sin\theta) = \epsilon$. Note that $\theta_i(\epsilon)$ is determined from the polar coordinates for the curves $(\xi_i(\epsilon), \eta_i(\epsilon))$, $i = 1, 2$, that are the solutions of the system $h(\xi,\eta) = \epsilon, k(\xi,\eta) = \pm k_*$. First, consider the properties of the curve $\hat{h}(\rho,\theta) = \epsilon$ in coordinates $(\rho,\theta)$. Let us write $\hat{h}(\rho,\theta)$ in the form

$$\begin{aligned}\hat{h}(\rho,\theta) &= \rho(\alpha\cos\theta - \beta\sin\theta) + \rho^2 f_1(\rho,\theta)\\ &= \gamma\rho\cos(\theta+\theta_0) + \rho^2 f_1(\rho,\theta) = \gamma[\rho\cos u + \rho^2 f(\rho,\theta)],\end{aligned}$$

where $\gamma = (\alpha^2+\beta^2)$, $u = \theta+\theta_0$, $f = f_1/\gamma$, $f(\rho,u) = f(\rho, u-\theta_0)$. On the plane $(\rho,\theta)$, the function $\hat{h}$ has singular points of the saddle type: $\rho = 0$, $\theta = \pi/2 + m\pi$, $m \in \mathbb{Z}$ (Figure 7.2). Separatrices of these points form the level set $\hat{h}(\rho,\theta) = 0$ and consist of the straight line $\rho = 0$ and the curves $u = \theta_k(\rho)$, where $\theta_k(\rho)$ is the solution of the equation $\cos u + \rho f(\rho,u) = 0$. For $\rho = 0$, this solution is equal to $\pi/2 + m\pi$, and is obtained from the implicit function theorem. Due to the $2\pi$-periodicity of $\hat{h}$ with respect to $u$, it is sufficient to consider the three curves corresponding to $m = -1, 0, 1$. Let us denote them by $\theta_-(\rho)$, $\theta_0(\rho)$, $\theta_+(\rho)$. Notice that $\theta_+(\rho) = \theta_-(\rho) + 2\pi$. Then, for sufficiently small $\rho_*$ and $\epsilon > 0$, the equation $\hat{h}(\rho,u) = \epsilon$ has a unique solution for $0 \le \rho \le \rho_*$. Namely, this solution is a connected smooth curve lying in the strip bounded by the straight lines $\rho = \rho_*$ and $\rho = 0$ between the curves $u = \theta_-(\rho)$ and $u = \theta_0(\rho)$. For $\epsilon < 0$ the solution of this equation is a curve lying in the same strip but between the curves $u = \theta_0(\rho)$ and $u = \theta_+(\rho)$. For example, let us choose $\epsilon > 0$, and divide the curve $\hat{h} = \epsilon$ into three parts. In order to determine these parts we consider, in the region $\hat{h} \ge 0$,

$0 \le \rho \le \rho_*$, the following intervals: $u = -\pi/2 + \nu\rho = u_-(\rho)$, $\nu + f(0, -\pi/2) > 0$ and $u = -\nu\rho = u_0(\rho)$, $\nu + f(0,0) > 0$ (see Figure 7.2). Then one can readily see that in the regions of the strip $0 \le \rho \le \rho_*$ where $\theta_-(\rho) \le u \le u_-(\rho)$ and $u_0(\rho) \le u \le \theta_0(\rho)$ the curve $\hat{h} = \epsilon$ has the form of the graphs of the functions $u_1(\rho, \epsilon)$ and $u_3(\rho, \epsilon)$, respectively. Notice that for a sufficiently small $\epsilon$, $0 \le \epsilon \le h_*$, and $\rho_*$ we have $|\partial u_i/\partial \rho| \le c_4$. In the region of the strip between the curves $u_-(\rho)$ and $u_0(\rho)$, the curve $\hat{h} = \epsilon$ has the form of the graph of the function $\rho = \rho(u, \epsilon)$ depending on $u$. Since $\partial^2\rho/\partial u^2 > 0$ in the region under consideration, this function has a unique minimum at the point $u = g(\epsilon)$, and takes its maximum value at one of the boundary points, i.e., at one of the intersection points of the graph with the curves $u_-(\rho)$ and $u_0(\rho)$. The minimum curve has the form $u = \phi(\rho)$, $\phi(0) = 0$. Therefore, the minimum value $\rho_{\min}(\epsilon)$ on the curve $\hat{h} = \epsilon$ has the form $\rho_{\min}(\epsilon) = \epsilon + O(\epsilon^2)$. On the curves $u_-(\rho)$ and $u_0(\rho)$, the function $\hat{h}$ has the form $\mu\rho^2 + O(\rho^3)$, $\mu > 0$. Therefore, the maximum value $\rho_{\max}(\epsilon)$ has the form $\rho_{\max}(\epsilon) = \epsilon^{1/2}/\mu + \cdots$. Let $\rho_1(\epsilon)$ and $\rho_3(\epsilon)$ denote the $\rho$-coordinates of the intersection points of the curve $\hat{h} = \epsilon$ with the curves $u_-(\rho)$ and $u_0(\rho)$. Besides, we denote the $u$-coordinates of the intersection points by $u_1(\epsilon)$ and $u_3(\epsilon)$, $u_1(0) = -\pi/2$, $u_3(0) = \pi/2$. We divide the integral $\tau_1(Q, \epsilon)$ into three integrals: $\tau_1 = I_1 + I_2 + I_3$. In $I_1, I_3$ we integrate over the graphs of the functions $u = u_1(\rho, \epsilon)$ and $u = u_3(\rho, \epsilon)$, respectively, and in $I_2$ we integrate over the graph of the function $\rho = \rho(u, \epsilon)$. Then we have the following estimates:

$$
\begin{aligned}
|I_{1,3}| &= \left| \frac{Q}{2} \int_{\hat{h}=\epsilon} \frac{(\partial u_{1,3}/\partial \rho) d\rho}{(Q^2 + \rho^2)^{1/2}} \right| \le \frac{|Q|}{2} c_4 \left| \int_{\rho_{1,3}(\epsilon)}^{\rho_*} \frac{d\rho}{(Q^2 + \rho^2)^{1/2}} \right| \\
&\le c_4 \frac{|Q|}{2} \ln \frac{\rho_* + (Q^2 + \rho_*)^{1/2}}{\rho_{1,3}(\epsilon) + (Q^2 + \rho_{1,3})^{1/2}}.
\end{aligned}
$$

Thus, we obtain that $\lim_{Q \to 0} I_{1,3} = 0$. Furthermore,

$$
I_2 = \frac{Q}{2} \int_{u_1(\epsilon)}^{u_3(\epsilon)} \frac{du}{(Q^2 + \rho^2(u, \epsilon))^{1/2}}.
$$

Therefore, for $Q > 0$, we have

$$
\frac{Q}{2} \frac{u_3(\epsilon) - u_1(\epsilon)}{(Q^2 + \rho^2_{\max}(\epsilon))^{1/2}} \le I_2 \le \frac{Q}{2} \frac{u_3(\epsilon) - u_1(\epsilon)}{(Q^2 + \rho^2_{\min}(\epsilon))^{1/2}}.
$$

For each $\epsilon$, we choose a neighborhood of zero in the interval $|Q| \le Q_0$ of the form $|Q| \le d(\epsilon)$, and such that as $\epsilon \to 0$, we have $\rho_{\min}(\epsilon)/d(\epsilon) \to 0$ and $\rho_{\max}(\epsilon)/d(\epsilon) \to 0$. We can do this since $\rho_{\min}(\epsilon) = O(\epsilon)$, $\rho_{\max}(\epsilon) = O(\epsilon^{\frac{1}{2}})$, i.e., it is enough to choose $d(\epsilon) = O(\epsilon^a)$, $0 < a < \frac{1}{2}$.

Thus, for $Q = d(\epsilon)$ and $\epsilon \to +0$, we have $I_2 \to \frac{\pi}{2}$. Similarly, it can be shown that $I_2 \to -\frac{\pi}{2}$ for $Q = -d(\epsilon)$. Calculating the same integrals for $\epsilon < 0$ one has to consider the strip between the curves $u = \theta_0(\rho)$ and $u = \theta_+(\rho)$, and choose a region of the form $|Q| < d(|\epsilon|)$. From now on by $d(\epsilon)$ we will mean $d(|\epsilon|)$.

Therefore, for a sufficiently small fixed $\epsilon$, the graph of the function $T(Q, \epsilon)$ has the shape of the letter Z. As $\epsilon \to \pm 0$, the topological limit of this family of curves is a curve formed by half of the circle $Q = 0$ on $C_0^-$ and two intervals adjacent to the boundary points of the half-circle. The difference between the cases $\epsilon \to +0$ and $\epsilon \to -0$ is that the limit curves contain different halves of the circle $Q = 0$.

The above properties of the function $T(Q,\epsilon)$ allow us to select on each annulus $C_\epsilon^-$ a more narrow annulus $\mu_\epsilon$ defined by the inequalities $|Q| < d(\epsilon)$. As $\epsilon \to +0$ this annulus contracts to the curve $\mu_0 : H = 0,\ K = -k_*,\ Q = 0$, which is the trace of the unstable manifold of the saddle singular point $p$ of the field $Y$. The annulus $C_\epsilon^-$ is foliated by the curves $\zeta(\tau_0,\epsilon)$, which are the graphs of the functions $\tau_0 + T(Q,\epsilon) (\operatorname{mod} 2\pi)$ (see Figure 7.3). Such a foliation will be called a $\zeta$-foliation. The coordinates on $C_\epsilon^-$ are $(Q, r\,(\operatorname{mod} 2\pi))$. Thus, as $\epsilon \to 0$, we have

$$T(d(\epsilon),\epsilon) - T(-d(\epsilon),\epsilon) \to \pi. \tag{7.4}$$

Therefore, the points $(d(\epsilon), T(d(\epsilon),\epsilon))$ and $(d(\epsilon), \pi + T(-d(\epsilon),\epsilon))$ in $\mu_\epsilon$ can be connected by the segment $n_0(\epsilon)$ which is transversal to the curves $\zeta(\tau_0,\epsilon)$, since, as $\epsilon \to 0$, the limit position of the tangent line at the point $Q = 0$ is $\tau = \text{const}$, and the tangent line to the graph of the function $T(Q,\epsilon)$ approaches the direction $Q = 0$. The foliation of the annulus $C_\epsilon^-$ by the curves $\zeta(\tau_0,\epsilon)$ is invariant with respect to the shift $\tau_0 \to \tau_0 + C (\operatorname{mod} 2\pi)$. Therefore, the segments $n_c(\epsilon)$ obtained by the shift of the segment $n_0(\epsilon)$ by $c$ along the coordinate $\tau$, form a foliation on $\mu_\epsilon$. Each fiber of this foliation is transversal to the curves $\zeta(\tau_0,\epsilon)$. Such a foliation will be called an $n$-foliation.

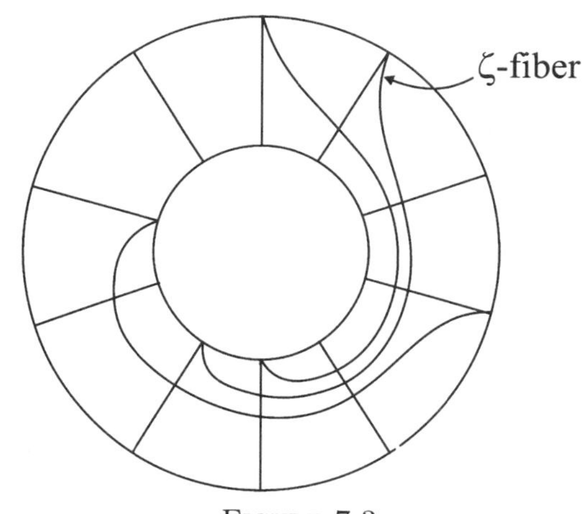


FIGURE 7.3

## 7.3. The construction of the conjugating homeomorphism in $U$

Let us construct a conjugating homeomorphism of the set $Z = \bigcup \mu_\epsilon$, $|\epsilon| \le h_*$ to the corresponding set $Z'$ of the second system $(X_{H'}, K')$. First we define a homeomorphism $q : \sigma \to \sigma'$, $q = (q_1, q_2)$ of rectangles $\sigma : |\epsilon| \le h_*$, $|k| \le k_*$ and $\sigma' : |\epsilon'| \le h'_*$, $|k'| \le k'_*$ such that it preserves the foliation $h = \text{const}$ and $q_1(0,k) = 0$. On each of the annuli $\mu_\epsilon$, $|\epsilon| \le h_*$, let us choose one of the boundary circles $l_1(\epsilon)$. For example, we may choose the circle for which $Q > 0$. Since each of the curves $\zeta(\tau_0,\epsilon)$ intersects the circle $l_1(\epsilon)$ transversally in a unique point, we may regard $\theta = \tau - T(d(\epsilon),\epsilon)$ as a parameter on $l_1(\epsilon)$. Moreover, for each point of $l_1(\epsilon)$ there is a unique segment $n_c(\epsilon)$ passing through it. Without loss of generality we may assume that if the curves $\zeta(\tau_0,\epsilon)$ and $n_c(\epsilon)$ are passing through the same point of $l_1(\epsilon)$, then $c = \theta$. This identification is possible due to the fact that the families of curves $n_c(\epsilon)$ and $\zeta(\tau_0,\epsilon)$ are invariant with respect to the shifts $c + c_0$ and $\tau + \tau_0$.

REMARK 7.1. Notice the following property of the foliations in $\mu_\epsilon$. Let us consider the curves $\zeta(\tau_0, \epsilon)$ and $\zeta(\tau_0 + \pi, \epsilon)$. Denote by $x$ the intersection point of the curve $\zeta(\tau_0, \epsilon)$ with $l_1(\epsilon)$ and by $y$ the intersection point of the same curve with the other boundary circle $l_2(\epsilon)$ of the annulus $\mu_\epsilon$. Consider the interval $n_\theta(\epsilon)$ passing through the point $x$, and let $y' = n_\theta(\epsilon) \cap l_2(\epsilon)$. Also, consider the interval $n_{\theta'}(\epsilon)$ passing through the point $y$, and let $x' = n_{\theta'}(\epsilon) \cap l_1(\epsilon)$. It follows from the construction of the foliations that the curve $\zeta(\tau_0 + \pi, \epsilon)$ connects the points $x'$ and $y'$, i.e., $\theta' = \theta + \pi$.

REMARK 7.2. Fix a value of $\theta$. Then there is an interval $n_\theta(\epsilon)$ defined for each value of $\epsilon > 0$ in $\mu_\epsilon$. The union over $\epsilon > 0$ of all these intervals forms a curvilinear triangle which has a vertex at some point of the curve $\mu_0$. Indeed, the intersection points of the interval $n_\theta(\epsilon)$ with $l_1(\epsilon)$ and $l_2(\epsilon)$ are the points with the angular coordinates $\tau$ equal to $\tau_1 = \theta + T(d(\epsilon), \epsilon)$ and $\tau_2 = \theta + T(-d(\epsilon), \epsilon)$, respectively (see the construction of the $n$-foliation). According to (7.4), the values of $\tau_1$ and $\tau_2$ converge to the same limit as $\epsilon \to +0$. The analogous triangle exists for $\epsilon \leq 0$ (for a fixed $\theta$).

Consider the arc on $l_1(\epsilon)$ corresponding to the coordinate $0 \leq \theta \leq \pi$. Also, consider the arc on $l_1'(\epsilon')$ corresponding to $0 \leq \theta \leq \pi$, where $l_1'(\epsilon')$ is the boundary curve with $Q' > 0$ for the annulus $\mu_{\epsilon'}$, $\epsilon' = q_1(\epsilon, -k_*)$ corresponding to the second system. Define a homeomorphism of these arcs which preserves the points 0 and $\pi$. We can extend this homeomorphism to a homeomorphism $g_\theta : l_1(\epsilon) \to l_1'(\epsilon')$ so that the points of $l_1(\epsilon)$ with coordinates $\theta$ and $\theta + \pi$ are mapped into the points $\theta'$ and $\theta' + \pi$, respectively.

Now, let $x \in Z$, $x \notin \mu_0$ be an arbitrary point. It determines the value $\epsilon = H(x)$ and the points $y$ and $y_1$ of intersection of the curves $\zeta(\tau_0, \epsilon)$ and $n_\theta(\epsilon)$ with $l_1(\epsilon)$. Let $y'$ and $y_1'$ be the images of these points under the map $g_\theta$. The leaf of the $\zeta'$-foliation passing through the point $y'$ and the leaf of the $n'$-foliation passing through the point $y_1'$ either have a single intersection point, or do not intersect at all. But then, according to Remark 7.1, this leaf of the $n'$-foliation intersects the leaf of the $\zeta'$-foliation passing through the point $y'$ obtained by the shift by $\pi$ along $\theta'$ at a single point $x'$ (see Figure 7.3). Thus, we obtain a map $g_Z : Z \setminus \mu_0 \to Z' \setminus \mu_0'$, $g_Z(x) = x'$. It follows from Remark 7.2 and the transversality of the $n$-foliation and the $\zeta$-foliation that this map can be uniquely extended to $\mu_0$. Therefore, we obtain a homeomorphism on $Z$.

Let us extend the homeomorphism $g_Z$ to the set $D = U \cap \{K = -k_*\} = \bigcup C_\epsilon^-$. To do that we first recall that $D \setminus Z$ is foliated by the annuli $H = \epsilon$ (to be more precise, for a fixed $\epsilon$, there are two disjoint annuli $Q > 0$ and $Q < 0$). Each of the annuli is foliated by the circles $Q = \text{const}$ and the curves that are the graphs of the functions $\tau_0 + T(Q, \epsilon)$ and are transversal to the circles. These curves are continuous in $\epsilon$. Therefore, there exist continuous coordinates $(\epsilon, Q, \tau_0)$ in $D \setminus Z$. Consider the rectangle $|H| \leq h_*$, $|Q| \leq Q_0$, and the set $|Q| \leq d(H)$, on the $(H, Q)$ plane. For $Q \geq 0$ the additional set $S$ defined by $d(H) < Q < Q_0$ is foliated by the intervals $H = \text{const}$. A similar set $S'$ exists for the second system as well. Notice that on the part of the boundary of the set $S$ that is the graph of the function $Q = d(H)$, the map $S \to S'$ is already defined by $\epsilon : \epsilon' = q_1(\epsilon, -k_*)$. We define the map $S \to S'$ so that it be a continuation of the map of the boundary $Q = d(H)$ and preserve the foliation by the intervals $H = \text{const}$. Now the map in $D \setminus Z$, $Q \geq 0$, is determined coordinatewise. Namely, once we have chosen a point $x \in D \setminus Z$ we can determine the values of $\epsilon = H(x)$, $Q = Q(x)$ and obtain the curve $\zeta(\tau_0, \epsilon)$ in the annulus $H = \epsilon$, $Q > 0$. The curve $\zeta(\tau_0, \epsilon)$ intersects the circle $l_1(\epsilon)$ in a single point $\theta_0$. Using the above-constructed map $S \to S'$ we determine $\epsilon'$, $Q'_0$. The homeomorphism $g_Z$ maps the point $\theta_0 \in l_1(\epsilon)$ to the point $\theta'_0 \in l'_1(\epsilon')$. Therefore, we have defined a curve $\zeta'(\tau'_0, \epsilon')$ on the annulus $H' = \epsilon'$, $Q > 0$ in $D' \setminus Z'$. This curve passes through the point $\theta'_0$, and there is a point on this curve with the given $Q'_0$. The correspondence $x \to x'$ is the desired map $D \setminus Z \to D' \setminus Z'$. A similar construction can be performed for $Q \leq 0$.

Hence, we have constructed a homeomorphism $g$ of the set $\Sigma^- = \{|H| \leq h_*$, $K = -k_*$, $|Q| \leq Q_0\}$ to the corresponding set $'\Sigma^-$ of the second system. Let us extend the homeomorphism $g$ to a homeomorphism $g_U$ of the whole neighborhood of the point $p$. In order to do that we first recall that any trajectory of the field $Y$ that does not belong to the stable manifold of the point $p$ and begins at some point of the set $\Sigma^+ = \{|H| \leq h_*, K = k_*, |Q| \leq Q_0\}$ intersects the set $\Sigma^-$. Taking a point $x$ that belongs to such a trajectory $L(x)$ we obtain the values $H(x)$, $Q(x)$ and $K(x)$. Let $y$ be the point of intersection of the trajectory $L(x)$ with $\Sigma^-$. The values of $H$ and $Q$ at the point $y$ are the same as at the point $x$. Let $g(y) = y'$ be the image of the point $y$ on $'\Sigma^-$. The functions $H'$ and $Q'$ have certain values at the point $y'$. There is a unique point $x'$ on the trajectory $L'(y')$ for which $K'(x') = q_2(H(x), K(x))$. Thus, we have constructed the homeomorphism $g_U$ everywhere in $U$, except for the points of the stable manifold of the point $p$. To finish the construction we will first show that $g_U$ can be extended to a homeomorphism $\Sigma^+ \to {}'\Sigma^+$.

At all points, except for the points of the trace $\mu_0^+$ of the stable manifold of the point $p$, this statement follows from the theorem on the continuous dependence on the initial conditions. Let $x \in \mu_0^+$. Then $H(x) = 0$, and we can determine the value $\tau = \tau(x)$. The interval $\tau = \tau(x)$ is mapped along the orbits of the vector field $Y$ into two curves in $\Sigma^-$ belonging to the annulus $C_0^-$. Moreover, as $Q \to \pm 0$, their boundary points are shifted by $\pi$ along the coordinate $\tau$. The images under the map $g$ of these curves are the curves lying inside the annulus $H' = 0$, and their boundary points (at $Q = 0$) are also shifted by $\pi$ along the coordinate $\tau'$. Therefore, the preimages along the vector field $Y$ of the two constructed curves are the two curves $'\Sigma^+$ which belong to $H' = 0$ and have a common boundary point $x' \in C_0^+$. The point $x'$ is assigned to correspond to the point $x$. It follows from the construction that such a correspondence extends the homeomorphism continuously from $\Sigma^+ \setminus \mu_0^+$ to the whole $\Sigma^+$. Now this homeomorphism can be extended along the coordinate $K$ to the stable manifold. Therefore, we have constructed the homeomorphism $g_U$ from a neighborhood of the point $p$ to a neighborhood of the point $p'$, which establishes the isoenergetic equivalence of the vector fields $(X_H, K)|_U$ and $(X_{H'}, K')|'_U$.

## 7.4. Proof of the isoenergetic equivalence theorem

THEOREM 7.4.1. *There exist extended neighborhoods $V$ and $V'$ of the points $p$ and $p'$, respectively, such that IHVFs $(X_H, K)$ and $(X_{H'}, K')$ are isoenergetically equivalent in $V$ and $V'$ if and only if they correspond to the same Case 1–3 in Section 1.*

Since for different cases the separatrix sets of garlands contain different types of SPTs and the isoenergetic isomorphism preserves the SPT types, the "only if" part of the theorem is obvious.

The proof of the "if" part is conducted for each of the three cases separately.

*Case* 1. The intersection of the separatrix set $W$ of the garland with the neighborhood $U$ consists of two two-dimensional disks (the local stable and unstable manifolds of the singular point $p$ of the system $X_H$) which intersect transversally at the point $p$. Outside $U$, the set $W$ is a two-dimensional cylinder. The boundary circles of this cylinder are determined by the equations $Q = \pm Q_0$, $H = 0$, $K = 0$. Therefore, all the components of the common level sets $H = \epsilon$, $K = k$, which intersect the sets $Q = \pm Q_0$ and lie outside $U$, are also cylinders for $|\epsilon|$ and $|k|$ small enough.

The extended neighborhood $V$ is obtained from $U$ by gluing a four-dimensional handle $R$ along two sets $Q = \pm Q_0$, which are diffeomorphic to $S^1 \times I_K \times I_H$, $I_K = [-k_*, k_*]$, $I_H = [-h_*, h_*]$. Here $R$ is diffeomorphic to $S^1 \times I \times I_K \times I_H$, where $S^1 \times I$ is the cylinder $W \setminus U$, and $R$ is glued along the two "soles" $S^1 \times \partial I \times I_K \times I_H$.

It is obvious from the construction of $V$ that $\rho(V) = \rho(U) = \sigma$, where $\rho$ is the moment map, and the bifurcation diagram consists of a single point $(0,0)$ since $V$ does not contain one-dimensional action orbits (see Figure 6.4 a)).

Recall that the level set $H = \epsilon$, $K = k_*$ is also a cylinder inside $U$. Thus, we have proved the following result.

PROPOSITION 7.4.1. *The set* $\partial V_\epsilon^+ = \{x \in V \mid H(x) = \epsilon, K = k_*\}$ *is a Liouville torus and the set* $\partial V^+ = \{x \in V \mid K = k_*\}$ *is diffeomorphic to* $T^2 \times [-h_*, h_*]$. *The similar statement is true for* $\partial V^-$.

As follows from the construction, the set $\partial V^+$ is divided into two parts by the submanifolds $Q = \pm Q_0$. One of them is $\Sigma_1^+ = \partial V^+ \cap \overline{U}$, the other is $\Sigma_2^+ = \partial V^+ \cap R$, and they both are homeomorphic to the direct product of an annulus and $I_H$. There are similar sets $'\Sigma_1^+$ and $'\Sigma_2^+$ for the second system $(X_{H'}, K')$.

Clearly, since $\partial V^+$ and $\partial' V^+$ are diffeomorphic to $T^2 \times I_H$ and there is a homeomorphism $g_U : \Sigma_1^+ \to {}'\Sigma_1^+$ preserving the leaves $H = \text{const}$, this homeomorphism can be extended to a homeomorphism $f^+ : \partial V^+ \to \partial' V^+$, preserving leaves $H = \text{const}$. The vector field $Y$ does not have singular points outside of $U$. Therefore, all the trajectories of $Y$ pass from $\Sigma_2^+$ to $\partial V^-$ in a bounded time. Hence, the homeomorphism on $V \setminus U$ can be extended along trajectories of the vector field $Y$.

*Case* 2. Let us construct the extended neighborhood $V$ of the point $p$ that consists of three parts:

a) a local part $U$, which is a neighborhood of the point $p$;

b) a neighborhood $U_R$ of the cylinder $R$ formed by the hyperbolic SPTs which belong to the level sets $H = \epsilon$, $|\epsilon| \leq h_*$;

c) a neighborhood of the rest of the separatrix set.

The local part $U$ is described above. Its boundaries are lateral sets defined by the equations $Q = \pm Q_0$ (the upper and the lower boundaries are defined by the equations $K = \pm k_*$).

Now let us pass to the set $U_R$. To describe its structure, let us first consider the cylinder $R$ for the values $H = \epsilon$, $|\epsilon| \leq h_*$. Denote by $W^s_{\text{loc}}$ and $W^u_{\text{loc}}$ local stable and unstable manifolds of the cylinder $R$ with respect to the vector field $X_H$. They both are diffeomorphic to $S \times I_H$, where $S$ is a circle. The manifold $W^s_{\text{loc}}$ is foliated by the level sets $M_\epsilon = \{H = \epsilon\}$ into the stable manifolds of the SPTs $R_\epsilon \subset M_\epsilon$. We denote $W^s_{\text{loc}}(\epsilon) = W^s_{\text{loc}} \cap M_\epsilon$. Similarly, we define $W^u_{\text{loc}}(\epsilon) = W^u_{\text{loc}} \cap M_\epsilon$.

On each $W^s_{\text{loc}}(\epsilon)$, $|\epsilon| \leq h_*$, we consider two closed curves $n_1^s(\epsilon)$ and $n_2^s(\epsilon)$ diffeotopic to the SPT $R_\epsilon$, lying on different sides of $R_\epsilon$ in $W^s_{\text{loc}}(\epsilon)$, and not intersecting $R_\epsilon$. These curves can be chosen so that they be transversal to trajectories of the field $X_H$ on $W^s_{\text{loc}}(\epsilon)$, and their union over $\epsilon$ form two smooth annuli. Denote these annuli by $N^s_1$ and $N^s_2$. Then $n^s_i = N^s_i \cap M_\epsilon$. Similarly we obtain the curves $n^u_i$ and annuli $N^u_i$, $i = 1, 2$. Let $\hat{U}_R$ be a neighborhood of the set $R$ that does not contain other SPTs of the vector field $X_H$, and $\hat{U}_R \cap U = \varnothing$. Choose a Riemannian metric in $\hat{U}_R$. Then we can consider the field $Y$. Consider the trajectories of the field $Y$ that pass through the points of curves $n^s_i(\epsilon)$, $n^u_i(\epsilon)$, $i = 1, 2$ (here and below when constructing $U_R(\epsilon)$ viewing Figure 7.4 is useful). The curves $n^s_i(\epsilon)$, $n^u_i(\epsilon)$ belong to the level set $M_\epsilon \cap \{K = K(R_\epsilon)\}$ and do not intersect $R_\epsilon$, which is the unique set of singular points of the field $Y$ on $M_\epsilon \cap \hat{U}_R$. Since the function $K$ varies monotonically along trajectories of the vector field $Y$, the trajectories passing through the points of $n^s_i(\epsilon)$ and $n^u_i(\epsilon)$, $i = 1, 2$, cannot tend to singular points in $R_\epsilon$. Thus, these trajectories reach the level sets $K = \pm k_*$. The four annuli $\mu^s_i(\epsilon)$, $\mu^u_i(\epsilon)$, $i = 1, 2$, formed by the pieces of the trajectories of the vector field $Y$ between $K = k_*$ and $K = -k_*$, together with the boundary surfaces $K = k_*$, $H = \epsilon$ and $K = -k_*$, $H = \epsilon$, cut out from $\hat{U}_R(\epsilon)$ the neighborhood $U_R(\epsilon)$ of $R_\epsilon$. The union of $U_R(\epsilon)$ over $\epsilon$, $|\epsilon| \le h_*$, forms a neighborhood $U_R$ of the cylinder $R$. Notice that the trajectories of the vector field $X_H$ on $M_\epsilon$ are transversal to the annuli $\mu^s_i(\epsilon)$, $\mu^u_i(\epsilon)$. The unions of the annuli $\mu^s_i(\epsilon)$ and $\mu^u_i(\epsilon)$ over $\epsilon$ form three-dimensional sets $\mathcal{M}^s_i$ and $\mathcal{M}^u_i$, which are parts of the boundary $\partial U_R$.

The intersection of $U_R(\epsilon)$ with $K = k_*$ forms two annuli $\nu^+_1(\epsilon)$, $\nu^+_2(\epsilon)$, which are neighborhoods of certain closed curves, namely the traces of the stable manifolds of the CSP $R_\epsilon$ of the vector field $Y$ (since $R_\epsilon$ is an orientable CSP, there are two such curves) (see Figure 7.4).

Outside the set $U \cup U_R$, the separatrix set of the garland consists of three cylinders. Two of them join $U$ and $U_R$, and one joins $U_R$ to itself. Therefore, for $h_*$, $k_*$ small enough, the level sets $H = \epsilon$, $K = k$, $|\epsilon| \le h_*$, $|k| \le k_*$, are cylinders outside $U \cup U_R$. The union of those cylinders (over $k$ and $\epsilon$) forms three connected sets diffeomorphic to $S^1 \times I \times I_H \times I_K$, which are called the the handles $G_i$, $i = 1, 2, 3$. Two of them, $G_1$ and $G_2$, are glued to $U$ along the sets $Q = \pm Q_0$ and to $U_R$ along the sets $\mathcal{M}^s_1$ and $\mathcal{M}^u_1$, respectively. Assume, for example, that $G_1$ is glued along $\mathcal{M}^s_1$ and contains a part of the global unstable manifold of the point $p$. Similarly, $G_2$ is glued to $U_R$ along $\mathcal{M}^u_1$ and contains a part of the global stable manifold of the point $p$. Then, the handle $G_3$ is glued along $\mathcal{M}^s_2$ and $\mathcal{M}^u_2$. Hence, we have constructed the neighborhood $V$. It follows from the construction that $\sigma = \rho(V) = \rho(U) = \rho(U_R)$, where $\rho$ is the moment map, and that the bifurcation diagram at $\sigma$ consists of the point $(0, 0)$ and a smooth curve $\rho(R)$, which is the graph of the function $k = \varphi(h)$, $\varphi(0) = 0$. Below we denote by $\partial V^+$ and $\partial V^-$ the sets $V \cap \{K = k_*\}$ and $V \cap \{K = -k_*\}$, respectively.

The trajectories of the vector field $X_H$ that start on $N^u_2$ reach $N^s_2$. In particular, the annulus between $R_0$ and $n^s_2(0)$ on $W^s_{\text{loc}}(0)$, and the annulus between $R_0$ and $n^u_2(0)$ on $W^s_{\text{loc}}(\epsilon)$, which are glued together along $R_0$, together with the annulus formed by the segments of trajectories of the vector field $X_H$ between $n^u_2(0)$ and $n^s_2(0)$, compose a torus $T_0$. The torus $T_0$ is smooth everywhere except for the points of $R_0$, and intersects the annuli $\mu^s_2(0)$, $\mu^u_2(0)$ transversally along the curves $n^s_2(0)$ and $n^u_2(0)$, respectively.

Now we consider the trajectories of the vector field $X_H$ passing through the points of the annulus $\mu^u_2(0)$. Due to the theorem on the continuous dependence on the initial conditions, these trajectories reach $\mu^s_2(0)$. It follows from the orientability

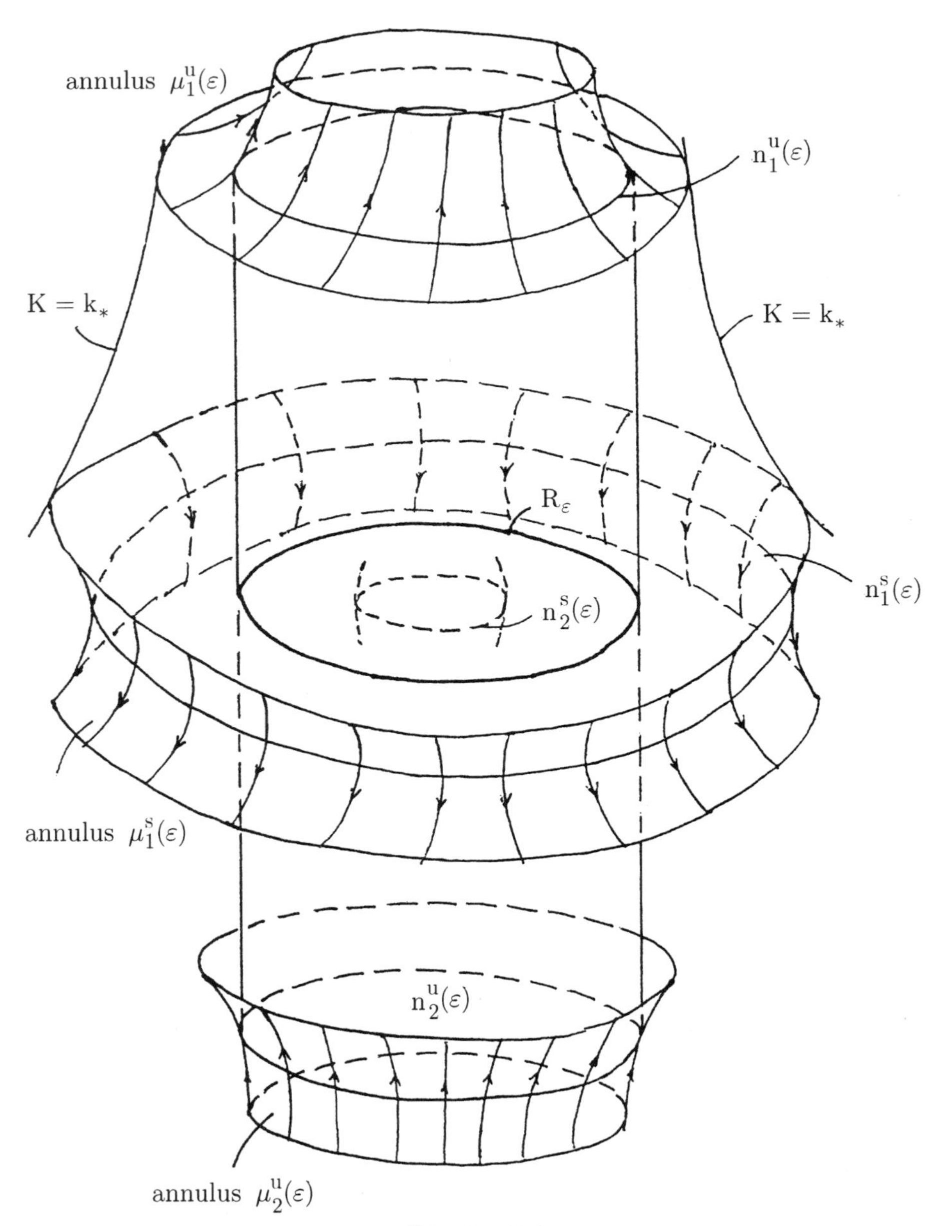


FIGURE 7.4

of the SPT $R_0$ that either for $k > 0$ or for $k < 0$, there is a torus $T_k$ close to $T_0$ on the level set $K = k$, for $k$ small enough. The tori $T_k$ do not intersect the neighborhood $U$. Let us assume that the tori $T_k$ correspond to $k > 0$. It is easy to see that the torus $T_{k_*} \subset \partial V_0^+$ contains one annulus, for example $\nu_2^+(0)$, from the boundary $\partial U_R(0)$ on which the trace of one half of the stable manifold of the CSP $R_0$ of the vector field $Y$ lies. The other half of this manifold belongs to the other annulus, $\nu_1^+(0)$, from the boundary $\partial U_R(0)$. This annulus adjoins the annuli $\mu_1^u(0)$ and $\mu_1^s(0)$ on $\partial U_R(0)$ (see Figure 7.4).

As follows from the proof of Proposition 7.4.1, the trajectories of the vector field $X_H$ passing through the curve $n_1^s \subset \mu_1^s(0)$ belong to the unstable manifold of the point $p$, and the trajectories of the vector field $X_H$ passing through the curve $n_1^u \subset \mu_1^u(0)$ belong to the stable manifold of the point $p$. Therefore, the trajectories of the vector field $X_H$ that start on $\mu_1^u(0)$ go into the neighborhood $U$ as $t$ increases, and they reach the boundary of $U$ where $H = 0$, $Q = Q_0$, or $Q = -Q_0$. Similarly, the trajectories that start on $\mu_1^s(0)$ go into the neighborhood $U$ as $t$ decreases, and they reach the second annulus $H = 0$, $Q = \pm Q_0$. The time for passing from $\partial U_R(0)$ to $\partial U$ is uniformly bounded. We assume, for example, that as time increases, the trajectories of the vector field $X_H$ inside $U$ go from $Q = -Q_0$ to $Q = Q_0$. Let us fix $K = k_*$ and consider the annulus $\nu_1^+(0)$ on $\partial U_R(0)$ that contains the trace of the half of the stable manifold of the CSP $R_0$. Denote the boundaries of this annulus by $l_1$ and $l_2$. Let $l_1 \subset \mu_1^u(0)$ and $l_2 \subset \mu_1^s(0)$. Consider the trajectories of the vector field $X_H$ that start at $l_1$. When the time increases they reach the annulus $H = 0$, $K = k_*$, which is a part of the boundary of $U$, and intersect it in the circle $Q = -Q_0$. When $t$ increases further, these trajectories, remaining on the annulus $H = 0$, $K = k_*$, intersect the circle $Q = Q_0$ and reach $l_2$. Hence, we obtain a Liouville torus $T_{k_*}^2$ different from $T_{k_*}$. Clearly, the level set $H = 0$, $K = k_*$ consists only of the tori $T_{k_*}^2$, $T_{k_*}$. We call the torus $T_{k_*}$ short, and the torus $T_{k_*}^2$ long. The same arguments show that the level set $H = 0$, $K = -k_*$ consists of a single torus, which is intersected by two smooth homotopic disjoint curves, the traces of the unstable manifold of the CSP $R_0$ of the vector field $Y$. Since the level set $K = k_*$ does not intersect the CSP we have the following result.

PROPOSITION 7.4.2. *In Case 2 one of the sets $\partial V^+$, $\partial V^-$ is diffeomorphic to $T^2 \times I$ and the other to $2(T^2 \times I)$.*

Let us pass to the proof of the isoenergetic equivalence theorem. Without loss of generality we may assume that $\partial V_\epsilon^+$ consists of two tori and $\partial V_\epsilon^-$ consists of one torus. First we construct the homeomorphism in each of the three sets $U$, $U_R$, $\hat{V} = V \setminus (U \cup U_R)$ separately. After that we formulate the compatibility conditions for those homeomorphisms. In $U$ the homeomorphism is already constructed. Let us construct the homeomorphism in $U_R$. First we do it without any agreement with the homeomorphism in $U$. Denote by $\nu_1^+(\epsilon)$ the annulus on $\partial V_\epsilon^+ \cap U_R(\epsilon)$ which contains the trace $W_1^s(\epsilon)$ of the stable manifold of the CSP $R_\epsilon$, and whose boundary circles belong to $\mu_1^s(\epsilon)$, $\mu_1^u(\epsilon)$. Respectively, $\nu_2^+(\epsilon)$ is the second annulus on $\partial V_\epsilon^+ \cap U_R(\epsilon)$, and it contains the second component $W_2^s(\epsilon)$ of the trace of the stable manifold of the CSP $R_\epsilon$ and its boundary circles belong to $\mu_2^s(\epsilon)$, $\mu_2^u(\epsilon)$ (see Figure 7.4).

Denote $\nu_1^+ = \bigcup \nu_1^+(\epsilon)$, $\nu_2^+ = \bigcup \nu_2^+(\epsilon)$, $|\epsilon| \leq h_*$. Similarly we construct the annuli $\nu_1^-(\epsilon)$, $\nu_2^-(\epsilon)$ on $\partial V_\epsilon^-$ and the sets $\nu_1^-$, $\nu_2^-$. Each of $\nu_1^-$, $\nu_2^-$ is diffeomorphic to $S^1 \times I \times I_H$, where $I$ is the interval $[-1, 1]$. Introduce the cyclic coordinate $\theta \in [0, 2\pi] (\operatorname{mod} 2\pi)$ on $S^1$ and the coordinate $Q$ on $I$. Thus, we obtain coordinates on $\nu_1^+$. Moreover, we will assume that $Q = 0$ is the union over $\epsilon$ of the traces $W_1^s(\epsilon)$ of the stable manifolds of the CSP $R_\epsilon$. Similarly, we construct the coordinates $(\theta_1, Q_1, \epsilon)$ on $\nu_2^+$. In this construction we may identify the coordinate $\theta_1$ with $\theta$, assuming that $\theta_1$ is equal to $\theta$ for those points of $W_1^s(\epsilon)$ and $W_2^s(\epsilon)$ which contain the trajectories of the vector field $Y$ that pass through the same point of $R_\epsilon$.

The stable and unstable manifolds of the cylinder $R$ are the unions over $\epsilon$ of the stable and unstable manifolds of the CSP $R_\epsilon$ with respect to the vector field $Y$. They divide $U_R$ into four closed subsets $U_R^i$, $i = 1, \dots, 4$. We introduce continuous coordinates $(Q, \theta, \epsilon, k)$ in $U_R^1$ (similarly in other sets). At any point $x \in U_R^1$ the values of $\epsilon$ and $k$ are simply $\epsilon = H(x)$ and $k = K(x)$. Let us assume, for definiteness, that $U_R^1$ intersects $\nu_1^+$ and $\nu_1^-$. To extend the coordinates $Q$, $\theta$ we use the Shoshitaishvili theorem [**62**]. It implies that in some neighborhood of the cylinder $R$ there exists a homeomorphism that maps the trajectories of the vector field $Y$ into the trajectories of the vector field

$$\theta = 0, \quad \dot{\epsilon} = 0, \quad \dot{u} = -a(\theta, \epsilon)u, \quad \dot{v} = -b(\theta, \epsilon)v, \tag{7.5}$$

where the functions $a(\theta, \epsilon)$ and $b(\theta, \epsilon)$ are continuous, positive, and $2\pi$-periodic with respect to $\theta$. It follows from (7.5) that, for a fixed $\epsilon$, the image on $\partial V^-$ of the curve $\theta = \theta_0$ on $\nu_1^+(\epsilon)$ under the mapping along trajectories of the vector field $Y$ splits into two parts. One of these parts belongs to $\nu_1^-(\epsilon)$, and the other belongs to $\nu_2^-(\epsilon)$. The part in $\nu_1^-(\epsilon)$ is everywhere smooth and has a limit point on $W_1^u(\epsilon)$ that satisfies the following condition: the $\alpha$-limit point (i.e., as $t \to -\infty$) of the trajectory of the vector field $Y$ passing through this point is the point of $R_\epsilon$ which is the $\omega$-limit point (i.e., as $t \to \infty$) for the trajectory passing through the point $\theta = \theta_0$ on $W_1^s(\epsilon)$. The $\theta$-foliation on $\nu_1^+$ is mapped into a continuous foliation of $\nu_1^- \cap U_R^1$. It follows from the same theorem that the $Q$-foliation on $\nu_1^+$, $Q > 0$ (we assume, for definteness, that the value of $Q$ is positive in $\nu_1^+ \cap U_R^1$) is mapped along the trajectories of the vector field $Y$ into a foliation of $\nu_1^-$ by smooth curves. Moreover, for a fixed $\epsilon$ and $Q \to +0$, the corresponding curve on $\nu_1^-$ tends to $W_1^u(\epsilon)$ so that $W_1^u(\epsilon)$ is its topological limit. Also, for a fixed $\epsilon$, the images of the curves $\theta = \theta_0$ and $Q = q$ on $\nu_1^-$ intersect transversally at a single point (the map along the trajectories of the vector field $Y$ is smooth outside $W_1^s(\epsilon)$).

Thus, we have constructed continuous coordinates in $U_R^1$. Namely, for an arbitrary point $x \in U_R^1$ that does not belong to the unstable manifold of the CSP $R$, we have $\epsilon = H(x)$, $k = K(x)$, and $\theta$ and $Q$ are determined by their values at the intersection point of the trajectory of the vector field $Y$ passing through $x$ and $\nu_1^+(\epsilon)$. For $x \in R$, we set $Q = 0$, and $\theta$ is determined by the intersection point of $\nu_1^+$ and the trajectory for which $x$ is the $\omega$-limit point. For a point $x$ from the unstable manifold, we set $Q = 0$, and $\theta$ is determined by the $\alpha$-limit point on $R$ of the trajectory passing through $x$. The above-constructed coordinates are smooth everywhere except for the points of the unstable manifold of the cylinder $R$.

REMARK 7.3. Below, for definteness, we denote by $U_R^2$ the part of $U_R$ that intersects both $\nu_1^+$ and $\nu_2^-$. We extend the coordinates $Q$, $\theta$ from $\nu_1^+$ so that the value of $Q$ satisfy the inequality $-1 \le Q \le 0$ in $U_R^2$. Also, by $U_R^3$ we denote the part of $U_R$ that intersects both $\nu_2^-$ and $\nu_2^+$, and by $U_R^4$ the remaining part of $U_R$. We assume that $Q_1 \ge 0$ on $U_R^3$, and that $Q_1 \le 0$ on $U_R^4$.

Constructing the homeomorphism $g_R : U_R \to U_{R'}$, we assume that $U_{R'}$ is also decomposed into $U_{R'}^i$, $i = 1, \dots, 4$, and that there are coordinate systems constructed in each $U_{R'}^i$, $i = 1, \dots, 4$. Let us define an arbitrary correspondence between $U_R^1$ and $U_{R'}^i$, and assume that $i = 1$. Then the correspondence of the other parts is uniquely determined. Namely, the boundary of $U_R^1$ contains $\nu_1^+$. There is another $U_R^i$, $i \ne 1$, adjacent to it. Similarly, for $U_{R'}^1$ there is a unique $U_{R'}^j$, $j \ne 1$, that is adjacent to ${}'\nu_1^+$. We set a correspondence between $U_R^i$ and $U_{R'}^j$, and so on. We will assume that the enumeration of the corresponding parts and the signs of the corresponding functions $Q$ and $Q_1$ in $U_R$ and $U_{R'}$ are the same.

The homeomorphism $g_R^1 : U_R^1 \to U_{R'}^1$ is constructed coordinatewise. Namely, we define an arbitrary homeomorphism $g_\theta$ of the circle $[0, 2\pi]\ (\mathrm{mod}\, 2\pi)$, and a homeomorphism $g_Q : [0,1] \to [0,1]$, $g_Q(0) = 0$ of the interval. We obtain two more homeomorphisms defining a homeomorphism $q : \sigma \to \sigma'$ on the moment plane that satisfies the properties $q = (q_1, q_2)$, $q_1(\epsilon, k) = q_1(\epsilon) = \epsilon'$, $q_2(\epsilon, k_*) = k'_*$, $q_2(\epsilon, k_0(\epsilon)) = {}'\varphi(\epsilon')$, where $(\epsilon, \varphi(\epsilon))$ $((\epsilon', {}'\varphi(\epsilon'))$, respectively) is the image of the cylinder $R$ $(R')$ under the moment map. The homeomorphisms $g_R^i$, $i = 2, 3, 4$, are constructed in the same way. Notice that in those parts where $Q$ (or $Q_1$) is negative the homeomorphism $g_Q$ is replaced by an arbitrary homeomorphism of the interval $[-1, 0]$ that preserves zero. It remains to verify the compatibility of constructed homeomorphisms on common boundaries. For definiteness, let $x \in U_R^1 \cap U_R^2$. Then $x$ belongs to the stable manifold of the cylinder $R$. The point $x$ has coordinates $(0, \theta, \epsilon, k)$ in $U_R^1$. According to the construction, it has the same coordinates in $U_R^2$. By the definitions of $g_R^1$ and $g_R^2$, the image of $x \in U_R^1$ (and $x \in U_R^2$) is the point $x' \in U_{R'}^1$ (and $x'' \in U_{R'}^2$) with the same coordinates $(0, g_\theta(\theta), q_1(\epsilon), q_2(\epsilon, k))$. Due to the way the coordinates were introduced, $x' = x''$ in $U_{R'}$. Compatibility conditions for the other parts are verified in the same way.

Now we construct a homeomorphism $g : V \to V'$ that coincides with $g_U$ on $U$. Also, $g$ will be compatible with $g_R$. Recall that for $g_R$ the homeomorphisms $g_\theta$, $g_Q$ and the correspondence between the parts $U_R^i$ and $U_{R'}^j$ were chosen arbitrarily. Now we use this freedom of choice. According to our assumptions the level set $\partial V_\epsilon^+$ consists of two Liouville tori $T_1^+(\epsilon)$ (long) and $T_2^+(\epsilon)$ (short), and $\partial V_\epsilon^-$ consists of a single torus $T^-(\epsilon)$. Extend the foliation $Q = \mathrm{const}$ from $U$ to $T_1^+(\epsilon)$ so that it coincide with the $Q$-foliation in $T_1^+(\epsilon) \cap U_R$. The leaves of the extended $Q$-foliation are closed curves. Also, we construct another foliation, called the $\theta$-foliation, of $T_1^+(\epsilon)$ by closed curves, which satisfies the following conditions: its leaves intersect the leaves of the $Q$-foliation transversally and each leaf contains exactly one segment of the local foliation $\tau = \mathrm{const}$ in $U$ and $\theta = \mathrm{const}$ in $U_R$. Below we assume that there are the following coordinates on $T_1^+ = \bigcup T_1^+(\epsilon)$, $|\epsilon| \le h_*$ : the cyclic coordinates $\theta$ and $\psi$ $(\mathrm{mod}\, 2\pi)$ and $\epsilon$, $\epsilon \in [-h_*, h_*]$, where $\psi$ parametrizes the $Q$-foliation. Notice that the $\theta$-foliation coincides with the $\tau$-foliation in $U$. Therefore, when we use the homeomorphism $g_U$, $\theta$ plays the role of $\tau$. Also, we introduce the coordinates $(\tau, \psi, \epsilon)$ on $T_2^+$, extending them from $U_R$. Now let us construct the homeomorphism $T_1^+ \to {}'T_1^+$. Let $x \in T_1^+$, and suppose that it has the coordinates $(\theta, \psi, \epsilon)$. The local homeomorphism $g_U$ defines the map $g_\theta : \theta \to \theta'$. The map with respect to $\epsilon$ is also defined above, $\epsilon' = q_1(\epsilon)$. Let us define the map with respect to the coordinate $\psi$. Let $s$ be the leaf of the $Q$-foliation that corresponds to the value $\psi = \psi_s$. The map $g_U$ defines the correspondence between the leaves $Q = \pm Q_0$ that belong to the boundary of $U$. Thus, we have determined the oriented (by the $Q$ coordinate) interval $J \subset [0, 2\pi](\mathrm{mod}\, 2\pi)$ and the interval $J'$ of the second system that correspond to $U$ and $U'$. Also, we have a homeomorphism from $J$ to $J'$. The set $U_R$ contains three leaves $Q = \pm 1$, $Q = 0$, and the corresponding values $\psi_\pm$, $\psi_0$ of the coordinate $\psi$ lie outside the interval $J$. In addition, $\psi_0$ is between $\psi_+$ and $\psi_-$. Similarly, we define the values $\psi'_\pm$, $\psi'_0$ for the second system. We define a homeomorphism with respect to the coordinate $\psi$ outside $J$ so that it extend the homeomorphism from $J$ and map the set of points $\{\psi_+, \psi_-, \psi_0\}$ into the set of points $\{\psi'_+, \psi'_-, \psi'_0\}$. Denote this homeomorphism by $g_\psi^1$. The homeomorphism $g_\psi^1$ defines the correspondence between the parts $U_R^i$ from $U_R$ and $U_{R'}^i$ from $U_{R'}$.

Outside $U \cap T_1^+$ the homeomorphism $g_1 : T_1^+ \to 'T_1^+$ is defined as a direct product of the homeomorphisms $g_\theta$, $g_\psi^1$ and $q_1$, and inside $U \cap T_1^+$, $g_1$ coincides with $g_U$.

The homeomorphisms $g_\theta$, $q_1$ are already defined for $T_2^+$. To define $g_\psi^2$ we will use the already defined correspondence between the parts $U_R^i$ and $U_{R'}^i$. We obtain the correspondence between the leaves $Q = \pm 1$ and $Q = 0$ intersecting $T_2^+$ and the similar leaves of the second system. There are values $\psi_\pm$, $\psi_0$ of the coordinate $\psi$ on $T_2^+$ and the values $\psi'_\pm$, $\psi'_0$ of the coordinate $\psi'$ on $'T_2^+$ corresponding to those leaves. We define an arbitrary homeomorphism $g_\psi^2$ of the coordinate $\psi$ that maps $\psi_0$ into $\psi'_0$, and the interval between $\psi_+$ and $\psi_-$ containing $\psi_0$ into the corresponding interval for the second system. The homeomorphism $g_2 : T_2^+ \to 'T_2^+$ is the direct product of the homeomorphisms $g_\theta$, $g_\psi^2$ and $q_1$.

The homeomorphisms $g_\theta$, $g_\psi^1$, $g_\psi^2$, $q$ completely determine the homeomorphism $g_R$. Now, it is possible to extend $g_U$ and $g_R$ to a homeomorphism $g : V \to V'$. In order to do that we notice that the trajectories of the vector field $Y$ that start at $t = 0$ on $\partial V^+$ outside $U$ and $U_R$ reach $\partial V^-$ in a uniformly bounded way. Thus, we can extend the homeomorphisms $g_1$ and $g_2$ along the trajectories of the vector field $Y$.

*Case* 3. We construct the extended neighborhood $V$ of the point $p$ out of three parts: the local neighborhood $U$, the neighborhood $U_R$ of the cylinder $R$ formed by the SPTs of the vector field $X_H$ (the CSP of the vector field $Y$), and the neighborhood $\hat{V}$ of the rest of the separatrix set. This division is carried out exactly like in Cases 1 and 2. The difference with Case 2 is that for the nonorientable SPT the boundary of the set $U_R$ consists of

a) one set $\nu^+ \subset \partial V^+$;

b) one set $\nu^- \subset \partial V^-$;

c) a set that consists of two components formed by the segments of the trajectories of the vector field $Y$ joining the boundaries of $\nu^+$ and $\nu^-$.

The sets $\nu^+$, $\nu^-$ are diffeomorphic to the direct product of an annulus and $I_H$, and each of the sets of the type c) is also diffeomorphic to the direct product of the annuli $\mu^s(0)$, $\mu^u(0)$ and $I_H$. As in Case 2, the annuli $\mu^s(\epsilon)$, $\mu^u(\epsilon)$ are transversal to the trajectories of the vector field $X_H$ on $M_\epsilon$. The trace of the stable manifold of the SPT $R_0$, which coincides with the trace of the unstable manifold of the point $p$ (Proposition 7.4.1), is the median of $\mu^s(0)$. Similarly, the trace of the unstable manifold of the SPT $R_0$, which coincides with the trace of the unstable manifold of the point $p$, is the median of $\mu^u(0)$. The segments of trajectories of the vector field $X_H$ between $\mathcal{M}^u = \bigcup \mu^u(\epsilon)$, $|\epsilon| \le h_*$, and one of the sets $Q = \pm Q_0 \subset \partial U$ form the handle $G_1$. Similarly, $G_2$ connects $\mathcal{M}^s = \bigcup \mu^s(\epsilon)$ and the second of the sets $Q = \pm Q_0$. The union $U \cup U_R \cup G_1 \cup G_2$ forms the neighborhood $V$. It follows from the construction that $\sigma = \rho(V) = \rho(U) = \rho(U_R)$, where $\rho$ is, as before, the moment map, and the bifurcation diagram in $\sigma$ is the graph of the smooth curve $k = \varphi(h)$, $\varphi(0) = 0$, which coincides with $\rho(R)$.

All the trajectories of the vector field $X_{H'}$ that pass through the closed curve $\mu^u(0) \cap \nu^+(0)$ must belong to $\partial V^+$. Their extensions starting from $U_R$ reach $U$ at the points of the circle $Q = Q_0$, $H = 0$, $K = k_*$ (or $Q = -Q_0$). Then they leave $U$ at the points of the circle $Q = -Q_0$, $H = 0$, $K = k_*$ (or $Q = Q_0$, respectively), and come back to $U_R$ at the points of $\mu^s(0) \cap \nu^+(0)$. Then, moving along $\nu^+(0)$, they reach $\mu^u(0) \cap \nu^+(0)$ forming a single Liouville torus $T^+(0) = \partial V_0^+$. Similarly, $\partial V_0^- = T^-(0)$. Thus, we have the following result.

PROPOSITION 7.4.3. *In Case 3, $\partial V_0^+$ and $\partial V_0^-$ are diffeomorphic to $T^2 \times I_H$.*

Now we use the double cover $\pi : \hat{U}_R \to U_R$ and the construction from Case 2 to construct the homeomorphism $g_R : U_R \to U_{R'}$. Doing that we have to make sure that the pairs of incident trajectories of the vector field $Y$ lying in the stable manifold of the CSP $R$ and having a common limit point on $R_\epsilon$ are mapped into similar pairs of incident trajectories of the second system. The same must be true for the pairs of trajectories of the unstable manifolds. Here, as in Case 2, we have freedom in choosing correspondence between the boundaries $\nu^+$ and $'\nu^+$, and in choosing the homeomorphisms $g_\theta$ and $g_\psi$. This freedom is used to make homeomorphisms $g_R$ and $g_U$ compatible.

Since annuli $\nu^+$ and $\partial V_\epsilon^+ \cap U$ lie on the same torus, the correspondence of the boundaries of the sets $\nu^+(\epsilon)$ and $'\nu^+(\epsilon')$ is uniquely determined by $g_U$. Extending the $\theta$-foliation from $\nu^+$ to $\partial V^+ \cap U$ it is necessary to ensure the compatibility of the local $\theta$-foliations on $U_R$ and $U$. Namely, by construction, $g_U$ maps the pair of leaves $\theta = \theta_0$ and $\theta = \theta_0 + \pi$ into the pair of leaves $\theta' = \theta_0'$ and $\theta' = \theta_0' + \pi$. This holds due to the fact that the map along trajectories of the vector field $Y$ from $\partial V_0^+$ to $\partial V_0^-$ is discontinuous. In turn, for every leaf $\theta = \theta_0$, there is an incident leaf (see earlier in this section). The $\theta$-foliation on $\partial V^+$ is constructed in such a way that the intersection of a leaf with $\partial V_\epsilon^+$ is a closed curve containing one segment of $\partial V_\epsilon^+ \cap U$ and one segment of $\nu^+(\epsilon)$. Moreover, a pair of leaves that contains the segments $\theta = \theta_0$ and $\theta = \theta_0 + \pi$ must contain a pair of incident leaves from $\nu^+(\epsilon)$. Then the homeomorphism $g_\theta$ is induced by the homeomorphism $g_U$, and $\psi_s$ is defined as an arbitrary extension from $U$ and $U_R$. Then, $g$ is constructed as in Case 2. Theorem 7.4.1 is proved.

## 7.5. The equivalence of actions

Let $\Phi$ and $\Phi'$ be two Poisson actions of the Morse type with singular points $p$ and $p'$, respectively, of the focus-focus type, and let $V$ and $V'$ be their extended neighborhoods.

THEOREM 7.5.1. *The actions $\Phi$ and $\Phi'$ are topologically equivalent if and only if the garlands of the corresponding singular points are the same.*

Theorem 7.5.1 follows immediately from Theorem 7.4.1, since, obviously, it is possible to choose the generators of the group $\mathbb{R}^2$ so that the IHVF generated by them satisfy the conditions of Theorem 7.4.1. Then the homeomorphism constructed in Theorem 7.4.1 realizes the topological equivalence of the actions.

## 7.6. The topology of $V$

In this section we will describe the topological structure of the neighborhood $V$. For this we will determine the structure of $V_\epsilon$ for all $\epsilon \neq 0$, the structure of $V_0$, and the deformations of $V_\epsilon$ as $\epsilon \to \pm 0$. Let us consider each case separately.

To describe $V_0$, we introduce the following standard three-dimensional set $P_0$. Consider the set $B$ on the $(x, y)$-plane determined by the following inequalities: $|xy| \leq 1/2$, $|x^2 - y^2| \leq 1/2$. Let $\hat{P}_0$ be the bundle over $S^1$ with the fiber $B$ and the structural group $\mathbb{Z}_2$ acting on $B$ by the antipodal involution $(x, y) \to (-x, -y)$. There is a unique section $\lambda$ of the bundle $\hat{P}_0$ defined by the point $(0, 0)$ of the fiber $B$. Then $P_0$ is obtained from $\hat{P}_0$ by identifying the section $\lambda$ with a point. Denote by $\Delta P_0$ the part of the boundary of $P_0$ determined by the equation $x \cdot y = \pm 1/2$. Let $U$ be a neighborhood of the point $p$ (see Section 7.2).

PROPOSITION 7.6.1. *$U_0 = U \cap V_0$ is homeomorphic to $P_0$.*

PROOF. There are three foliations in $U_0$, $K = k$, $Q = q$, $\tau = \tau_0 \pmod{2\pi}$ (see Section 7.2). We define similar foliations in $P_0$. In order to do that consider the foliation $x \cdot y = c$ in $B$. If $c \neq 0$, then the set $x \cdot y = c$ consists of two pieces of the hyperbola. Then there is a corresponding connected fiber of the bundle $\hat{P}_0$ diffeomorphic to a cylinder. The resulting foliation plays the role of the foliation $Q = q$ in $P_0$. The foliation in $P_0$ similar to the $\tau$-foliation is obtained from the foliation $s = \text{const}$, $s \in S^1$. The foliation that plays the role of the $K$-foliation is obtained as the natural extension to $P_0$ of the foliation defined by the function $G = x^2 - y^2$ in $B$. Then the proof is carried out similarly to Section 7.3. □

In what follows, we denote by $Z$ the direct product of the annulus $C$ and the interval $I = [-1, 1]$, so that $\Delta Z = \partial C \times I$. There is a natural foliation of $Z$ by the annuli defined by the leaves $C \times \{x\}$, $x \in I$. In other words, it is defined by the function $F(c, x) = x$, $x \in I$, $c \in C$. Let us proceed with the description of the topology of $V_\epsilon$ in each of the Cases 1–3.

*Case* 1. If $\epsilon \neq 0$, then there are no SPTs in $V_\epsilon$, and $\partial V_\epsilon^+$ and $\partial V_\epsilon^-$ consist of one torus each (Proposition 7.4.2). Thus, $V_\epsilon$ is diffeomorphic to $T^2 \times I_K$. The point $p$ belongs to $V_0$, which is formed by gluing to $U_0$ a set diffeomorphic to $Z$ along the two annuli $Q = \pm Q_0$, this set (diffeomorphic to $Z$) being a neighborhood in $V_0$ of the part of the separatrix set $W$ lying outside $U_0$. Thus, $V_0$ is homeomorphic to the set obtained by gluing to $P_0$ the set $Z$, via the diffeomorphism $g : \Delta Z \to \Delta P_0$ defined by the following construction. Fix the orientation of $Z$ and $P_0$. This determines the orientation of $\Delta Z$ and $\Delta P_0$. Each of the sets $\Delta Z$ and $\Delta P_0$ consists of two components $\Delta Z^1$, $\Delta Z^2$ and $\Delta P_0^1$, $\Delta P_0^2$, respectively. Define the diffeomorphisms $g_i : \Delta Z^i \to \Delta P_0^i$, $i = 1, 2$, with the following properties.

a) The orientations on $\Delta P^i$ induced by the diffeomorphisms $g_i$ are either both the same as, or both opposite to the fixed orientations of $\Delta P_0^i$.

b) The diffeomorphisms preserve the circle foliations induced on $\Delta Z^i$ and $\Delta P_0^i$ by the functions $F$ and $G$, respectively. Moreover, the circles on $\Delta Z^1$ and $\Delta Z^2$ that correspond to the same value of $F$ are mapped to two circles on $\Delta P_0^1$ and $\Delta P_0^2$ that correspond to the same value of $G$.

Under these conditions the $F$-foliation and the $G$-foliation form a single foliation which is determined by a single function on the glued manifold. The leaves of this foliation (except for the leaf $G = 0$) are two-dimensional tori. Clearly, the sets obtained by this construction are homeomorphic for all $g_1$, $g_2$ that satisfy the conditions a)–b). In particular, it follows that $V_0$ is homeomorphic to the set $P_4$ (see Section 6.8 in Chapter 6) in which the axial circle is identified with a point.

Let us describe the deformations of the submanifolds $V_\epsilon$ as $\epsilon \to \pm 0$. As $\epsilon \to +0$ for a fixed $k \neq 0$, the torus $T^k(\epsilon) \subset V_\epsilon$, which forms the level set $H = \epsilon$, $K = k$, $|k| \leq k_*$, merges into the corresponding torus $T^k(0)$ on $V_0$. For $k = 0$, the topological limit of the tori $T_0(\epsilon)$ is the separatrix set $W$, which is obtained from $T_0(\epsilon)$ by contracting to a point the homotopically nontrivial closed curve. For simplicity, let us assume that the vector field $Y_H$ constructed in (7.1), with the roles played by $H$ and $K$ reversed, has the same properties as the vector field $Y$. Then we see that the curve that is contracted to a point as $\epsilon \to 0$, is the trace of the stable (or unstable) manifold of the point $p$ of the vector field $Y_H$ on $K = 0$.

*Case* 2. Consider the set $Z_1$ which is the direct product of a circle and the flat set $B_1$ with the corresponding foliation generated by the smooth function $F$ (see Figure 7.5). Due to the construction in Case 1, $V_0$ is homeomorphic to $Z_1$ glued with $P_0$. Since $V_\epsilon$ is obtained by gluing $U_\epsilon$ (diffeomorphic to $Z$) to $V_\epsilon \setminus U_\epsilon$ (diffeomorphic to $Z_1$), the manifolds $V_\epsilon$, $\epsilon \neq 0$, are homeomorphic to $P_1$ (see Section 6.8 in Chapter 6).

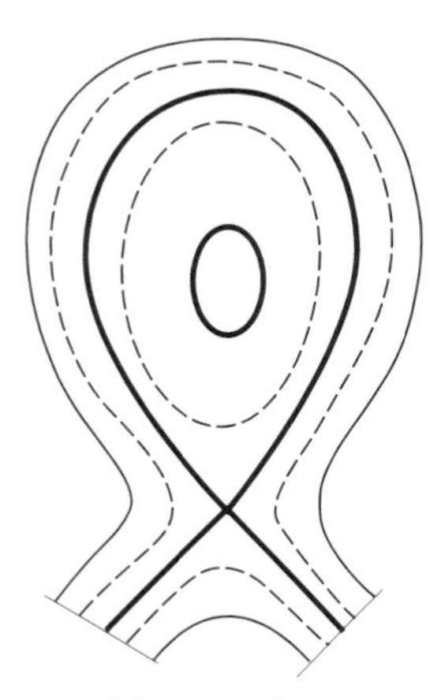

FIGURE 7.5

To describe the deformation process $V_\epsilon \to V_0$ as $\epsilon \to 0$, we consider the foliation of the set $\sigma$ on the moment plane by the smooth curves $k = \psi(h, \lambda)$. They form a foliation on $\sigma$ among whose leaves are, in particular, the bifurcation curve (for $\lambda = 0$) and segments $K = \pm k_*$. Then, moving along a single leaf $k = \psi(h, \lambda)$, $\lambda \neq 0$, we obtain a deformation in $V$ of the surface $V_{hk} = \{H = h, K = k\}$, $k = \psi(h, \lambda)$, which does not bifurcate as $\epsilon \to 0$ and deforms into the surface $V_{0k_0}$, $k_0 = \psi(0, \lambda)$, that consists of a single torus (when $k_0 < 0$), or a pair of tori (when $k_0 > 0$). When we move along the bifurcation curve we obtain a deformation of the set formed by the merging stable and unstable manifolds of the SPT $R_\epsilon$. It is homeomorphic to the direct product of a "figure eight" and a circle. The set $W$ is obtained by contracting (to a point) a closed curve that is homopotic to the axial line and does not intersect it.

*Case* 3. The set $V_0$ is constructed from $P_0$ and $\hat{P}_0$, as in Case 1. After the gluing we obtain the set $P_5$ (see Section 6.8 in Chapter 6) in which one SPT is contracted to a point.

Since there is a single nonorientable SPT on $V_\epsilon$, it is easy to see that the manifold $V_\epsilon$, $\epsilon \neq 0$, is diffeomorphic to $P_4$.

To describe the deformation $V_\epsilon \to V_0$, we foliate $\sigma$ by the smooth curves $k = \psi(h, \lambda)$ as was done in Case 2. In particular, the bifurcation curve (when $\lambda = 0$) and the intervals $K = \pm k_*$ are leaves of this foliation. Then, changing $\epsilon$ on a fixed leaf, we obtain a deformation (when $\lambda \neq 0$) of the Liouville torus $V_{\epsilon k}$ to the torus $V_{0k_0}$, $k_0 = \psi(0, \lambda)$. For $\lambda = 0$, $\epsilon \neq 0$, we obtain a separatrix surface $W_\epsilon$ of the SPT $R_\epsilon$ in $V_\epsilon$ formed by the merging stable and unstable manifolds. Topologically this is a fiber bundle of the "figure eight" $x^2 - y^2 + x^4 = 0$ on the $(x, y)$-plane over the circle, where the structural group $\mathbb{Z}_2$ acts on the "figure eight" by rotation by the angle $\pi$. As $\epsilon \to 0$, this surface is deformed into the separatrix set $W$ of the garland. Namely, a homotopically nontrivial circle that belongs to $W_\epsilon$ and does not intersect $R_\epsilon$, is contracted to a point.

CHAPTER 8

# Realization

In this chapter we construct various IHVFs such that in extended neighborhoods of their singular points the situations studied in the previous chapters are realized (see [**73**]). Some of the examples have illustrative meaning only; others are richer in content and appear in applications.

## 8.1. Elliptic points

Let $(\mathbb{R}^4, \Omega)$ be the standard linear symplectic space with coordinates $(x_1, x_2, y_1, y_2)$ and 2-form $\Omega = dx_1 \wedge dy_1 + dx_2 \wedge dy_2$. Both cases of Theorem 4.1.1 are realized by the IHVF $(X_H, K)$, where

$$H = \frac{\omega_1}{2}(x_1^2 + y_1^2) + \frac{\omega_2}{2}(x_2^2 + y_2^2),$$
$$K = \frac{\nu_1}{2}(x_1^2 + y_1^2) + \frac{\nu_2}{2}(x_2^2 + y_2^2),$$

and $\Delta = \omega_1\nu_2 - \omega_2\nu_1 \neq 0$. For $\omega_1\omega_2 > 0$ we have Case A of Theorem 4.1.1, and for $\omega_1\omega_2 > 0$ we have Case B.

## 8.2. Singular point of the saddle-center type

As in the case of an elliptic point, consider the following IHVF $(X_H, K)$ in $(\mathbb{R}^4, \Omega)$:

$$H = \sigma\left(x_1y_1 - \frac{1}{4}(x_1^2 + y_1^2)^2\right) + \frac{1}{2}(x_2^2 + y_2^2), \quad K = x_2^2 + y_2^2,$$

where $\sigma = \pm 1$. The point $O = (0,0,0,0)$ is a simple singular point of the saddle-center type, and the points $O_1 = (\frac{\sqrt{2}}{2}, 0, \frac{\sqrt{2}}{2}, 0)$, $O_2 = (-\frac{\sqrt{2}}{2}, 0, -\frac{\sqrt{2}}{2}, 0)$ are simple elliptic points. The degeneracy set consists of two planes $\Sigma_1 = \{x_2 = y_2 = 0\}$ and $\Sigma_2 = \{x_1 = y_1 = 0\}$ containing the point $O$ and the planes $\Sigma_3 = \{x_1 = y_1 = \frac{\sqrt{2}}{2}\}$ and $\Sigma_4 = \{x_1 = y_1 = -\frac{\sqrt{2}}{2}\}$. Therefore, the bifurcation diagram looks like shown in Figure 8.1. The phase portrait of the restriction of the IHVF to $\Sigma_1$ is shown in Figure 8.2. The level $H = 0$ in the portrait is a "figure eight" formed by the separatrices of the saddle $(0,0)$. As was pointed out in Chapter 5, Cases A and B of Theorem 5.2.1 differ in the number of SPTs from $\Sigma_1$ belonging to those levels $H \neq 0$ that contain a SPT from $\Sigma_2$. In our example, SPTs from $\Sigma_2$ belong to levels $H > 0$. It is easy to see that in our case, for $\sigma = 1$, two periodic trajectories from $\Sigma_1$ lie on the level $H = h$ if $0 < h < \frac{1}{16}$, and there is only one such trajectory if $h > 0$ (Case A). If $\sigma = -1$, there is only one periodic trajectory from $\Sigma_1$ on each level set $H < 0$ (Case B).

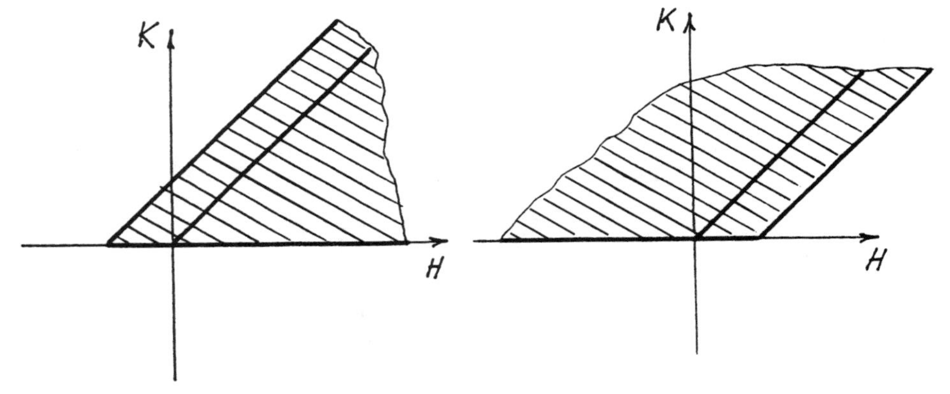


FIGURE 8.1

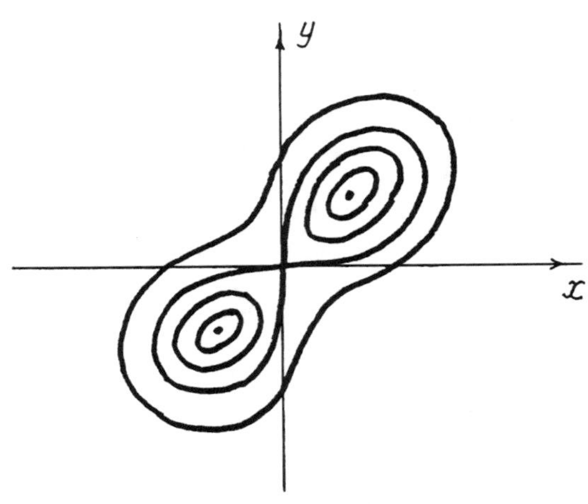


FIGURE 8.2

## 8.3. Singular point of the saddle type

Integrable systems with singular points of the saddle type quite often appear in applications (see, for example, [**23, 64, 69**]). Nevertheless, since we are mostly interested in the geometric structure of IHVF, we will construct our examples explicitly. The other reason is that, as far as we know, not every possible situation described in Chapter 6 have been found in concrete systems. Below we will describe all possible cases which were studied in Chapter 6.

**8.3.1. Cases 1A and 1B.** Consider two symplectic planes $(\mathbb{R}_i^2, dx_i \wedge dy_i)$ with Hamilton functions $H_i = \frac{1}{2}(y_i^2 - x_i^2) + \frac{1}{4}x_i^4$, $i = 1, 2$. On each of the planes consider the region $D_i$ determined by inequality $|H_i| \leq \frac{1}{4}$. The phase portrait of the vector field in $D_i$ is shown in Figure 8.3. Consider now the Hamilton functions $H = H_1 + \sigma H_2, \sigma = \pm 1$, on the direct product $D_1 \times D_2$.

If $\sigma = 1$, we have Case 1A, and if $\sigma = -1$, we have Case 1B. Indeed, the symplectic submanifolds $\Sigma_1, \Sigma_2$ are merely $D_1 \times \{0\}$ and $\{0\} \times D_2$ (see Section 6.1). All the loops are orientable since the local manifolds which determine orientability of the loop $\gamma$ on $\Sigma_i$ (see Section 6.2 in Chapter 6) are merely the direct product of the loop and an interval of the unstable manifold of the point $(0, 0)$ on $D_j$, $i \neq j$, $j = 1, 2$. Case 1A occurs if $\sigma = 1$, because there are two SPTs on each level set

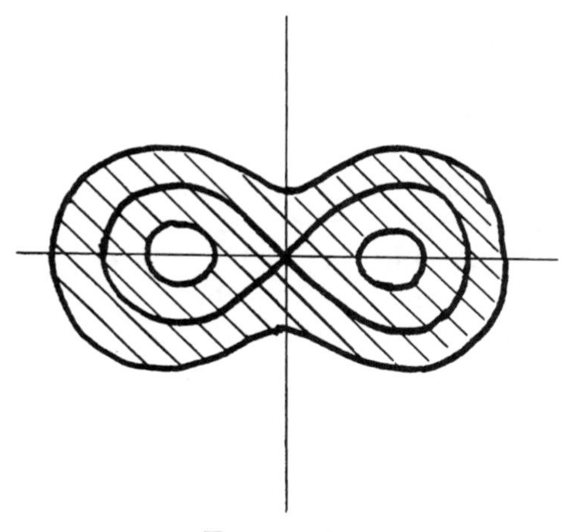

FIGURE 8.3

$H < 0$ on both $\Sigma_i$. If $\sigma = -1$, there are two SPTs on $\Sigma_1$ and one on $\Sigma_2$; therefore, here we have Case 1B.

**8.3.2. Cases 2A and 2B.** To obtain these cases we use an IHVF on the phase manifold $T^*S^2$ (which is, as usual, the cotangent bundle over the two-dimensional sphere), which is invariant with respect to an involution, has no fixed points, and has two saddles that are symmetric with respect to the involution. We will show that on the quotient manifold with respect to the involution, which can be viewed as the cotangent bundle over the real projective space $\mathbb{RP}^2$, the induced quotient system is an IHVF with the necessary properties. As a system on $T^*S^2$ we take the well-known Neumann system [**63**], which describes the motion of a point on the sphere in the square potential field and, simultaneously, describes stationary solutions of the Landau–Lifshits equation [**18, 19, 20, 28, 29, 64, 65**]. In fact, without analyzing orientability types of the loops, this example was considered in [**65**]. Our considerations are based on the results of [**66**].

To define the system, consider the space $T^*\mathbb{R}^3$ with coordinates $(M_1, M_2, M_3, m_1, m_2, m_3)$ and constraints $m^2 = 1$, $M \cdot m = 0$, where $M = (M_1, M_2, M_3)$, $m = (m_1, m_2, m_3)$, $M \cdot m = M_1 m_1 + M_2 m_2 + M_3 m_3$, $m^2 = m_1^2 + m_2^2 + m_3^2$. In this space, consider the vector field defined by

$$M' = [JM, M], \quad m' = [M, m], \tag{8.1}$$

where the square brackets denote the standard vector product in $\mathbb{R}^3$, and $J = \operatorname{diag}(J_1, J_2, J_3)$ is a diagonal $3 \times 3$-matrix such that $J_1 < J_2 < J_3$. System (8.1) is a Hamiltonian system of the form $\dot{u} = A\nabla H(u)$ with the skew-symmetric matrix $A$ of the following type (see the example in Section 1.2 in Chapter 1):

$$A = \begin{pmatrix} L(M) & L(m) \\ L(m) & 0 \end{pmatrix}, \qquad L(x) = \begin{pmatrix} 0 & x_3 & -x_2 \\ -x_3 & 0 & x_1 \\ x_2 & -x_1 & 0 \end{pmatrix},$$

$(x_1, x_2, x_3)^T \in \mathbb{R}^3$ and $H = \frac{1}{2}(M^2 + m \cdot Jm)$. The matrix $A$ defines a Poisson structure in $T^*\mathbb{R}^3$ with the Poisson bracket of functions $H$ and $K$ determined by $\{H, K\} = A\nabla H \cdot \nabla K$ (where $\cdot$ denotes the standard inner product in $\mathbb{R}^6$). The Poisson structure is degenerate since it has two invariant functions (the Casimir functions) $m^2$ and $M \cdot m$. On common level sets of these functions, which are smooth four-dimensional manifolds diffeomorphic to $T^*S^2$, the Poisson structure becomes nondegenerate and defines a symplectic manifold. Since the vector field (8.1) has the integrals $m^2$ and $M \cdot m$, a Hamiltonian system is induced on each common level set (in particular, on the manifold $N = \{m^2 = 1, M \cdot m = 0\}$). The Hamilton function of the system is the restriction of the quadratic function $H$ to $N$. Moreover, the system has an additional integral $Q = \frac{1}{2}(M \cdot JM - J_1J_2J_3(m \cdot J^{-1}m))$. It is easy to check that the functions $H$ and $Q$ are independent almost everywhere on the manifold $N$, i.e., they define an IHVF which we will denote by $(X_H, Q)$. The algebraic structures fundamental for the construction are described in more detail in [**28**], [**29**].

The involution on $N$ mentioned above is $\sigma : (M, m) \to (M, -m)$. The action of $\sigma$ on $N$ has no fixed points. Therefore, since the quotient of the sphere $m^2 = 1$ by the map $m \to -m$ is $\mathbb{RP}^2$, the quotient manifold $\tilde{N} = M/\sigma$ is a smooth manifold which is nothing else but $T^*\mathbb{RP}^2$. It is easy to see that the restriction of the field (8.1) to $N$ is invariant with respect to $\sigma$. Thus, there is an integrable Hamiltonian vector field induced on $\tilde{N}$.

Following [**66**] (see also the references there), we describe the structure of the solutions of the vector field $X_H$ on $N$. The vector field $X_H$ on $N$ has two symmetric singular points $P^{\pm} = (0,0,0;0,0,\pm 1)$ of the saddle type. The degeneracy set in $N$ consists, first, of the three cylindrical two-dimensional submanifolds, $C_1 = \{m_1 = M_2 = M_3 = 0\}$, $C_2 = \{m_2 = M_1 = M_3 = 0\}$, $C_3 = \{m_3 = M_1 = M_3 = 0\}$ such that the points $P^{\pm}$ belong to the intersection of $C_1$ and $C_2$. The degeneracy set also contains one more two-dimensional manifold, diffeomorphic to the torus $\mathbb{T}^2$, but it is irrelevant for our considerations (see [**66**]). The phase portrait of the restriction of the IHVF to $C_1$ and $C_2$ is shown in Figure 8.4, where one of the saddles is $P^+$ and the other is $P^-$.

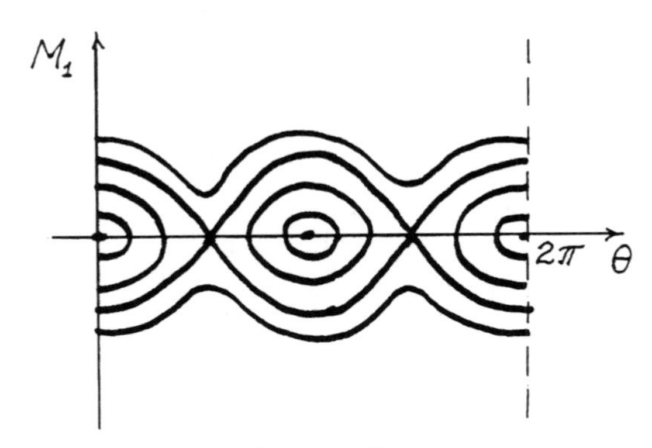


FIGURE 8.4

The bifurcation diagram of the IHVF $(X_H, Q)$ is shown in Figure 8.5, where the letters stand for the images of the corresponding symmetric pairs of singular points $P^{\pm}$, $R^{\pm} = (0,0,0;0,\pm 1,0)$ and $S^{\pm} = (0,0,0;\pm 1,0,0)$. The diagram consists of three rays: $l_1 : J_1H - Q = \frac{1}{2}J_1(J_2 + J_3)$, $l_2 : J_2H - Q = \frac{1}{2}J_2(J_1 + J_3)$, $l_3 : J_3H - Q = \frac{1}{2}J_3(J_2 + J_1)$, and the arc $H = \gamma + \frac{1}{2}\mathrm{tr}J$, $Q = -\frac{1}{2}\gamma^2$, $\gamma \in [-J_2, -J_1]$. In a neighborhood of the point $P$ the diagram consists of two pieces of the lines $l_1$ and $l_2$.

After factorizing $N$ with respect to the action of the involution $\sigma$ the points $P^+$ and $P^-$ merge into a single point $P \in \tilde{N}$ such that the stable manifolds of $P^+$

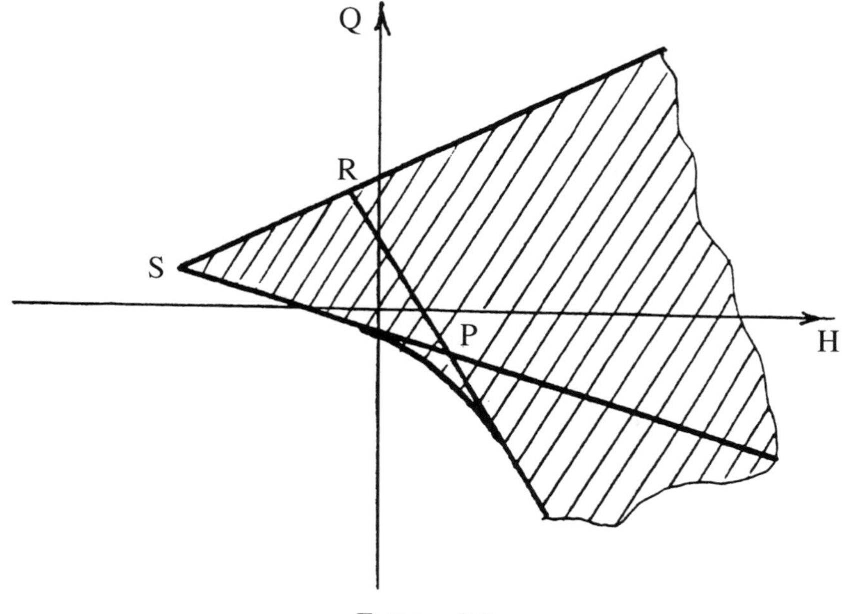


FIGURE 8.5

and $P^-$ identify with the stable manifold of the point $P$. The same holds for the unstable manifolds. After this identification the heteroclinic trajectories joining $P^+$ and $P^-$ and going into $N$ (and the other way around) become homoclinic trajectories (loops). These heteroclinic trajectories belong to the cylinders $C_1$, $C_2$, and are one-dimensional orbits of the induced Poisson action. It becomes clear from the formulas below that the other trajectories of the filed $X_H$, which belong, for instance, to the unstable manifold of the point $P^-$, are homoclinic, and return to the stable manifold of the point $P^-$. After identification these trajectories remain homoclinic. To define the type of the loop after identification, according to the construction from Chapter 6, we proceed as follows.

Fix a point $P^-$ and a heteroclinic trajectory $\gamma$ starting at this point. Near $P^-$ the trajectory $\gamma$ belongs to $D_-$, one of the two halves of the local unstable manifold $W^u(P^-)$ into which the two neighboring $\gamma$ heteroclinic trajectories divide the manifold $W^u(P^-)$ (this neighboring heteroclinic orbits make up (locally, near $P^-$) the segment $d$ of Section 6.2 in Chapter 6; if $\gamma$ belongs to $C_1$, then the neighboring trajectories belong to $C_2$, and vice versa). The trajectory $\gamma$ comes to the point $P^+$. Moreover, when continuing along trajectories of the field $X_H$, the half-disk $D_-$ merges with the local saddle manifold $W^c(P^+)$, or, to be more precise, with one of its halves, $D^c_+$, that appears as a result of dividing $W^c(P^+)$ by the heteroclinic trajectories neighboring $\gamma$ (see details in Chapter 6). Therefore, we get a strip which consists of $D_-$, $D^c_+$, and intervals of trajectories going from $D_-$ to $D^c_+$ near $\gamma$. The identification with respect to $\sigma$ turns the strip either into a cylinder or a Möbius band. Now we show that in fact it is always a Möbius band. In order to do so, we use the formulas which describe those solutions of (8.1) on $W^u(P^-)$ for which $m_1 < 0$. They were obtained in [**67**], and can be written in the following convenient form:

$$\begin{aligned} m_1 &= -\frac{2\omega_1^2\sqrt{1-d_2^2}(1+d_1\tanh(\rho_1\xi))}{\cosh(\rho_2\xi)D(\xi)}, \\ m_2 &= -\frac{2\omega_1^2\sqrt{1-d_1^2}(d_2+\tanh(\rho_1\xi))}{\cosh(\rho_1\xi)D(\xi)}, \\ m_3 &= (D(\xi))^{-1}((1-\omega_1^2)(1+d_2\tanh(\rho_2\xi))(1+d_1\tanh(\rho_1\xi)) \\ &\qquad - (1+\omega_1^2)(d_1+\tanh(\rho_2\xi))(d_2+\tanh(\rho_2\xi))), \end{aligned}$$

where

$$\begin{aligned} \rho_1 = \sqrt{J_3 - J_2}, \quad \rho_2 &= \sqrt{J_3 - J_1}, \\ \omega_1 = (1-k_1)/(1+k_1), \quad k_1^2 &= (J_3-J_2)/(J_3-J_1), \\ D(\xi) = (1+\omega_1^2)(1+d_2&\tanh(\rho_2\xi))(1+d_1\tanh(\rho_1\xi)) \\ &+ (\omega_1^2-1)(d_1+\tanh(\rho_1\xi))(d_2+\tanh(\rho_2\xi)), \end{aligned}$$

and the parameters $d_1$, $d_2$ satisfy the inequality $|d_i| < 1$. The functions $M_i(\xi)$ are given by the relation $M = m \times m'$. Heteroclinic trajectories are derived from the above formulas in the limit as $d_1 \to \pm 1$ or $d_2 \to \pm 1$. Let us choose a heteroclinic trajectory which corresponds to $d_1 \to -1$. It has the form $m_1 = M_2 = M_3 = 0$, $m_2 = \cosh^{-1}(\rho_1\epsilon)$, $m_3 = \tanh(\rho_1\epsilon)$, $M_1 = \rho_1 m_2$. It is easy to check that the tangent plane to $W^u(P^-)$ is determined by the equations $M_2 = -\rho_2 m_1$, $M_1 = \rho_1 m_2$, $m_3 = -1$, $M_3 = 0$, and the plane tangent to $W^c(P^+)$ is determined by the equations $M_1 = \rho_1 m_2$, $M_2 = \rho_2 m_1$, $m_3 = 1$, $M_3 = 0$. Evidently, it is possible to choose $(m_1, m_2)$ as coordinates on the manifolds $W^u(P^-)$, $W^c(P^+)$. Moreover, the equation for the trajectory $\gamma$ in these coordinates is $m_1 = 0$, in $W^u(P^-)$ as well as in $W^c(P^+)$. Thus the side of the strip connecting $D_-$ and $D_+^c$ is determined by the sign of the coordinate $m_1$. Since gluing identifies the points $(M, m)$ and $(M, -m)$, it is clear that the identification produces a Möbius band provided that trajectories close to $\gamma$ on $W^u(P^-)$ with $m_1 < 0$ get to $W^c(P^+)$ with $m_1 < 0$. Since $D(\xi) > 0$, from the expression for $m_1$ we can see that $m_1$ is negative for all $\xi$. Therefore, we have an IHVF with a singular point of the saddle type $P$ which has four nonorientable homoclinic loops.

We now determine which of the Cases A and B takes place for our system. In order to do it, we need to calculate, for sufficiently small $\epsilon > 0$, the number of SPTs on the level set $H = H(P) + \epsilon$ of the quotient system. First, consider the original system. All the SPTs belong to the degeneracy set. It is evident from the bifurcation diagram that if $H > H(P^+)$, then near the separatrix of the saddle point, all the SPTs we are interested in belong either to $C_1$ or to $C_2$. The restriction of the field $X_H$ to $C_1$ has the form $\theta' = M_1$, $M_1' = -(J_3 - J_2)\sin 2\theta$, where $m_2 = \cos\theta$, $m_3 = \sin\theta$ (we recall that $m_1 = M_2 = M_3 = 0$ on $C_1$). The involution $\sigma$ in these coordinates has the form $(\theta, M_1) \to (\theta + \pi, M_1)$. Since the saddles are the points $(\pi/2, 0)$ and $(3\pi/2, 0)$, it is clear that for $\epsilon > 0$ we have two periodic trajectories. One of them encircles the cylinder $C_1$ in the region $M_1 > 0$, and the other encircles it in the region $M_1 < 0$ (see Figure 8.4). The restriction of the field $X_H$ to the cylinder $C_2$ has similar properties. If $H > H(P^+)$, then the field has two periodic trajectories. Therefore, the quotient system realizes Case 2A.

According to the results of Chapter 6, since the lines $Q = \text{const}$ intersect the bifurcation diagram transversally, the IHVF $(X_Q, H)$ realizes Case 2B provided that $J_1 \neq 0$, $J_2 \neq 0$.

**8.3.3. Cases 3, 4.** To realize these cases we present a special construction, similar to the symplectic skew product, that allows us to produce IHVF with the necessary properties.

On the symplectic plane $(x_1, y_1)$ with the 2-form $dy_1 \wedge dx_1$, consider the Hamiltonian vector field with the Hamilton function

$$H_1 = \frac{y_1^2 - x_1^2}{2}\sigma(x_1) + (1 - \sigma(x_1))(y_1 \cdot \operatorname{sgn}(y) - 1),$$

where $\sigma(x_1)$ is an even $C^\infty$-smooth function with the following properties: $\sigma(x_1) = 1$ for $|x_1| \leq \frac{3}{4}$, $0 < \sigma(x_1) < 1$ and $\sigma(x_1)$ is monotone for $\frac{3}{4} < |x_1| < 1$, and $\sigma(x_1) \equiv 0$ for $|x_1| \geq 1$. In the region $D_1 = \{(x_1, y_1) \mid |H_1| \leq \frac{1}{8}, |x_1| \leq 2\}$ the Hamilton function is a $C^\infty$-smooth function that coincides with $y_1 - 1$ in the rectangles $\Pi_1^+ = \{(x_1, y_1) \mid |y_1 - 1| \leq \frac{1}{8}, 1 \leq x \leq 2, y_1 > 0\}$ and $\Pi_2^+ = \{(x_1, y_1) \mid |y_1 - 1| \leq \frac{1}{8}, -2 \leq x \leq -1, y_1 > 0\}$, and in the rectangles $\Pi_1^-$ and $\Pi_2^-$ symmetric to $\Pi_1^+$ and $\Pi_2^+$ with respect to the $x_1$ axis, it coincides with the function $-y_1 - 1$. There is a saddle at the point $(0, 0)$. The phase portrait of the vector field in $D_1$ is shown in Figure 8.6.

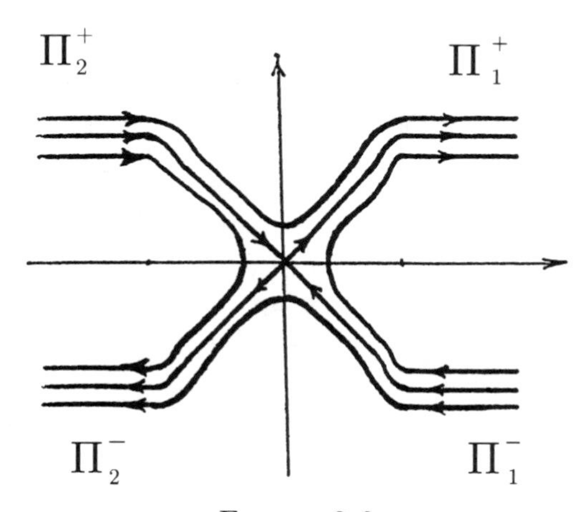


FIGURE 8.6

On the plane $(x_2, y_2)$ with the 2-form $dy_2 \wedge dx_2$, consider the Hamiltonian field $X_2$ with the Hamilton function

$$H_2 = \frac{y_2^2 - x_2^2}{2} - \frac{1}{4}x_2^4$$

and take it in the region $D_2 : |H| \leq \frac{1}{8}$ (see Figure 8.3). The Hamilton function $H = H_1 + H_2$ generates the direct product of two Hamiltonian systems with the Hamilton functions $H_1$ (in $D_1$) and $H_2$ (in $D_2$) in the region $D_1 \times D_2$. The following properties of the obtained system in $D$ are obvious.

1. It is an IHVF and the second integral is $K = H_2$.
2. The degeneracy set of the system consists of the sets $\Sigma_2 = \{(0, 0)\} \times D_2$, which contains two loops $\gamma_1, \gamma_3$, and $\Sigma_1 = D_1 \times \{(0, 0)\}$ where the separatrix intervals of the saddle $O = (0, 0)$ form a "cross".

As in Subsection 8.3.1, the loops $\gamma_1, \gamma_3$ are orientable, and our problem here is to complete the construction of the system in order to obtain two loops on $\Sigma_1$ such that either one of them is orientable and the other is not (Case 3) or both are nonorientable (Case 4).

We define the map $g_1$ of the set $\Pi_1^+ \times D_2$ to the set $\Pi_1^- \times D_2$ by $g_1(x_1, y_1, x_2, y_2) = (3 - x_1, -y_1, k_1 x_2, k_1 y_2), k_1 = \pm 1$. The map is symplectic regardless of the sign of $k_1$. Moreover, $g_1$ maps the field $X_H$ to itself, $Dg_1 X_H(\xi) = X_H(g_1 \xi)$, $\xi \in \mathbb{R}^4$. The symplectic diffeomorphism $g_2$ is defined similarly, $g_2 : \Pi_2^+ \times D_2 \to \Pi_2^- \times D_2$, $g_2(x_1, y_1, x_2, y_2) = (-x_1 - 3, -y_1, k_2 x_2, k_2 y_2)$, $k_2 = \pm 1$. We identify the points $\xi$ and $g_1(\xi)$, $\xi \in \Pi_1^+ \times D_2$, and the points $\eta$ and $g_2(\eta)$, $\eta \in \Pi_2^+ \times D_2$, in $D$. It is clear from the construction that the quotient manifold we have obtained is a smooth symplectic manifold $V$ with boundary, the functions $H$ and $K$ can be uniquely extended to $V$ and define an IHVF on $V$ (for more information about symplectic gluing, see also [**70**]). If we choose opposite signs of $k_1$ and $k_2$ in the formulas for $g_1$, $g_2$, then we obtain Case 3, and if we choose $k_i = -1, i = 1, 2$, then we obtain Case 4. Naturally, if we put $k_i = 1, i = 1, 2$, then we obtain Case 1. Since in the constructed example there is only one SPT on each $\Sigma_1$ and $\Sigma_2$ in the region $H > 0$, we have Cases 3A or 4A. In order to get Cases 3B and 4B it is enough to consider the system with the Hamilton function $-H_2$ on $D_2$.

## 8.4. Singular point of the saddle-focus type

Let us proceed to the construction of IHVFs that realize Cases 1–3 studied in Chapter 7. In order to construct these examples we will use symplectic polar coordinates in a neighborhood of the singular point. They were applied in the works [**56, 57**] to study stability of Hamiltonian systems with two degrees of freedom in neighborhoods of singular points with eigenvalues $\pm i\omega_1$, $\pm i\omega_2$ in presence of resonances of different orders between the frequencies $\omega_1$ and $\omega_2$.

Let $(p_1, p_2, q_1, q_2)$ be the standard symplectic coordinates in $(\mathbb{R}^4, dp_1 \wedge dq_1 + dp_2 \wedge dq_2)$. It is easy to check that the substitution

$$\begin{aligned} p_1 &= P\cos(\phi) - \frac{k}{r}\sin(\phi), \qquad & q_1 &= r\cos(\phi), \\ p_2 &= P\sin(\phi) + \frac{k}{r}\cos(\phi), \qquad & q_2 &= r\sin(\phi), \end{aligned} \tag{8.2}$$

is symplectic ($dp_1 \wedge dq_1 + dp_2 \wedge dq_2 = dP \wedge dr + dk \wedge d\phi$), and it is invertible in the region $r^2 = q_1^2 + q_2^2 \neq 0$. The inverse map has the form

$$r = \sqrt{q_1^2 + q_2^2}, \quad k = p_2 q_1 - p_1 q_2, \quad P = (p_1 q_1 + p_2 q_2)/\sqrt{q_1^2 + q_2^2},$$

and $\phi \in [0, 2\pi)$ can be found from the equations $\cos(\phi) = q_1/r$, $\sin(\phi) = q_2/r$. The coordinates $(P, k, r, \phi)$ will be called the symplectic polar coordinates.

Consider the Hamilton function

$$H = \frac{p_1^2 + p_2^2}{2} + p_2 q_1 - p_1 q_2 - (q_1^2 + q_2^2) + H_1(p_1, p_2, q_1, q_2)$$

such that the function $H_1$ is infinitesimally small of order three or higher at the origin. It is easy to check that the Hamiltonian field $X_H$ has a singular point of the saddle-focus type at the origin. In the symplectic polar coordinates $H$ has the form

$$H = \frac{1}{2}(P^2 + k^2/r^2) + k - \frac{1}{2}r^2 + \tilde{H}_1. \tag{8.3}$$

If $H_1$ is an arbitrary smooth function of the homogeneous quadratic polynomials $p_1^2 + p_2^2 = P^2 + k^2/r^2$, $p_2q_1 - p_1q_2 = k$, $q_1^2 + q_2^2 = r^2$, $p_1q_1 + p_2q_2 = Pr$, then the function $K = p_2q_1 - p_1q_2$ is an integral of the field $X_H$. Taking different $H_1$, we obtain fields with different structures. From now on we choose $H_1$ to be a function of only $q_1^2 + q_2^2 = r^2$.

Let us notice that in the symplectic polar coordinates the Hamilton function does not depend on the angle $\phi$. Hence the coordinate $k$ conjugate to $\phi$ (i.e., the function $K$) is an integral. Therefore, these coordinates reduce our Hamilton function to another one which depends on parameter $k$ and has only one degree of freedom. At this point we would like to make some comments about the correspondence between trajectories of the reduced system and trajectories of the initial one. Note that trajectories of the Hamiltonian vector field $X_K$ are circles (except for the singular point $O = (0,0,0,0)$). The singular point $O$ belongs to the level set $K = 0$. Thus, to study trajectories on the level set, one has to put $k = 0$ in the reduced Hamilton function (8.3). The stable $W^s(O)$ and unstable $W^u(O)$ manifolds of the saddle-focus point $O$ belong to the same level set $K = 0$. Moreover, if the closure of $W^s(O)$ or $W^u(O)$ contains a hyperbolic SPT as well, then it also belongs to the level $K = 0$. It is obvious from (8.2) that if $k = 0$, then the reduced system has to be studied for all $r \geq 0$. Such trajectories of the reduced system correspond to two-dimensional invariant surfaces of the initial vector field $X_H$ that appear as a result of twist by $\phi$. This follows from the fact that $\dot{\phi} = 1$ for $K = 0$ since $H_1$ does not depend on $p_1, p_2$. In particular, SPTs of the field $X_H$ correspond to those singular points of the reduced system for which $r \neq 0$.

**8.4.1. Case 1.** Let $H_1 = r^4/4$; then for $k = 0$ we have

$$H_0 = (P^2 - r^2)/2 + r^4/4.$$

For $r = 0$, the level set $H_0 = 0$ is a separatrix loop of the saddle $(0,0)$. In the initial system, the stable and unstable manifolds of the saddle-focus point $O$, merged together, correspond to this loop. Therefore, Case 1 is realized by the corresponding IHVF $(X_H, K)$ with $H = H_0 + k^2/r^2 + k$, or, in the original coordinates,

$$H = \frac{p_1^2 + p_2^2}{2} + p_2q_1 - p_1q_2 - (q_1^2 + q_2^2) + \frac{1}{4}(q_1^2 + q_2^2)^2, \quad K = p_2q_1 - p_1q_2.$$

**8.4.2. Case 2.** For $k = 0$, let

$$H_0 = \frac{P^2 - r^2}{2} - \frac{1}{9\sigma^2}\int_0^r \xi(1-\xi^2)(\sigma^2 - \xi^2)(9 - \xi^2)d\xi + \frac{r^2}{2} = \frac{P^2 - r^2}{2} + f(r^2),$$

where $1 < \sigma^2 < 9$, $\sigma > 0$. Then it is easy to check that the singular points in the half-plane $r \geq 0$ (except the saddle $(0,0)$) are two centers at the points $(0,1)$, $(0,3)$, and the saddle at the point $(0,\sigma)$. The number $\sigma$ is uniquely determined by the condition $H_0(0,\sigma) = 0$. This guarantees that the separatrices of the saddles $(0,0)$ and $(0,\sigma)$ merge, i.e., the phase portrait of the reduced system in the half-plane $r \geq 0$ has the form shown in Figure 8.7. This means that the IHVF $(X_H, K)$ with

$$H = \frac{p_1^2 + p_2^2}{2} + p_2q_1 - p_1q_2 - (q_1^2 + q_2^2) + f(q_1^2 + q_2^2), \quad K = p_2q_1 - p_1q_2$$

has the necessary structure, i.e., it has a one-parameter family (depending on $H$) of hyperbolic orientable SPTs. For $H = K = 0$ the stable manifold of the saddle-focus merges with the half of the unstable manifold of the same SPT, the unstable manifold of the saddle-focus merges with the half of the stable manifold of the SPT, and the remaining half of the stable manifold and half of the unstable manifold merge with each other.

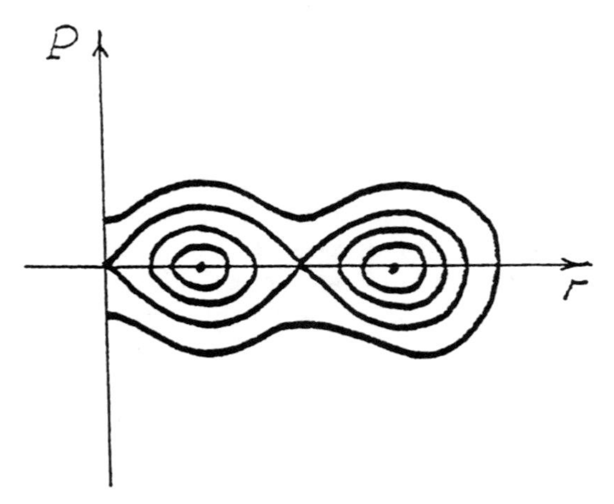


FIGURE 8.7

REMARK 8.1. The case when the stable and unstable manifolds of a saddle-focus merge with the unstable and stable manifolds of a parabolic SPT can be realized similarly. We will give a corresponding example, although we have not considered this case in Chapter 7.

For $k = 0$ consider the following Hamilton function:

$$H_0 = \frac{P^2 - r^2}{2} + f(r^2), \quad f(r^2) = \frac{3}{8}r^4 - \frac{3}{32}r^6 + \frac{1}{128}r^8.$$

The system with the Hamilton function $H_0$ has three singular points for $r \geq 0$: the saddle $(0,0)$, the center $(1,0)$, and the parabolic point $(2,0)$. The phase portrait of the system is shown in Figure 8.8. Thus, for

$$\begin{aligned} H &= \frac{p_1^2 + p_2^2}{2} + p_2 q_1 - p_1 q_2 - (q_1^2 + q_2^2) + \frac{3}{8}(q_1^2 + q_2^2)^2 \\ &\quad - \frac{3}{32}(q_1^2 + q_2^2)^3 + \frac{1}{128}(q_1^2 + q_2^2)^4, \\ K &= p_2 q_1 - p_1 q_2, \end{aligned}$$

the IHVF $(X_H, K)$ has a saddle-focus on the level set $H = K = 0$ such that its stable and unstable manifolds merge with the unstable and stable manifolds of the parabolic SPT, respectively.

**8.4.3. Case 3.** In order to construct the IHVF that realizes Case 3, we glue an extended neighborhood of a saddle-focus from two pieces; one of them, $N_1$, contains a cylinder formed by nonorientable hyperbolic SPTs, and the other, $N_2$, is a neighborhood of the saddle-focus.

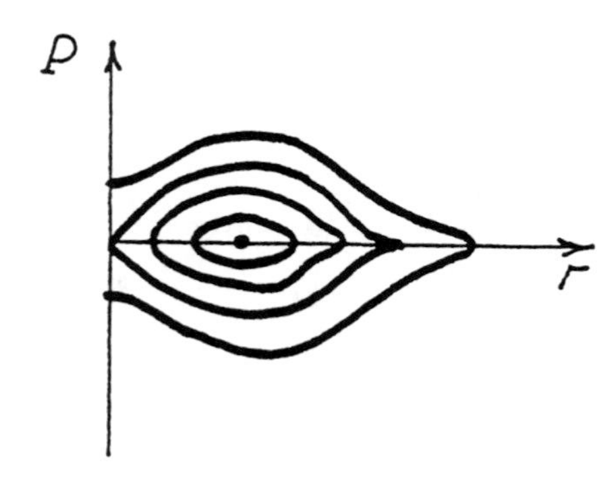


FIGURE 8.8

Consider the direct product of two two-dimensional symplectic manifolds with boundary: the annulus $D_1 = \{(\rho, \phi) \mid 1 \le \rho \le 2,\ 0 \le \phi \le 2\pi \,(\mathrm{mod}\, 2\pi)\}$ with 2-form $d\rho \wedge d\phi$, and the symplectic plane $\mathbb{R}^2$ with 2-form $dx \wedge dy$. Consider the Hamiltonian vector field on $D_1$ with the Hamilton function $H_1 = \rho$. On $\mathbb{R}^2$ the vector field is defined by the $C^\infty$-Hamilton function

$$H_2 = (xy\sigma(x) + y \cdot \mathrm{sgn}(x)(1 - \sigma(x)))\sigma(y) + x \cdot \mathrm{sgn}(y)(1 - \sigma(y)),$$

where $\sigma(t)$ is an even function that was used in Subsection 8.3.3. The vector field constructed has the following properties.

1. At the point $(0, 0)$, the vector field has a singular point of the saddle type such that its stable and unstable manifolds are the lines $y = 0$ and $x = 0$, respectively.

2. For sufficiently small $\delta > 0$, there are no other singular points in the region $D_2 = \{(x, y) \mid |x| \le 2,\ |y| \le 2,\ |H_2| \le \delta\}$.

3. In the rectangle $\Pi_1^+ = \{1 \le x \le 2,\ |y| \le \delta\}$ the Hamilton function coincides with $H_2 = y$, and in the symmetric rectangle $\Pi_2^+ = \{-2 \le x \le -1,\ |y| \le \delta\}$ the Hamilton function coincides with $H_2 = -y$; analogously, in the rectangles $\Pi_1^- = \{1 \le y \le 2,\ |x| \le \delta\}$ and $\Pi_2^- = \{-2 \le y \le -1,\ |x| \le \delta\}$ the Hamilton function $H_2$ coincides with $x$ and $-x$, respectively.

4. In the constructed region $D_2$ the vector field is symmetric with respect to the linear involution $(x, y) \to (-x, -y)$, i.e., if $\mu$ is the linear involution, $\mu^2 = \mathrm{id}$, and $v$ is a vector field, then $v(\mu\xi) = \mu v(\xi)$.

Consider the Hamilton function $H = H_1 + H_2$ on the direct product $N_1 = D_2 \times D_1$. The IHVF $(X_H, H_1)$ is defined on $N_1$ so that it has the invariant symplectic cylinder $x = y = 0$, which is foliated by the orientable hyperbolic SPTs. An SPT $\rho = c$, $x = y = 0$ lies on the level set $H = c$ and its stable manifold is $y = 0$, $\rho = c$ and the unstable one is $x = 0$, $\rho = c$. Henceforth we denote by $S_i^+ \subset N_1$ ($S_i^- \subset N_1$) the sets $\Pi_i^+ \times D_1$ (respectively, $\Pi_i^- \times D_1$).

Notice that the constructed vector field in $N_1$ is symmetric with respect to the symplectic involution without fixed points $(x, y, \rho, \phi) \to (-x, -y, \rho, \phi + \pi)$. Therefore, the involution induces a $C^\infty$-smooth integrable Hamiltonian vector field on the quotient of the manifold $N_1$, which admits a cylinder formed by nonorientable hyperbolic SPTs.

Let us turn to $N_2$. In $(\mathbb{R}^4, \Omega)$, $\Omega = dp_1 \wedge dq_1 + dp_2 \wedge dq_2$, consider the IHVF with the quadratic Hamilton function $H = \xi + \eta$, where $\xi = p_1 q_1 + p_2 q_2$, $\eta = p_2 q_1 - p_1 q_2$, and with an additional integral $K = \eta$. Remark that the function $\xi$ is also an integral of the vector field. The IHVF $(X_H, \eta)$ has a singular point of the saddle-focus type at the point $(0,0,0,0)$. Its stable manifold is the plane $q_1 = q_2 = 0$, and its unstable manifold is the plane $p_1 = p_2 = 0$. Consider the set determined by the inequalities $\epsilon \le p_1^2 + p_2^2 \le \epsilon^2$, $q_1^2 + q_2^2 \le Q^2$, where the number $Q > 0$ will be defined later. In this set we introduce symplectic coordinates $(R_1, \eta, \xi, \psi)$, where

$$R_1 = \ln r_1, \quad p_1 = r_1 \cos(\psi),$$
$$p_2 = r_1 \sin(\psi), \quad q_1 = (\xi \cos(\psi) + \eta \sin(\psi))/r_1, \quad q_2 = (\xi \sin(\psi) - \eta \cos(\psi))/r_1.$$

In these coordinates the form $\Omega$ is $dR_1 \wedge d\xi + d\eta \wedge d\psi$, and the vector field $X_H$ can be written as

$$\dot{R}_1 = -1, \quad \dot{\eta} = 0, \quad \dot{\xi} = 0, \quad \dot{\psi} = 1.$$

Consider the domain $B^+ = \{(R_1, \eta, \xi, \psi) \mid 1 \le R_1 \le 2,\ |\xi| \le \delta,\ |\eta| \le 1/2,\ 0 \le \psi \le 2\pi (\operatorname{mod} 2\pi)\}$, where the number $\delta$ is the same we used for the construction of the set $N_1$. Since $q_1^2 + q_2^2 = (\xi^2 + \eta^2)/r_1^2$, we can choose such a domain provided that the inequality $Q^2 \ge (4 + \delta^2)\epsilon^{-2}$ holds. Analogously, using the symplectic coordinates $(\xi, \eta, R_2, \chi)$, we construct the domain $B^-$ in the set $\epsilon \le q_1^2 + q_2^2 \le \epsilon^2$, $p_1^2 + p_2^2 \le Q^2$, where

$$R_2 = \ln r_2, \quad q_1 = r_2 \cos(\chi), \quad q_2 = r_2 \sin(\chi),$$
$$p_1 = (\xi \cos(\chi) - \eta \sin(\chi))/r_2, \quad p_2 = (\xi \sin(\chi) + \eta \cos(\chi))/r_2,$$

and $d\xi \wedge dR_2 + d\eta \wedge d\chi$. Here $B^- = \{(\xi, \eta, R_2, \chi) \mid 1 \le R_1 \le 2,\ |\xi| \le \delta,\ |\eta| \le 1/2,\ 0 \le \chi \le 2\pi (\operatorname{mod} 2\pi)\}$. In these symplectic coordinates the vector field $X_H$ has the form

$$\dot{\xi} = 0, \quad \dot{\eta} = 0, \quad \dot{R}_2 = 1, \quad \dot{\chi} = 1.$$

Now we define the set $N_2$ by the following inequalities: $|\xi| \le \delta$, $|\eta| \le 1/2$, $0 \le r_1 \le \epsilon^2$, $0 \le r_2 \le \epsilon^2$. This set is a neighborhood of stable and unstable manifolds of the saddle-focus with $B^+, B^- \subset N_2$.

To complete the construction of the example, let us glue two copies $N_2^{(1)}$, $N_2^{(2)}$ of the set $N_2$ to the set $N_1$. We denote by $B_i^+, B_i^-$ the corresponding sets $B^+, B^-$ in $N_2^{(i)}$, by $(R_1^i, \eta^i, \xi^i, \psi^i)$ the coordinates in $B_i^+$, and by $(\xi^i, \eta^i, R_2^i, \chi^i)$ the coordinates in $B_i^-$. Each copy of $N_2^{(i)}$ is glued to $N_1$ in such a way that $B_i^+$, $B_i^-$ in $N_2^{(i)}$ are identified by a symplectic diffeomorphism with the sets $S_i^+, S_i^-$ in $N_1$, respectively. Then, obviously, the manifold obtained after identifications is symplectic. Moreover, we will show that it is possible to define an IHVF on the manifold, which is the union of the IHVFs on each of the pieces.

To begin with, we introduce the gluing diffeomorphism for $N_1$ and $N_2^{(1)}$. For $B_1^+$ and $S_1^-$ the gluing map has the form $y = 3 - R_1^1$, $x = \xi^1$, $\rho = \eta^1$, $\phi = \psi^1 (\operatorname{mod} 2\pi)$. For $B_1^-$ and $S_1^+$ the gluing map is $y = 3 - R_2^1$, $x = \xi^1$, $\rho = \eta^1$, $\phi = \chi^1 (\operatorname{mod} 2\pi)$. To glue $N_2^{(2)}$ and $N_1$ we use the following diffeomorphisms: $y = R_1^2 - 3$, $x = -\xi^2$, $\rho = \eta^2$, $\phi + \pi = \psi^2 (\operatorname{mod} 2\pi)$ for $B_2^+$ and $S_2^-$; and $x = R_2^2 - 3$, $x = -\xi^2$, $\rho = \eta^2$, $\phi + \pi = \chi^2 (\operatorname{mod} 2\pi)$ for $B_2^-$ and $S_2^+$. As the result of the gluing we obtain a $C^\infty$-smooth symplectic manifold $\tilde{M}$ with boundary. Moreover, we can and will assume that $N_1, N_2^{(1)}, N_2^{(2)}$ are submersed symplectic submanifolds. According to the construction, functions $H$, $K$, hence also the corresponding Hamiltonian vector fields, can be smoothly extended to $\tilde{M}$. Note that the stable and unstable manifolds of the SPT $x = y = \rho = 0$ on $N_1$ are glued with the unstable and stable manifolds of the saddle-focus points in $N_2^{(1)}$ and $N_2^{(2)}$. We define a symplectic involution without fixed points on $\tilde{M}$. Namely, in $N_1$ the involution coincides with the symplectic diffeomorphism $\tau : (x, y, \rho, \phi) \to (-x, -y, \rho, \phi + \pi)$. We define a map from $N_2^{(1)}$ to $N_2^{(2)}$ in the following way. Since $N_2^{(1)}$ and $N_2^{(2)}$ are two identical copies of $N_2$ with global coordinates $(p_1, p_2, q_1, q_2)$, in order to distinguish different copies, we attach the index $i$ to the coordinates in the $i$th copy, i.e., the coordinates in $N_2^{(i)}$ are denoted by $(p_1^i, p_2^i, q_1^i, q_2^i)$. Then we define the symplectic diffeomorphism $\nu : N_2^{(1)} \to N_2^{(2)}$ in the following way: a point $m \in N_2^{(1)}$ is mapped into the point $m' = \nu(m) \in N_2^{(2)}$ with the same coordinates in $N_2^{(2)}$ as $m$ has in $N_2^{(1)}$. Now we have to check that this map agrees with the involution $\tau$ on $N_1$, i.e., that if $m \in N_1 \cap N_2^{(1)}$, then $\tau(m) = \nu(m)$.

To be specific, let us assume that $m \in B_1^+ \cap N_2^{(1)}$, hence $m \in S_1^- \cap N_1$. Then, if $m = (a_1, a_2, b_1, b_2)$ in the coordinates on $N_2^{(1)}$, then $\nu(m) = (a_1, a_2, b_1, b_2)$ in the coordinates on $N_2^{(2)}$. Let us show that the point $\tau(m) \in N_2^{(2)}$ has the same coordinates. The coordinates of the point $m$ in the set $B_1^+$ in $N_2^{(1)}$ are $R_1^1 = \frac{1}{2}\ln(a_1^2 + a_2^2)$, $\xi^1 = a_1 b_1 + a_2 b_2$, $\eta^1 = a_2 b_1 - a_1 b_2$, and $\psi^1 \in [0, 2\pi)$ is uniquely defined by the equations $\sin\psi^1 = a_2/\sqrt{a_1^2 + a_2^2}$, $\cos\psi^1 = a_1/\sqrt{a_1^2 + a_2^2}$. On the other hand, this point has the following coordinates on $S_1^-$ :

$$y_0 = 3 - R_1^1, \quad x_0 = \xi^1, \quad \rho_0 = \eta^1, \quad \phi_0 = \psi^1.$$

Then the point $\tau(m) \in S_2^-$ has the coordinates

$$(x_1, y_1, \rho_1, \phi_1) = (-x_0, -y_0, \rho_0, \phi_0 + \pi (\mathrm{mod}\, 2\pi)).$$

By construction, the same point belongs to $B_1^+$ and has the following coordinates:

$$R_1^2 = y_1 + 3 = -y_0 + 3 = R_1^1, \quad \xi^2 = -x_1 = x_0 = \xi^1,$$
$$\eta^2 = \rho_1 = \rho_0 = \eta^1, \quad \psi^2 = \phi_1 + \pi = \phi_0 + 2\pi = \phi_0 (\mathrm{mod}\, 2\pi) = \psi^1.$$

Therefore, we have proved that the coordinates of the point $\tau(m)$ in $N_2^{(2)}$ are equal to the coordinates of the point $m$ in $N_2^{(1)}$, i.e., $\tau(m) = \nu(m)$. Analogously, one can check the coincidence of $\tau(m)$ and $\nu(m)$, for $m \in B_1^-$. Thus, the symplectic involution $\kappa$, $\kappa^2 = \mathrm{id}$, $\kappa|_{N_1} = \tau$, $\kappa|_{N_2^{(1)}} = \nu$, $\kappa|_{N_2^{(2)}} = \nu^{-1}$ acts without fixed points on the symplectic manifold $\tilde{M}$. The quotient manifold $M = \tilde{M}/\kappa$ is symplectic, and IHVF on $M$ is induced by the IHVF $(X_H, H_1)$ on $\tilde{M}$. There is only one singular point of the saddle-focus type in $M$. Moreover, its stable and unstable manifolds merge with the unstable and stable manifolds of a nonorientable hyperbolic SPT, respectively.

APPENDIX A

# Normal Forms of Quadratic Hamilton Functions and Their Centralizers in $sp(4, \mathbb{R})$

The aim of this Appendix is to give the list of normal forms of all possible quadratic Hamilton functions in the case of two degrees of freedom, and to present for each normal form all the quadratic functions that are the additional integrals of the corresponding linear Hamiltonian system. To begin with, we formulate Williamson's results about the normal form of a quadratic Hamilton function in a linear symplectic space as it appears in Appendix 6 of [**21**] (see also [**72**]).

Let $V$ be a real $2n$-dimensional linear symplectic space, $[\cdot,\cdot]$ a skew-symmetric inner product, and let $C : V \to V$ be a linear operator that satisfies the equality

$$[Cx, y] + [x, Cy] = 0.$$

Then, as was pointed out in Section 2.1 of Chapter 2, the function $H(x) = \frac{[x,Cx]}{2}$ is a quadratic form. The related linear Hamiltonian vector field $v$ has the form $v(x) = Cx$ (see Section 2.1 in Chapter 2). Introducing a symplectic basis in $V$ one can write the linear Hamiltonian vector field in the form

$$\dot{\xi} = IA\xi,$$

where $A$ is a symplectic matrix, and $I$ is the standard matrix (see Section 2.1). It is known ([**21**], [**22**]) that for the matrix $IA$ the following statement is true.

PROPOSITION. *The eigenvalues of the matrix $IA$ split into pairs of real numbers $(-\lambda, \lambda)$, pairs of imaginary numbers $(i\omega, -i\omega)$, quadruples of complex conjugate numbers $\pm\alpha \pm i\beta$, $\alpha\beta \neq 0$, and the zero eigenvalue. Multiplicity of the zero eigenvalue is always even, and multiplicities of each element of one pair or quadruple are equal. Jordan blocks which correspond to the elements of one pair or quadruple have the same structure, and Jordan blocks of odd dimension which correspond to the eigenvalue zero always appear in pairs.*

THEOREM (Williamson). *A linear real symplectic space $V$ with a quadratic form $H$ is decomposed into a direct sum of mutually skew-orthogonal symplectic subspaces such that $H$ is represented as a sum of standard (quadratic) forms, each corresponding to a given subspace. Each of the subspaces is invariant with respect to the corresponding Hamiltonian system and corresponds to a pair of real eigenvalues $\pm\lambda$, $\lambda \in \mathbb{R}$, $\lambda \neq 0$, to a pair of imaginary eigenvalues $\pm i\omega$, $\omega \in \mathbb{R}$, $\omega \neq 0$, to a quadruple of complex conjugate eigenvalues $\pm\alpha \pm i\beta$, $\alpha\beta \neq 0$, $\alpha$, $\beta \in \mathbb{R}$, or to the zero eigenvalue; it is determined by the Jordan form of the corresponding matrix.*

Let us present the standard forms for the case $n = 2$ assuming that there are symplectic coordinates in $V$ such that $V$ is identified with the coordinate space $\mathbb{R}^4$ with coordinates $(p_1, p_2, q_1, q_2)$, and the skew-symmetric inner product of vectors

$\xi = (p_1, p_2, q_1, q_2)$ and $\eta = (p_1', p_2', q_1', q_2')$ has the form

$$[\xi, \eta] = p_1 q_1' - q_1 p_1' + p_2 q_2' - q_2 p_2'.$$

If $n = 2$ there are 15 different possible cases for $H$ depending on the types of the eigenvalues and Jordan forms of the matrices of the linear Hamiltonian system. Writing down the corresponding normal forms of the Hamilton function $H$, we will also show all possible quadratic functions which are in involution with $H$, i.e., the centralizer $\mathcal{C}$ of the function $H$ in the algebra $sp(4, \mathbb{R})$ of quadratic functions with Poisson bracket as an operation. For each of the cases we will also present the form of the corresponding discriminant $\Delta$ of the subalgebra which is defined by the pair $(H, K)$, $K \in \mathcal{C}$. For every case under consideration we assume $H$ to be fixed.

1. The eigenvalues are real and different, $(\pm\lambda_1, \pm\lambda_2), \lambda_1 \neq \lambda_2, \lambda_1\lambda_2 \neq 0$,

$$H = \lambda_1 p_1 q_1 + \lambda_2 p_2 q_2, \quad K = \mu_1 p_1 q_1 + \mu_2 p_2 q_2.$$

The condition for the algebra to be two-dimensional is $\lambda_1\mu_2 - \lambda_2\mu_1 \neq 0$. In this case, for fixed $\mu_1, \mu_2$ the centralizer $\mathcal{C}$ coincides with the algebra generated by the pair $H, K$. Here the discriminant $\Delta$ does not vanish identically.

2. The eigenvalues are $(\pm\lambda, \pm i\omega)$, $\lambda$, $\omega \in \mathbb{R}$, $\lambda$, $\omega \neq 0$,

$$H = \lambda p_1 q_1 + \frac{\omega}{2}(p_2^2 + q_2^2), \quad K = \mu p_1 q_1 + \nu(p_2^2 + q_2^2).$$

The condition for the algebra to be two-dimensional is $\lambda\nu - \omega\mu \neq 0$. In this case, for fixed $\mu, \nu$, the centralizer $\mathcal{C}$ coincides with the algebra generated by the pair $H, K$. Here the discriminant $\Delta$ does not vanish identically.

3. The eigenvalues are $(\pm i\omega_1, \pm i\omega_2)$, $\omega_1 \neq \omega_2$, $\omega_1, \omega_2 \in \mathbb{R}$, $\omega_1, \omega_2 \neq 0$,

$$H = \frac{\omega_1}{2}(p_1^2 + q_1^2) + \frac{\omega_2}{2}(p_2^2 + q_2^2), \quad K = \frac{\nu_1}{2}(p_1^2 + q_1^2) + \frac{\nu_2}{2}(p_2^2 + q_2^2).$$

The condition for the algebra to be two-dimensional is $\omega_1\nu_2 - \omega_2\nu_1 \neq 0$. In this case, for fixed $\nu_1, \nu_2$ the centralizer $\mathcal{C}$ coincides with the algebra generated by the pair $H, K$. Here the discriminant $\Delta$ does not vanish identically.

4. The eigenvalues are $(\pm\alpha \pm i\beta)$, $\alpha\beta \neq 0$, $\alpha$, $\beta \in \mathbb{R}$,

$$H = \alpha(p_1 q_1 + p_2 q_2) + \beta(p_1 q_2 - p_2 q_1),$$
$$K = \gamma(p_1 q_1 + p_2 q_2) + \delta(p_1 q_2 - p_2 q_1).$$

The condition for the algebra to be two-dimensional is $\alpha\delta - \beta\gamma \neq 0$. In this case, for fixed $\gamma, \delta$ the centralizer $\mathcal{C}$ coincides with the algebra generated by the pair $H, K$. Here the discriminant $\Delta$ does not vanish identically.

5. The eigenvalues $(\pm\lambda, \pm\lambda)$, $\lambda \in \mathbb{R}$, $\lambda \neq 0$, have multiplicity two, the Jordan block which corresponds to $\lambda$ is two-dimensional,

$$H = \lambda(p_1 q_1 + p_2 q_2) + p_1 q_2, \quad K = \mu_1(p_1 q_1 + p_2 q_2) + \tau p_1 q_2.$$

The condition for the algebra to be two-dimensional is $\lambda r - \mu \neq 0$. In this case, for fixed $\mu, \tau$ the centralizer $\mathcal{C}$ coincides with the algebra generated by the pair $H, K$. The discriminant $\Delta$ vanishes identically in this case.

6. The eigenvalues $(\pm\lambda, \pm\lambda)$, $\lambda \in \mathbb{R}$, $\lambda \neq 0$, have multiplicity two, the corresponding Jordan blocks are one-dimensional,

$$H = \lambda(p_1 q_1 + p_2 q_2), \quad K = d_{11} p_1 q_1 + d_{12} p_2 q_1 + d_{21} p_1 q_2 + d_{22} p_2 q_2.$$

The condition for the algebra to be two-dimensional is $(d_{11}-d_{22})^2+d_{12}^2+d_{21}^2 \neq 0$. Put

$$\Delta = (\lambda^2\alpha_1^2 + \lambda\sigma\alpha_1\alpha_2 + d\alpha_2^2)^2\alpha_2^2(\sigma^2-4d),$$

where $\sigma = d_{11}+d_{22}$, $d = d_{11}d_{22}-d_{12}d_{21}$. If $\Delta \neq 0$, we have simple LPA with a point of the saddle-saddle type if $(\sigma^2-4d)>0$, or of focus-focus type if $(\sigma^2-4d)<0$.

7. The eigenvalues $(\pm i\omega, \pm i\omega)$, $\omega \in \mathbb{R}$, $\omega \neq 0$, have multiplicity two, the corresponding Jordan blocks are two-dimensional,

$$H = \frac{\kappa}{2}(q_1^2+q_2^2) + \omega(p_2q_1 - p_1q_2), \quad K = \frac{\tau}{2}(q_1^2+q_2^2) + \sigma(p_2q_1 - p_1q_2),$$

$\kappa = \pm 1$. The condition for the algebra to be two-dimensional is $\kappa\sigma - \tau\omega \neq 0$. For fixed $\tau, \sigma$, the centralizer $\mathcal{C}$ coincides with the algebra spanned by the pair $H, K$. The discriminant $\Delta$ is equal to zero.

8. The eigenvalues $(\pm i\omega, \pm i\omega)$, $\omega \in \mathbb{R}$, $\omega \neq 0$ have multiplicity two, the corresponding Jordan blocks are one-dimensional,

$$\begin{aligned} H &= \frac{\omega}{2}(p_1^2+q_1^2) + \kappa\frac{\omega}{2}(p_2^2+q_2^2), \\ K &= \frac{d_{11}}{2}(p_1^2+q_1^2) + \frac{d_{22}}{2}(p_2^2+q_2^2) \\ &\quad + d_{12}(p_2q_1 - \kappa p_1q_2) + d_{21}(\kappa p_1p_2 + q_1q_2), \end{aligned}$$

$\kappa = \pm 1$. The condition for the algebra to be two-dimensional is

$$(d_{11}-\kappa d_{22})^2 + \omega d_{12}^2 + d_{21}^2 \neq 0.$$

Put $\Delta = P(\alpha_1,\alpha_2)((d_{11}-kd_{22})^2 + 4k(d_{12}^2+d_{21}^2))$, where

$$\begin{aligned} P(\alpha_1,\alpha_2) &= -\frac{\alpha_2^2}{2}[(\omega\alpha_1 + d_{11}\alpha_2)(k\omega\alpha_1 + d_{22}\alpha_2) - (d_{12}^2+d_{21}^2)\alpha_2^2)]^2 \\ &\quad \times [2\omega\alpha_1 + (d_{11}+kd_{22})\alpha_2)]^2. \end{aligned}$$

It is obvious that $P$ does not vanish identically. If $\Delta \neq 0$, we obtain a simple LPA. Moreover, if $\delta = (d_{11}-kd_{22})^2 + 4k(d_{12}^2+d_{21}^2) \geq 0$, then we have an elliptic point, and if $\delta < 0$, a point of the focus-focus type.

9. The eigenvalues are $(\pm\lambda, \pm 0)$, $\lambda \in \mathbb{R}$, $\lambda \neq 0$, the corresponding Jordan block of zero eigenvalue is two-dimensional,

$$H = \lambda p_1q_1 + \frac{\kappa}{2}q_2^2, \quad K = \mu p_1q_1 + \frac{\sigma}{2}q_2^2,$$

$\kappa = \pm 1$. The condition for the algebra to be two-dimensional is $\lambda\sigma - k\mu \neq 0$. For fixed $\mu, \sigma$ the centralizer $\mathcal{C}$ coincides with the algebra generated by the pair $H, K$. The discriminant $\Delta$ is equal to zero in this case.

10. The eigenvalues are $(\pm i\omega, \pm 0)$, $\omega \in \mathbb{R}$, $\omega \neq 0$, Jordan block of zero eigenvalue is two-dimensional,

$$H = \frac{\omega}{2}(p_1^2+q_1^2) + \frac{\kappa}{2}q_2^2, \quad K = \nu(p_1^2+q_1^2) + \frac{\sigma}{2}q_2^2.$$

The condition for the algebra to be two-dimensional is $\omega\sigma - \kappa\mu \neq 0$. For fixed $\nu, \sigma$, the centralizer $\mathcal{C}$ coincides with the algebra generated by the pair $H, K$. The discriminant $\Delta$ is equal to zero in this case.

11. The eigenvalues $(\pm\lambda, \pm\lambda)$, $\lambda \in \mathbb{R}$, $\lambda \neq 0$, have multiplicity two, the Jordan blocks corresponding to zero eigenvalues are one-dimensional,

$$H = \lambda p_1q_1, \quad K = \mu p_1q_1 + \frac{1}{2}(ap_2^2 + 2bp_2q_2 + cq_2^2).$$

The condition for the algebra to be two-dimensional is $a^2 + b^2 + c^2 \neq 0$. $\Delta = P(\alpha_1, \alpha_2)(b^2 - ac)$, where

$$P(\alpha_1, \alpha_2) = -\frac{1}{4}\alpha_2^2(\lambda\alpha_1 + \mu\alpha_2)^2((b^2 - ac)\alpha_2^2 - (\lambda\alpha_1 + \mu\alpha_2)^2)^2$$

does not vanish identically. Note that if $b^2 - ac \neq 0$ one has a simple LPA. Moreover, if $b^2 - ac > 0$, we have a point of the saddle-saddle type, if $b^2 - ac < 0$, we have a point of the saddle-center type.

12. The eigenvalues $(\pm i\omega, \pm 0)$, $\omega \in \mathbb{R}$, $\omega \neq 0$, have multiplicity two, Jordan blocks of zero eigenvalues are one-dimensional,

$$H = \frac{\omega}{2}(p_1^2 + q_1^2), \quad K = \nu(p_1^2 + q_1^2) + \frac{1}{2}(ap_2^2 + 2bp_2q_2 + cq_2^2).$$

The condition for the algebra to be two-dimensional is $a^2 + b^2 + c^2 \neq 0$. $\Delta = P(\alpha_1, \alpha_2)(b^2 - ac)$, where

$$P(\alpha_1, \alpha_2) = \frac{1}{4}\alpha_2^2(\omega\alpha_1 + \nu\alpha_2)^2((b^2 - ac)\alpha_2^2 + (\omega\alpha_1 + \nu\alpha_2)^2)^2$$

does not vanish identically. Note that if $b^2 - ac \neq 0$ we have a simple LPA. Moreover, if $b^2 - ac > 0$, we have a point of the saddle-center type, and if $b^2 - ac < 0$, we have an elliptic point.

13. Zero eigenvalue with multiplicity four forming a four-dimensional Jordan block,

$$H = \frac{k}{2}(p_1^2 - 2q_1q_2) - p_1q_2, \quad K = \frac{\nu}{2}(p_1^2 - 2q_1q_2) - (\kappa\nu p_1q_2 + \frac{1}{2}bq_2^2),$$

$\kappa = \pm 1$. The condition for the algebra to be two-dimensional is $b \neq 0$. For fixed $\nu, b$, the centralizer $\mathcal{C}$ coincides with the algebra generated by $H, K$. The discriminant $\Delta$ vanishes identically in this case.

14. Zero eigenvalue with multiplicity four forming two two-dimensional Jordan blocks

$$H = \frac{1}{2}(q_1^2 + \kappa q_2^2), \quad K = \frac{a}{2}q_1^2 + \frac{b}{2}q_1^2 + c(p_2q_1 - \kappa p_1q_2) + dq_1q_2,$$

$\kappa = \pm 1$. The condition for the algebra to be two-dimensional is $(a\kappa - b)^2 + c^2 + d^2 \neq 0$. The discriminant $\Delta$ vanishes identically in this case.

15. Zero eigenvalue with multiplicity four and with one two-dimensional Jordan block and two one-dimensional ones,

$$H = \frac{\kappa}{2}q_1^2, \quad K = \frac{a}{2}p_2^2 + p_2(b_1q_1 + b_2q_2) + \frac{1}{2}(c_{11}q_1^2 + 2c_{12}q_1q_2 + c_{22}q_2^2),$$

$\kappa = \pm 1$. The condition for the algebra to be two-dimensional is $a^2 + b_1^2 + b_2^2 + c_{12}^2 + c_{22}^2 \neq 0$. The discriminant $\Delta$ vanishes identically in this case. If $b_2^2 - ac_{22} \neq 0$, we have Cases 9 or 10.

APPENDIX B

# The Gradient System on $M$ Compatible with the Hamiltonian

In the previous chapters we often used a special vector field constructed by means of the pair of functions $H, K$ on the manifold $M$. Here we give the general construction of such a field and consider some of its properties.

Suppose that $M$ is equipped with a Riemannian metric. One of possible definitions of a Riemannian metric is the following ([**59**]). Let $D^1(M)$ be the space of $C^\infty$-smooth vector fields on $M$ with $C^\infty$-topology, and $C^\infty(M)$ the space of $C^\infty$-smooth real functions on $M$. The Riemannian metric $g$ on $M$ is a $C^\infty$ bilinear map

$$g : D^1(M) \times D^1(M) \to C^\infty(M)$$

which has the following properties:

1. locality: $g(\phi u, \psi v) = \phi\psi g(u, v)$, for every $u, v \in D^1(M), \phi, \psi \in C^\infty(M)$;
2. symmetry: $g(u, v) = g(v, u)$;
3. positive definiteness: $g(u, u) \geq 0$ and $g(u, u) = 0 \iff u \equiv 0$.

It is known that there is a smooth Riemannian metric on every smooth paracompact manifold.

As soon as one has a Riemannian metric on $M$, given a smooth function $F \in C^\infty(M)$ it is possible to determine uniquely the smooth vector field $\nabla F$ called the gradient field of the function $F$ or the gradient vector field constructed by the function $F$. Let us define the isomorphism $R$ between $D^1(M)$ and the space $D_1(M)$ of smooth 1-forms on $M$ by setting $R(v)$ to be the 1-form which is defined as $g(\cdot, v)$.

If $F \in C^\infty(M)$, then $R^{-1}dF$ is the required vector field $\nabla F$. The definition implies that

$$g(u, \nabla F) \equiv dF(u), \tag{B.1}$$

for every $u \in D^1(M)$.

The field $\nabla F$ has the property that the function $F$ does not decrease along the trajectories of $\nabla F$. Moreover, if $\nabla F(x) \neq 0$, then the function $F$ strictly increases in a neighborhood of the point $x$. In particular, in a region which has a compact closure the only possible $\omega$- and $\alpha$-limit points of the trajectories of the field $\nabla F$ are the singular points of the field $\nabla F$, i.e., the critical points of the function $F$.

Suppose now that there are two smooth functions $H$ and $K$ on the manifold $M$. Then it is possible to consider the vector field

$$Y = -\nabla K + \frac{g(\nabla K, \nabla H)}{g(\nabla H, \nabla H)} \nabla H. \tag{B.2}$$

This vector field is defined everywhere except for the points where $\nabla H = 0$. Moreover, since

$$\|Y\| = \max_X \sqrt{g(Y,Y)} \leq 2\|\nabla K\|, \tag{B.3}$$

the vector field (B.1) is bounded in any compact region. It is easy to see that the field $Y$ is a projection to the tangent space to the level submanifold $H = \text{const}$ of the vector field $-\nabla K$ along the direction of the field $\nabla H$.

LEMMA B.1. *The function $H$ is a first integral of the field $Y$.*

PROOF. Obviously, we have $L_Y H = dH(Y) = g(Y, \nabla H) = -g(\nabla K, \nabla H) + g(\nabla K, \nabla H) = 0$. □

It follows from the lemma that the level sets $H = c$ are invariant with respect to the flow of the vector field $Y$. At all points such that $\nabla H \neq 0$, these sets are smooth submanifolds.

It follows from (B.3) that critical points of the function $K$ are also singular points of the field $Y$, regardless of whether they are critical for the function $H$ or not. Therefore, at any point $x_0$ such that $\nabla H(x_0) = \nabla K(x_0) = 0$, the field $Y$ can be extended by continuity to be equal to zero. Later we use the field $Y$ in situations where either $\nabla H \neq 0$ or if $\nabla H(x_0) = 0$, then $\nabla K(x_0) = 0$ as well.

Suppose $dH \neq 0$ at a point $x_0$. Then $\nabla H(x_0) \neq 0$ and the field $Y$ is defined in a neighborhood of the point $x_0$. Consider the trajectory of the field $Y$ through the point $x_0$. Similarly to the case of a gradient vector field the following statement holds.

LEMMA B.2. *If the fields $\nabla K$ and $\nabla H$ are noncollinear at the point $x_0$, then, in a neighborhood of the point $x_0$ the function $K$ decreases along the trajectories of the field $Y$.*

PROOF. Let $x(t)$ denote the solution of the equation $\dot{x} = Y$ passing through the point $x_0$ at $t = 0$, and $\alpha = g(\nabla K, \nabla H)/g(\nabla H, \nabla H)$. Then, due to the Cauchy–Bunyakovskii inequality,

$$\begin{aligned} \frac{d}{dt} K(x(t))|_{t=0} &= dK(Y) = dK(-\nabla K) + \alpha dK(\nabla H) \\ &= -g(\nabla K, \nabla K) + \alpha g(\nabla H, \nabla K) \\ &= \frac{-g(\nabla K, \nabla K) g(\nabla H, \nabla K) + g(\nabla H, \nabla K)^2}{g(\nabla H, \nabla H)} < 0. \end{aligned}$$

The lemma is proved. □

LEMMA B.3. *If $\nabla H \neq 0$ at the point $x_0$, then $x_0$ is a singular point of the field $Y$ if and only if the vectors $\nabla K(x_0)$ and $\nabla H(x_0)$ are collinear.*

PROOF. If $Y(x_0) = 0$ and $\nabla H(x_0) \neq 0$, then $\nabla K(x_0) = \alpha \nabla H(x_0)$. Conversely, if $\nabla K(x_0) = \rho \nabla H(x_0)$, then $Y(x_0) = 0$. The lemma is proved. □

Now we give another characterization of the field $Y$.

Let $V_c$ be a nonsingular level hypersurface $H = c$, i.e., $dH \neq 0$ on $V_c$. By virtue of Lemma B.1, $V_c$ is an invariant submanifold of the field $Y$.

PROPOSITION B.1. *The restriction of the field $Y$ to the invariant hypersurface $V_c$ coincides with the gradient field constructed by means of the restriction of the function $K$ to $V_c$ with respect to the Riemannian metric $g_c$ which is the restriction of the metric $g$ to $V_c$.*

PROOF. Since the property we consider is a local one, we conduct the proof using local coordinates $(x_1, \dots, x_n)$ such that $H$ is equal to $x_1$. The existence of such coordinates follows from the condition $dH \neq 0$.

Recall that in any local coordinates $(x_1, \dots, x_n)$, for any two vector fields $u = \sum a_i(x)\partial/\partial x_i$ and $v = \sum b_i(x)\partial/\partial x_i$ the Riemannian metric $g$ has the form

$$g\Big(\sum_{i=1}^{n} a_i(x)\partial/\partial x_i, \sum_{j=1}^{n} b_i(x)\partial/\partial x_i\Big) = \sum_{i,j=1}^{n} g_{ij}(x)a^i(x)b^j(x),$$

where $g_i(x) = g(\frac{\partial}{\partial x_i}, \frac{\partial}{\partial x_j})$, and $G = (g_{ij}(x))$ is a symmetric, positive definite matrix whose elements are $C^\infty$ functions. Denote by $\hat{G} = (h_{ij}(x))$ the inverse matrix $G^{-1}$. It is also symmetric and positive definite. Using the definition of the gradient of the function $H$,

$$dH_x(\xi) = g(\xi, \nabla H),$$

for any $x \in M$, $\xi \in T_xM$ we obtain the representation of this gradient in local coordinates. Namely,

$$\sum g_{ij}\xi^i(\nabla H)^j = \sum \frac{\partial H}{\partial x_i} dx_i(\xi).$$

Hence

$$\sum_i \Big(\sum_j g_{ij}(\nabla H)^j\Big)\xi^i = \sum \frac{\partial H}{\partial x_i}\xi^i.$$

Since the vector $\xi$ is arbitrary, we have

$$\frac{\partial H}{\partial x_i} = \sum_{j=1}^{n} g_{ij}(\nabla H)^j, \quad i = 1, \dots, n,$$

or, using the matrix representation, $\nabla_e H = G\nabla H$, i.e., $\nabla H = G^{-1}\nabla_e H$, where $\nabla H = (\frac{\partial H}{\partial x_1}, \dots, \frac{\partial H}{\partial x_n})^T$ is the Euclidean gradient of the function $H$ written in the coordinates $(x_1, \dots, x_n)$. In these coordinates the vector field $Y$ has the following form:

$$Y = \hat{G}\left(-\begin{pmatrix} \frac{\partial K}{\partial x_1} \\ \vdots \\ \frac{\partial K}{\partial x_n} \end{pmatrix} + \frac{\sum_{j=1}^{n} h_{1j}\frac{\partial K}{\partial x_j}}{h_{11}}\begin{pmatrix} 1 \\ 0 \\ \vdots \\ 0 \end{pmatrix}\right). \tag{B.4}$$

Indeed,

$$\begin{aligned} g(\nabla K, \nabla H) &= dH(\nabla K) \\ &= \sum_{i=1}^{n} \frac{\partial H}{\partial x_i} dx_i(\nabla K) = \sum_i \frac{\partial H}{\partial x_i} dx_i\Big(\sum_k \Big(\sum_j h_{kj}\frac{\partial K}{\partial x_j}\Big)\frac{\partial}{\partial x_k}\Big) \\ &= \sum_{i=1}^{n} \frac{\partial H}{\partial x_i}\Big(\sum_{j=1}^{n} h_{ij}\frac{\partial K}{\partial x_j}\Big) = (\hat{G}\nabla_e K, \nabla_e H). \end{aligned}$$

Here $(\cdot,\cdot)$ denotes the standard Euclidean inner product with respect to the coordinates $(x_1,\dots,x_n)$. It follows from (B.4) that $\dot{x}_1 = 0$, which is an expression of the fact that the function $H$ is an integral of the field $Y$. The other property of the field $Y$ is that the coefficients of $\frac{\partial K}{\partial x_1}$ in the equations for $\dot{x}_i$, $i \geq 2$, all vanish. This means that the equations for $x_i$, $i \geq 2$, depend on $x_1$ parametrically, i.e., via entries of the matrix $\hat{G}$, where $x_i = \text{const}$. Thus, the system of equations for $\dot{x}_2,\dots,\dot{x}_n$ has the form

$$\dot{x}_i = -\hat{G}_c \nabla_e K_c, \quad \hat{G}_c = (d_{ij}), \tag{B.5}$$

where

$$d_{ij} = \frac{h_{11}h_{ij} - h_{i1}h_{1j}}{h_{11}}, \quad i,j = 2,\dots,n, \qquad K_c = K(c, x_2, \dots, x_n).$$

Now let us find the form of the matrix $G_c$ of the Riemannian metric $g_c$ in our coordinates. It is obvious that the vector $\xi \in T_xM$ belongs to the tangent space to $V_c$ at the point $x$ if and only if it has zero first coordinate (recall that in our coordinates the hypersurface $V_c$ is given by the equation $x_1 = c$).

Let $\xi$ be a tangent vector to $V_c$ at some point $x \in V$. Then its coordinates are $(0, \xi_2, \dots, \xi_n)$. For two such vectors $\xi$, $\eta$ from $T_xV$ one has

$$(G_c\xi_c, \eta_c) = g_c(\xi_c, \eta_c) = g(\xi, \eta) = (G\xi, \eta).$$

Since the first coordinates of the vectors $\xi$ and $\eta$ are equal to zero, it is easy to see that

$$(G\xi, \eta) = \sum_{i,j=2}^{n} g_{ij}\xi^i\eta^j,$$

where we have substituted $c$ instead of $x_1$ in the functions $g_{ij}$. Hence the matrix $G_c$ is obtained from the matrix $G$ by deleting the first column and the first row. It follows from the symmetry and positive definiteness of the matrix $G$ that the matrix $G_c$ is also symmetric and positive definite.

Let us show now that the matrix $\hat{G}_c$ in (B.5) is the inverse to $G_c$, which means that the field (B.4) is the gradient field on $V_c$ constructed by the restriction of the function $K$ with respect to the matrix $g_c$. Recall that the matrices $G$ and $\hat{G}$ are mutually inverse, i.e., we have the equalities $\sum g_{ij}h_{jk} = \delta_{ik}$, where $\delta_{ik}$ is the Kronecker symbol. Therefore, for the matrices $G_c$ and $\hat{G}_c$ and for any $i,k = 2,\dots,n$ we have

$$\begin{aligned}
\sum_{j=2}^{n} g_{ij}d_{jk} &= \sum_{j=2}^{n} g_{ij}\frac{-h_{j1}h_{1k} + h_{jk}h_{11}}{h_{11}} = -\frac{h_{1k}}{h_{11}}\sum_{j=2}^{n} g_{ij}h_{j1} + \sum_{j=2}^{n} g_{ij}h_{jk} \\
&= -\frac{h_{1k}}{h_{11}}\Big(\sum_{j=1}^{n} g_{ij}h_{j1} - g_{i1}h_{11}\Big) + \sum_{j=1}^{n}(g_{ij}h_{jk} - g_{i1}h_{1k}) \\
&= -\frac{h_{1k}}{h_{11}}\delta_{i1} + h_{k1}g_{i1} + \delta_{ik} - g_{i1}h_{1k} = \delta_{ik},
\end{aligned}$$

as $i \geq 2$. The statement is proved. □

There are two corollaries of the last proposition which we use.

PROPOSITION B.2. *Let $D \subset M$ be a compact region in which either $\nabla H \neq 0$ or if $\nabla H(x_0) = 0$, then $\nabla K(x_0) = 0$. Then any positive (negative) semitrajectory $L$ of the field $Y$ through a point $a \in D$ either reaches the boundary of the region as* *time increases (decreases) or its $\omega$-limit ($\alpha$-limit) set consists of singular points of the field $Y$ or of points $x_0$ such that $\nabla H(x_0) = 0$.*

PROOF. Consider, for example, the case of a positive semitrajectory. It follows from the conditions of the proposition that the field $Y$ is defined in the whole region $D$. Moreover, we extend the field $Y$ by continuity to be equal to zero at all points $x_0$ such that $\nabla H(x_0) = 0$. Hence through a given point of $D$ passes a unique trajectory of the field $Y$. Suppose a semitrajectory $L$ stays in $D$ for all $t > 0$. Then its $\omega$-limit set also belongs to $D$. It follows from the invariance of the level sets $V_c = \{H = c\}$ that the field is the gradient field on $V_c$ (see Proposition B.1), so that the $\omega$-limit set of the semitrajectory $L$ consists of the singular points of the field $Y$ (see [**60**]). The proposition is proved. □

Taking into account the proposition above, our next problem is to show that the linearization operator of the vector field (B.5) has the same spectrum structure at a singular point as the linearization operator at the singular point of the vector field $\nabla_e K_c$. In order to do this we prove the following algebraic statement.

LEMMA B.4. *Let $A$ and $B$ be two real symmetric $n \times n$ matrices, and let $B$ be positive definite. Then the numbers of positive, negative, and zero eigenvalues of the matrix $A$ are equal to those of the matrix $BA$.*

PROOF. It follows from the equality $\det(BA - \lambda E) = \det B \det(A - \lambda B^{-1})$ that the eigenvalues of the matrix $BA$ coincide with the eigenvalues of the regular bundle of the matrices $A - \lambda B^{-1}$ ([**39**]), i.e., with the solutions of the equation $\det(A - \lambda B^{-1}) = 0$. It follows from the condition of the lemma that the matrix $BA$ has only real eigenvalues ([**39**]). Consider the quadratic form $(A\xi, \xi)$, where, as before, $(\cdot, \cdot)$ is the standard inner product. An orthogonal transformation $\xi = C\xi$ reduces the form to the principal axes, i.e., to the form $\lambda_1 \eta_1^2 + \cdots + \lambda_n \eta_n^2$, where $\lambda_1, \ldots, \lambda_n$ are the eigenvalues of the matrix $A$. On the other hand, by a nondegenerate linear transformation $\xi = Z\zeta$, the quadratic forms $(A\xi, \xi)$ and $(B\xi, \xi)$ can be reduced to the forms $\mu_1 \zeta_1^2 + \cdots + \mu_n \zeta_n^2$ and $\zeta_1^2 + \cdots + \zeta_n^2$, respectively. Here $\mu_1, \ldots, \mu_n$ are the eigenvalues of the bundle of the matrices $A - \lambda B^{-1}$, i.e., the eigenvalues of the matrix $BA$ (see [**39**], Chapter 10). Finally, we obtain the statement of the lemma from the inertia law for quadratic forms. □

Now we return to the field $Y$. Let $x_0$ be a singular point for the field $Y$ and satisfy the condition that $\nabla H(x_0) \neq 0$. Let $K_c$ be the restriction of the function $K$ to $V_c$. Then the point $x_0$ is critical for the function $K_c$. Thus, the Hessian of the function $K_c$ is well defined at the point $x_0$. In the local coordinates the Hessian is the matrix $A$ of second derivatives of the function $K_c$ calculated at the point $x_0$. This matrix is symmetric. It is easy to see that the linearization of the field (B.5) at the point $x_0$ has the form

$$\dot{\xi} = \hat{G}_c(x_0) A \xi. \tag{B.6}$$

Recall that the matrix $\hat{G}_c$ is symmetric and positive definite. Thus, Lemma B.4 implies the following result.

PROPOSITION B.3. *The structures of the eigenvalues of the matrix $A$ and the matrix $\hat{G}_c(x_0)A$ coincide.*

Now let $M$ be a smooth four-dimensional symplectic manifold, and $(X_H, K)$ an IHVF. Fix a Riemannian metric on $M$. Then the isoenergetic submanifolds

$V_c = \{H = c\}$ are invariant subsets of the field $Y$. The following statement about the singular points of the field $Y$ is true.

PROPOSITION B.4. *Suppose IHVF $(X_H, K)$ is given. A point $x_0 \in M$ such that $dH_{x_0} \neq 0$ is a singular point of the field $Y$ if and only if $x_0$ belongs to a one-dimensional orbit of the induced action.*

The statement follows from Lemma B.3.

Let us now study the behavior of trajectories of the field $Y$ in a neighborhood of a point of a one-dimensional orbit $\gamma$ of the induced action. It follows from the results of Chapter 2 that, in a neighborhood of a point $a \in \gamma$ such that $dH_a \neq 0$, there are symplectic coordinates $(x_1, x_2, y_1, y_2)$ such that $H = x_1$, $K = k(x_1, x_2, y_2)$. In these coordinates points of one-dimensional orbits satisfy the system of equations $k_{x_2} = 0$, $k_{y_2} = 0$. In the case of elliptic or hyperbolic one-dimensional orbit the solution of the system forms a two-dimensional surface given by the functions $x_2 = \phi(x_1)$, $y_2 = \psi(x_1)$. This solution is foliated by segments of one-dimensional orbits, the coordinate $y_1$ serving as a parameter on a fixed orbit $\gamma_c$. This orbit is obtained when $x_1 = c = H(a)$. Moreover, the determinant $\Gamma_0 = k_{x_2x_2}k_{y_2y_2} - k^2_{x_2y_2}$ is positive for an elliptic orbit and negative for a hyperbolic one. In a neighborhood of the point $a$ on the level set $V_c$, for $|c - H(a)|$ sufficiently small, the restriction of the field $Y$ has the form (B.4). As was pointed out above, singular points of the restriction of the field $Y$ to the level set $V_c$ are the points of the curve $\gamma_c$.

It follows from Proposition B.3 and the form of the matrix $A$,

$$A = \begin{pmatrix} k_{x_2x_2} & 0 & k_{x_2y_2} \\ 0 & 0 & 0 \\ k_{x_2y_2} & 0 & k_{y_2y_2} \end{pmatrix},$$

that for an elliptic orbit, the restriction $Y_c$ of the field $Y$ to $V_c$ has one zero and a couple of real eigenvalues of the same sign at each singular point, and for a hyperbolic orbit it has one zero and a couple of real eigenvalues of different signs.

Consider now the field $Y$ in a four-dimensional region $U$ in $M$. We assume that there is a smooth two-dimensional symplectic submanifold $\Sigma$ in $U$ which is foliated by one-dimensional orbits of the action $\Phi$. Let us assume that all orbits are either elliptic (Case 1) or hyperbolic (Case 2). If all the orbits are closed, we assume that $\Sigma$ is compact. If there is a nonclosed orbit in $\Sigma$, we assume that there is a curvilinear rectangle foliated by segments of one-dimensional orbits. Moreover, two of its boundary segments are transversal to the orbit and the remaining two are segments of orbits.

The structure of the vector field $Y$ in a neighborhood $V \subset U$ which contains $\Sigma$ is described as follows.

PROPOSITION B.5. *In Case 1, there exists a neighborhood $V$ of the set $\Sigma$ such that for every point $a \in \Sigma$ there is a unique smooth two-dimensional disk $D_a$ which belongs to the level set $H = H(a)$ and is a strong stable (if the nonzero eigenvalues of the linearized system are negative) or strong unstable (if the nonzero eigenvalues of the linearized system are positive) manifold of the vector field $Y$ at the point $a$. The neighborhood $V$ is smoothly foliated by the disks $D_a$ and is diffeomorphic to the direct product $D \times \Sigma$.*

*In Case 2, there exists a neighborhood $V$ of the set $\Sigma$ such that every point $a \in \Sigma$ has smooth one-dimensional strong stable manifold $W^s(a)$ and strong unstable manifold $W^u(a)$, and $W^s(a) \cap W^u(a) = \{a\}$. The sets $W^s_\Sigma = \bigcup_{a\in\Sigma} W^s(a)$ and*

*$W^u_\Sigma = \bigcup_{a\in\Sigma} W^u(a)$ form two smooth three-dimensional submanifolds with boundary. They are diffeomorphic to $I \times \Sigma$ if $\Sigma$ is a curvilinear rectangle and in the case of the orientable hyperbolic closed orbits. In the case when $\Sigma$ is foliated by nonorientable closed one-dimensional hyperbolic orbits, $\Sigma$ is diffeomorphic to a fiber bundle with an annulus as the base and segment as the fiber. Moreover, $W^s_\Sigma$ and $W^u_\Sigma$ intersect transversally along $\Sigma$.*

The proof of the part of this proposition related to the structure of the strong stable and strong unstable manifolds follows from the results of Hirsch, Pugh, and Shub [**61**]. The topological description of the manifolds follows from Proposition 2.3.2 (for one-dimensional hyperbolic orbits) and the general theory of fiber bundles in case when the base is an annulus or a disk [**41**].

The last result of this appendix, which is used in Chapters 6 and 8 to construct foliations near one-dimensional hyperbolic orbits, concerns with the properties of the vector field $Y$ in a neighborhood of a segment of a one-dimensional hyperbolic orbit. Let $\gamma$ be a segment of such an orbit which belongs to the level set $V_c$. It follows from Proposition B.5 that the curve $\gamma$ on $V_c$ is a transversal intersection of two smooth two-dimensional manifolds $W^s_\gamma$ and $W^u_\gamma$ (the manifolds $W^s_\gamma$ and $W^u_\gamma$ are the intersections of $W^s_\Sigma$ and $W^u_\Sigma$ with the invariant submanifold $V_c$). The manifolds $W^s_\gamma$ and $W^u_\gamma$ are foliated by the one-dimensional strong stable (respectively, strong unstable) manifolds of the singular points in $\gamma$. For definiteness, fix a point $m \in W^s_\gamma\backslash\gamma$. Then there is a semitrajectory $L^+$ from $W^s_\gamma$ that passes through the point $m$. The $\omega$-limit set of this trajectory is a point $a \in \gamma$. The point $a$ is the $\alpha$-limit set of two semitrajectories in $W^u_\gamma$. We will denote them by $L_1^-$ and $L_2^-$. Let us choose a sufficiently small neighborhood $\sigma \subset V_c$ of the point $m$. The neighborhood $\sigma$ can be chosen in such a way that it is divided by the submanifolds $W^s_\gamma$ into two half-neighborhoods. We denote these half-neighborhoods by $\sigma_l$ and $\sigma_r$. Also, choose a submanifold $N^s \subset V_c$ transversal to $L^+$ at a point $m_1^+ \in L^+$, such that the motion from $m_1^+$ to $m$ along the trajectory $L^+$ be accomplished in a positive time $T$. Now let us fix two points $m_1^- \in L_1^-$ and $m_2^- \in L_2^-$ and two manifolds $N_1^u$ and $N_2^u$ transversal to $L_1^-$ and $L_2^-$ at the points $m_1^-$ and $m_2^-$, respectively. It follows from Lemma B.2 that the function $K$ decreases along each trajectory of the vector field $Y$. In particular, the same is true for the combined curves $L^+ \cup L_1^- \cup \{a\}$ and $L^+ \cup L_2^- \cup \{a\}$. Thus, such a curve (a trajectory or a combined curve) can be parametrized by $K$.

LEMMA B.5. *Let $\{r_n\}$ be a sequence of points that belongs to $\sigma_r$ and converges to the point $m$. Then, for $n$ large enough each trajectory $L(r_n)$ of the vector field $Y$ that passes at $t = 0$ through the points $r_n$, intersects $N^s$ and the submanifold $N_1^u$. Moreover, as $n \to \infty$, the segment of the trajectory $L(r_n)$ between $N^s$ and $N_1^u$ converges, uniformly with respect to the parameter $K$, to the combined curve $L_r = L^+ \cup L_1^- \cup \{a\}$. A similar statement is true for a sequence of points in $\sigma_l$, the limit being the curve $L_l = L^+ \cup L_2^- \cup \{a\}$.*

PROOF. To prove the lemma we use the Soshitaishvili theorem [**62**] which implies that in a neighborhood of the point $a$ in $V_c$ the vector field $Y$ is topologically equivalent to the following vector field in $\mathbb{R}^3$ in a neighborhood of the origin:

$$\dot{x} = -x, \quad \dot{y} = y, \quad \dot{z} = 0. \tag{B.7}$$

Indeed, the central manifold of the restriction of the vector field $Y$ to $V_c$ is the curve of singular points $\gamma$, and the two nonzero eigenvalues are real and have different signs.

Let $h$ be the homeomorphism that realizes this equivalence. Then $h(a) = (0,0,0)$ and the $C^0$-surfaces $h(N^s)$ and $h(N^u)$ are surfaces of one-time intersection of the vector field (B.7), and $h$ maps the stable and unstable manifolds of the field $Y$ onto stable and unstable manifolds of the field (B.7), respectively. Since $r_n \to m$ and $m \in W^s_\gamma$, we have that $h(r_n) \to h(m) = (x_0, 0, 0)$, $h(r_n) = (x_n, y_n, z_n)$, $x_n \to x_0$, $y_n \to 0$, $z_n \to 0$, where, for definiteness, we assume that $x_0 > 0$. The condition $r_n \in \sigma_r$ means that the $y$-coordinate of each point $h(r_n)$ has the same sign, say positive. This clearly implies that as $n \to \infty$, the trajectories of the vector field (B.7) passing through the points $h(r_n)$ converge to the curve that consists of two segments $y = z = 0$, $0 \leq x \leq h(m_1^+)$ and $x = z = 0$, $0 \leq y \leq h(m_1^-)$. Due to the fact that the trajectories of the vector field $Y$ are transversal to the surfaces $K = \text{const}$ on $V_c$, the trajectories of the vector field (B.7) can also be parametrized by the parameter $K$ transferred by the homeomorphism $h$. The lemma is proved. $\square$

# Bibliography


[1] V. I. Arnold, *A theorem of Liouville concerning integrable problems of dynamics*, Sibirsk. Mat. Zh. **4** (1963), no. 2, 471–474; English transl. in Siberian Math. J. **4** (1963).

[2] S. Smale, *Topology and mechanics*, 1, 2, Invent. Math. **10** (1970), 305–331; **11** (1970) 45–64.

[3] J. R. Marsden and A. Weinstein, *Reduction of symplectic manifolds with symmetry*, Reports Math. Phys. **5** (1974), no. 1, 121–130.

[4] L. M. Lerman and Ya. L. Umanskii, *The structure of a Poisson action of* $\mathbb{R}^2$ *on a four-dimensional symplectic manifold*, Selecta Math. Sovietica **6** (1987), no. 4, 365–396.

[5] ———, *The structure of a Poisson action of* $\mathbb{R}^2$ *on a four-dimensional symplectic manifold.* 2, Selecta Math. Sovietica **7** (1988), no. 1, 39–48.

[6] ———, *On the classification of four-dimensional Hamiltonian systems in extended neighborhoods of simple singular points*, Methods of the Qualitative Theory of Differential Equations (E. A. Leontovich-Adronova, ed.), Gorkov. Gos. Univ., Gorki, 1988, pp. 67–75. (Russian)

[7] ———, *Classification of four-dimensional Hamiltonian systems and Poisson actions of* $\mathbb{R}^2$ *in extended neighborhoods of simple singular points*, 1, Mat. Sb. **183** (1992), no. 12, 141–176; English transl., Russian Acad. Sci. Sb. Math. **77** (1993), 511–542.

[8] ———, *Classification of four-dimensional Hamiltonian systems and Poisson actions of* $\mathbb{R}^2$ *in extended neighborhoods of simple singular points*, 2, Mat. Sb. **184** (1993), no. 4, 105–138; English transl., Russian Acad. Sci. Sb. Math. **78** (1994), 479–506.

[9] A. T. Fomenko, *The topology of surfaces of constant energy of integrable Hamiltonian systems and obstructions to integrability*, Izv. Akad. Nauk SSSR Ser. Mat. **50** (1986), no. 6, 1276–1307; English transl., Math USSR-Izv. **29** (1987), 629–658.

[10] ———, *Topological invariants of Hamiltonian systems that are integrable in the sense of Liouville*, Funktsional. Anal. i Prilozhen. **22** (1988), no. 4, 38–51; English transl. in Functional Anal. Appl. **22** (1988).

[11] A. T. Fomenko and H. Zieschang, *A criteria for topological equivalence of integrable Hamiltonian systems with two degrees of freedom*, Izv. Akad. Nauk SSSR Ser. Mat. **54** (1990), no. 3, 546–575; English transl. in Math USSR-Izv. **36** (1991).

[12] A. T. Fomenko, *Integrability and nonintegrability in geometry and mechanics*, Kluwer, Dordrecht, 1988.

[13] L. Marcus and K. S. Meyer, *Generic Hamiltonian dynamical systems are neither integrable nor ergodic*, Mem. Amer. Math. Soc., no. 144 (1974).

[14] V. I. Arnold, V. V. Kozlov, and A. I. Neishtadt, *Mathematical aspects of classical and celestial mechanics*, Encyclopedia of Mathematical Sciences, vol. 3, Springer-Verlag, Berlin–Heidelberg–New York, 1988.

[15] A. I. Neishtadt, *The separation of motions in systems with rapidly rotating phase*, Prikl. Mat. Mekh. **48** (1982) no. 2, 197–204; English transl. in J. Appl. Math. Mech. **48** (1982).

[16] J. Milnor, *Lectures on the h-cobordism theorem*, Princeton Univ. Press, Princeton, NJ, 1965.

[17] D. V. Anosov, *Geodesic flows on closed Riemannian manifolds of negative curvature*, Trudy Mat. Inst. Steklov. **90** (1967); English transl. in Proc. Steklov Math. Inst. **1969**.

[18] V. M. Eleonskii, N. N. Kirova, and N. E. Kulagin, *New conservation law for the Landau–Lifshitz equations*, Zh. Eksp. Teoret. Fiz. **77** (1979), 409–419; English transl. in Soviet Physics JETP **50** (1979).

[19] ———, *On random degeneration of the self-localized solutions of the Landau–Lifshitz equations*, Zh. Eksp. Teoret. Fiz. **75** (1978), no. 6, 2210–2219; English transl. in Soviet Physics JETP **48** (1978).

[20] A. P. Veselov, *The Landau–Lifshits equation and integrable systems of classical mechanics*, Dokl. Akad. Nauk SSSR **270** (1983), no. 5, 1094–1097; English transl. in Soviet Math. Dokl. **27** (1983).
[21] V. I. Arnold, *Mathematical methods of classical mechanics*, Springer-Verlag, Berlin–Heidelberg, 1978.
[22] R. Abraham and J. Marsden, *Foundations of mechanics*, Benjamin, New York–Amstredam, 1978.
[23] A. T. Fomenko, *Symplectic geometry. Methods and applications*, Moskov. Gos. Univ., Moscow, 1988. (Russian)
[24] V. I. Arnold and A. B. Givental, *Symplectic geometry*, Encyclopedia of Mathematical Sciences, vol. 4, Springer-Verlag, Berlin–Heidelberg, 1988.
[25] A. Weinstein, *Lectures on symplectic manifolds*, CBMS Conference Ser., vol. 29, Amer. Math. Soc., Providence, RI, 1977.
[26] S. Helgason, *Differential geometry and symmetric spaces*, Academic Press, New York and London, 1962.
[27] B. A. Dubrovin, S. P. Novikov, and A. T. Fomenko, *Modern geometry*, vol. 1, "Nauka", Moscow, 1979; English transl., Springer-Verlag, New York, Berlin, 1984.
[28] S. P. Novikov, *Hamiltonian formalism and multivalued analog of Morse theory*, Uspekhi Mat. Nauk **37** (1982), no. 5, 3–49; English transl. in Russian Math. Surveys **37** (1982).
[29] B. A. Dubrovin, I. M. Krichever, and S. P. Novikov, *Integrable systems.* 1, Encyclopedia of Mathematical Sciences, vol. 3, Springer-Verlag, Berlin–Heidelberg–New York, 1988.
[30] N. N. Nekhoroshev, *The action-angle variables and their generalizations*, Trudy Moskov. Mat. Obshch. **26** (1972), 181–198; English transl. in Moscow Math. Soc. Trans. **26** (1974).
[31] V. I. Arnold, *Lectures on bifurcations and versal families*, Uspekhi Mat. Nauk **27** (1972), no. 5, 119–184; English transl. in Russian Math. Surveys **27** (1972).
[32] N. K. Gavrilov, *On some bifurcations of the equilibrium states of codimension two for divergence-free fields*, Methods of the Qualitative Theory of Differential Equations (E. A. Leontovich-Adronova, ed.), Gorkov. Gos. Univ., Gorki, 1985, pp. 46–55. (Russian)
[33] V. V. Nemytskii and V. V. Stepanov, *Qualitative theory of differential equations*, Gostehizdat, Moscow–Leningrad, 1947; English transl., Princeton Univ. Press, Princeton, NJ, 1960.
[34] A. A. Andronov, E. A. Leontovich, I. I. Gordon, and A. G. Maier, *Qualitative theory of dynamical systems of the second order*, "Nauka", Moscow, 1966. (Russian)
[35] S. Smale, *Differentiable dynamical systems*, Bull. Amer. Math. Soc. **73** (1967), 747–817.
[36] W. H. Gottschalk and G. A. Hedlund, *Topological dynamics*, Amer. Math. Soc., Providence, RI, 1955.
[37] V. V. Kalashnikov, *Bott property and general position property of integrable Hamiltonian systems*, Uspekhi Mat. Nauk **48** (1993), no. 6, 151–152; English transl. in Russian Math. Surveys **48** (1993).
[38] ______, *On the generic character of Bott integrable Hamiltonian systems*, Mat. Sb. **185** (1994), no. 1, 107–120; English transl. in Russian Acad. Sci. Sb. Math. **79** (1995).
[39] F. R. Gantmakher, *Theory of matrices*, "Nauka", Moscow, 1988.
[40] J. Williamson, *On an algebraic problem concerning the normal forms of linear dynamical systems*, Amer. J. Math. **58** (1936), 141–163.
[41] D. Husemoller, *Fibre bundles*, McGraw-Hill, New York, 1966.
[42] A. D. Bruno, *Local methods in nonlinear differential equations*, Springer-Verlag, Berlin, 1989.
[43] D. Birkhoff, *Dynamical systems*, Amer. Math. Soc., New York, 1927.
[44] K. L. Siegel, *On the integrals of canonical systems*, Ann. Math. **42** (1941), 806–822.
[45] J. Moser, *On generalization on a theorem of Liapounoff*, Comm. Pure Appl. Math. **11** (1958), 257–271.
[46] H. Russmann, *Über das Verhalten analytisher Hamiltonscher Differentialgleichungen in der Nähe einer Gleichgewichtslösung*, Math. Ann. **154** (1964), 285–300.
[47] J. Vey, *Sur certains systèmes dynamiques séparables*, Amer. J. Math. **100** (1978), 591–614.
[48] H. Ito, *Convergence of Birkhoff normal forms for integrable systems*, Comment. Math. Helv. **64** (1989), 412–461.
[49] L. H. Eliasson, *Hamiltonian systems with Poisson commuting integrals*, Thesis, Univ. of Stockholm, 1984.
[50] ______, *Normal forms for Hamiltonian systems with Poisson commuting integrals*, Comment. Math. Helv. **65** (1990), 4–35.

[51] A. G. Kushnirenko, *Analytic action of a semisimple Lie group near fixed point is equivalent to a linear action*, Funktsional. Anal. i Prilozhen. **1** (1967), no. 1, 89–90; English transl. in Functional Anal. Appl. **1** (1967).

[52] T. Poston and I. Stewart, *Catastrophe theory and its applications*, Pitman, London–San Francisco–Melbourn, 1978.

[53] G. Seifert and W. Trelfall, *Lehrbuch der Topologie*, Teubner, Leipzig, 1934.

[54] V. S. Medvedev and Ya. L. Umanskii, *On decomposition of $n$-dimensional manifolds into simple manifolds*, Izv. Vyssh. Uchebn. Zav. Mat. **1979**, no. 1, 46–50; English transl. in Soviet Math. (Iz. Vuz.) **23** (1979).

[55] G. M. Goluzin, *Geometric theory of functions of a complex variable*, "Nauka", Moscow, 1966; English transl., Amer. Math. Soc., Providence, RI, 1969.

[56] A. G. Sokolskii, *On stability of autonomous Hamiltonian systems with two degrees of freedom in the case of equal frequencies*, Prikl. Mat. Mekh. **38** (1974), no. 5, 791–799; English transl. in J. Appl. Math. Mech. **38** (1974).

[57] ———, *On stability of autonomous Hamiltonian systems with two degrees of freedom in the case of first order resonance*, Prikl. Mat. Mekh. **41** (1977), no. 1, 24–33; English transl. in J. Appl. Math. Mech. **41** (1977).

[58] J.-C. Van der Meer, *The Hamiltonian Hopf bifurcation*, Lecture Notes in Math., vol. 1160, Springer-Verlag, Berlin, 1985.

[59] J. T. Schwartz, *Differential geometry and topology*, New York, 1968.

[60] S. Smale, *On gradient dynamical systems*, Ann. Math. **74** (1961), 199–206.

[61] M. W. Hirsch, C. C. Pugh, and M.Shub, *Invariant manifolds*, Lecture Notes in Math., vol. 583, Springer-Verlag, Berlin, 1977.

[62] A. N. Shoshitaishvili, *Bifurcations of the topological type of vector field near a singular point*, Trudy Sem. Petrovsk., Vyp. 1 (1975), 279–309. (Russian)

[63] C. Neumann, *De problemate quodam mechanico, quod ad primam integralium ultraellipticorum classem revocatur*, J. Reine. Angew. Math. **56** (1858), 46–63.

[64] M. P. Kharlamov, *Topological analysis of integrable problems in dynamics of solid body*, Leningrad. Gos. Univ., Leningrad, 1988. (Russian)

[65] R. Devaney, *Transversal homoclinic orbits in an integrable system*, Amer. J. Math. **100** (1979), 632–642.

[66] L. M. Lerman, *Once again on the structure of integrable stationary waves for the Landau–Lifshitz equation*, Selecta Math. Sovietica **12** (1993), 333–351.

[67] A. I. Bobenko, *Explicit integration of nonlinear equations by the inverse problem method with the parameter on an elliptic curve*, Ph. D. Thesis, Leningrad, 1985.

[68] P. J. Hilton and S. Wylie, *Homology theory. An introduction to algebraic topology*, Cambridge Univ. Press, Cambridge, 1960.

[69] A. V. Bolsinov, A. T. Fomenko, and S. V. Matveev, *Topological classification of the integrable Hamiltonian systems with two degrees of freedom. The list of systems of little complexity*, Uspekhi Mat. Nauk **45** (1990), no. 2, 49–77; English transl. in Russian Math. Surveys **45** (1990)

[70] V. F. Lazutkin, *KAM-theory and semi-classical approximation to eigenfunctions*, Springer-Verlag, New York, 1993.

[71] A. T. Fomenko and D. B. Fuks, *A course in homotopic topology*, "Nauka", Moscow, 1989. (Russian)

[72] D. M. Galin, *Versal deformations of linear Hamiltonian systems*, Trudy Sem. Petrovsk., Vyp. 1 (1975), 63–74. (Russian)

[73] L. M. Lerman and Ya. L. Umanskii, *Classification of four-dimensional Hamiltonian systems and Poisson action of $\mathbb{R}^2$ in extended neighborhoods of simple singular points. 3. Realization*, Mat. Sb. **186** (1995), no. 10, 89–102; English transl. in Russian Acad. Sci. Sb. Math. **80** (1996).

# Selected Titles in This Series

(*Continued from the front of this publication*)



(See the AMS catalog for earlier titles)